An Introduction to Economic Geology and Its Environmental Impact

To Jo, Nick, Caroline and Jason

An Introduction to Economic Geology and Its Environmental Impact

ANTHONY M. EVANS

BSc, PhD, MIMM, FGS
Honorary Research Fellow in Geology,
University of Leicester

b

**Blackwell
Science**

© 1997 by
Blackwell Science Ltd
Editorial Offices:
Osney Mead, Oxford OX2 0EL
25 John Street, London WC1N 2BL
23 Ainslie Place, Edinburgh EH3 6AJ
350 Main Street, Malden
    MA 02148 5018, USA
54 University Street, Carlton
    Victoria 3053, Australia

Other Editorial Offices:
Blackwell Wissenschafts-Verlag GmbH
Kurfürstendamm 57
10707 Berlin, Germany

Blackwell Science KK
MG Kodenmacho Building
7–10 Kodenmacho Nihombashi
Chuo-ku, Tokyo 104, Japan

First published 1997

Set by Excel Typesetters Co., Hong Kong
Printed and bound in Great Britain
by Hartnolls Ltd, Bodmin, Cornwall

The Blackwell Science logo is a
trade mark of Blackwell Science Ltd,
registered at the United Kingdom
Trade Marks Registry

DISTRIBUTORS

    Marston Book Services Ltd
    PO Box 269, Abingdon
    Oxon OX14 4YN
    (*Orders*: Tel: 01235 465500
              Fax: 01235 465555)

USA
    Blackwell Science, Inc.
    Commerce Place
    350 Main Street
    Malden, MA 02148 5018
    (*Orders*: Tel: 800 759 6102
                    617 388 8250
              Fax: 617 388 8255)

Canada
    Copp Clark Professional
    200 Adelaide St, West, 3rd Floor
    Toronto, Ontario M5H 1W7
    (*Orders*: Tel: 416 597-1616
                    800 815-9417
              Fax: 416 597-1617)

Australia
    Blackwell Science Pty Ltd
    54 University Street
    Carlton, Victoria 3053
    (*Orders*: Tel: 3 9347 0300
              Fax: 3 9347 5001)

A catalogue record for this title
is available from the British Library

ISBN 0-86542-876-X

Library of Congress
Cataloging-in-publication Data

Evans, Anthony M.
    An introduction to economic geology
and its environmental impact /
Anthony M. Evans.
        p.   cm.
    Includes bibliographical references
(p.   ) and index.
    ISBN 0-86542-876-X
    1. Geology, Economic.
    2. Mines and mineral resources—
    Environmental aspects. I. Title.
TN260.E93   1997
553—dc21   97-6494

Contents

Part 3: Mineralization in Time and Space

Preface

There is still little choice in the field of introductory textbooks in economic geology suitable for undergraduates, particularly those pursuing combined studies or general degrees, and this volume is an attempt to widen the choice. Much of its content is drawn from courses that I taught at Leicester University and from my books on ore geology. However, in this volume I have tried to draw the reader's attention to the great need in mineral exploration and exploitation to do all we can to preserve the environment, remembering that pollution and contamination can be as bad or even worse after mining operations terminate as during their productive phase. Chapter 3 is therefore an introduction to environmental implications which are then at least touched on in most of the chapters in Part 2 and often considered at some length.

I must take this opportunity to emphasize my hope that, if this book should be adopted as a set course book, the students concerned will be given adequate practical periods in which to handle and examine hand specimens and thin and polished sections of the common ore types and their typical host rocks. Too often one gains the impression that fresh graduates have gleaned their ideas of what an orebody looks like from being given a few mineralogical and museum specimens to gaze at. It is essential that lecturers build up typical ore suites from working mines in abundance to allow students to handle as many run-of-the-mill ores and their wall rocks as possible. This should be accompanied by developing such skills as visual assaying, recognizing wall rock alteration, using textural evidence to decide on the mode of genesis and so on.

As in previous books, I have attempted to keep the reader aware of financial realities, in particular by including a chapter on mineral economics and by including many statistics on product price, grades of ore, ore reserves and tonnage produced; with special emphasis on similar matters for the energy commodities.

For ease of reference, emphasis of important points and help with revision I have used bold type for certain words and phrases to draw the eye.

I am very grateful to Mr David Highley of the British Geological Survey for his help in some of the more difficult areas of mineral statistics. To

Mr Simon Rallison of Blackwell Science I offer my deep thanks for his constant help and encouragement.

As with all my writings, this book would not have seen the light of day without my loving wife's constant encouragement and her editorial and typing skills. For this stimulation and invaluable help I am truly grateful.

Anthony M. Evans
Burton on the Worlds
January 1997

Units and Abbreviations

Note on units

With few exceptions, the units used are all S.I. (Système International), which has been in common use by engineers and scientists since 1965. The principal exceptions are: (a) for commodity prices still quoted in old units, such as troy ounces for precious metals and the short ton (= 2000 lb); (b) when there is uncertainty about the exact unit used, tons in certain circumstances might be short or long (2240 lb); (c) degrees Celsius (centigrade)—geologists do not seem to be able to envisage temperatures in degrees Kelvin! (neither do meteorologists!); and (d) centimetres (cm), which like °C refuse to die because they are so useful!

S.I. prefixes commonly used in this text are k = kilo-, 10^3; M = mega-, 10^6 (million); G = giga-, 10^9 (billion is rarely used as it has different meanings on either side of the Atlantic. On the few occasions when it appears it means 10^9).

Some abbreviations used in the text

A.S.T.M.	American Society for Testing Materials	e	estimated
B.S.	British Standard	E.U.	European Union, sometimes still referred to as E.E.C.
B.T.u.	British thermal Unit; see Section 25.3.2	F.O.B.	Freight on board
C.I.F.	Carriage, insurance and freight	I.T.C.	International Tin Council
		L.N.G.	Liquid natural gas
C.I.P.E.C.	Conseil Inter-governmental des Pays Exportateurs de Cuivre (Intergovernmental Council of Copper Exporting Countries)	L.P.G.	Liquid petroleum gas
		M.E.C.	Market economy countries
		O.P.E.C.	Organization of Petroleum Exporting Countries
		P.G.M.	Platinum group metals
C.I.S.	Commonwealth of Independent States (includes many of the countries formerly in the U.S.S.R.)	R.E.E.	Rare earth elements
		R.F.	Russian Federation
		t p.a.	Tonnes *per annum*
		t p.d.	Tonnes *per diem*
d.a.f.	dry, ash-free (coal)	W.W.1	World War One
d.m.m.f.	dry, mineral matter-free (coal)	W.W.2	World War Two

Part 1
Principles

The poor tread lightest on the earth. The higher our income, the more resources we control and the more havoc we wreak.

Paul Harrison (b. 1936). U.S. playwright, director.
Guardian (London, 1 May 1992)

Is a park any better than a coal mine? What's a mountain got that a slag pile hasn't? What would you rather have in your garden—an almond tree or an oil well?

Jean Girandoux (1882–1944), French diplomat, author.
A prospector in *The Madwoman of Chaillot*, Act 1

1: The Importance and History of Mining

1.1 Beginnings

Why is economic geology essential to our progress? It has been said that agriculture is the most basic of man's activities, but in this time of **overpopulation** and indeed in the preceding centuries it would have been impossible to feed the world's teeming millions without recourse to mining and the use of minerals and metals for ploughing, fertilization, harvesting, food preparation and so on. Indeed, of these two basic industries—agriculture and mining—mining is the older. It started some 500 000 years ago when man commenced exploring for tool-making materials — **flint, obsidian, quartzite** were chipped and ground to sharp cutting edges. Early man found that some localities were richer than others and began to search for these richer deposits, to mine them, establish **axe factories** and to trade in the finished products. He soon developed an appreciation of their geological occurrence. For example, by 3 000 B.C. **large underground flint mines** were in operation at Grime's Graves in Norfolk, U.K. and it is clear that the miners had noted that particular horizons in the chalk host rock carried the best and most numerous flints.

Stone tools continued to be refined and more uses found for this material, particularly when agriculture became important and milling equipment was required leading to the development of **querns** during the Neolithic era. In this period, clay and other materials began to be exploited for **pottery manufacture**. But older than these events was the exploitation of **pigments** both for painting cave walls and body decoration. Natural iron oxides (ochres) and iron-rich clays (umbers) were among the first to be used and mine workings in Swaziland have been carbon-dated at 40 000 B.C. Cinnabar (red mercury sulphide used as vermilion) was mined in central Turkey 8 000 years ago. Wad (manganese oxides) was used for black, galena for grey (and as eye shadow by the ancient Egyptians). In St Jerome's Bible, Jezebel is recorded as painting her face with the black cosmetic stibio (stibnite). It has been suggested that efforts to purify pigments led to the discovery of smelting as Neolithic beads (*c*. 8 000 years old) of smelted lead have been found in Turkey apparently predating any other smelting by a significant period.

Quern—a stone hand mill for grinding cereal grain.

3

1.2 The importance of metals

Materials scientists are restricted in what they can make by the raw materials, like iron ore, that can be mined from the earth. The chemical, oil and metallurgical industries produce valuable substances such as fertilizers, fuels, metals and plastics from raw materials explored for by geologists and exploited by miners in or on the earth. Table 1.1 shows the six most important sources of raw materials and the products that are obtained from them. For the mining and manufacture of these products, **metals are essential** for mining machinery, power supply, transport and so on. Thus, although many of the materials we use such as stone, cement, clay, limestone, salt are not essentially metallic, metals are required **for their extraction and manufacture**.

The true facts concerning the initial discovery and use of metals are still being elucidated by archaeologists. The most abundant native metal by far is copper and it is often found in large and jagged pieces — perhaps Neolithic man found it by cutting his feet on it when fording streams. One source of native copper in eastern Turkey was the Ergani Maden Copper Mine on the Tigris—copper needles and scrapers 9 000 years old have been found 20 km downstream. But the first evidence of the **mining and smelting of copper ores** comes from the Timna region of Israel and the nearby Feinan area of Jordan where it occurred about 5–6 000 B.C. Copper production then spread through Turkey, Iraq and Iran to central and south-western Europe. When **tin** and copper minerals were (probably accidentally) smelted together with charcoal about 4 600 years ago, the **Bronze Age** came into being. Not until another 2 000 years or so had elapsed did **iron smelting** become widespread.

In Roman times, **bronze** and **iron** were used for weaponry, tools and farming implements; **copper** for jugs and ornaments; **lead** for pewter, plumbing and coffins; **gold** and **silver** for ornaments and, together with

Table 1.1 Important raw materials and their products.

Raw material	Substances obtained from the raw material
Air	Oxygen, nitrogen, argon
Coal	Fuels (e.g. coke, coal gas), dyes and plastics from coal tar
Oil and natural gas	Fuels, chemicals (e.g. plastics, pesticides, perfumes)
Plants	Foods (e.g. flour, oats, sugar, cooking oil), timber, clothing (e.g. cotton, linen), rubber
Rocks	Metals (e.g. iron, copper, gold), limestone, aggregates for building, brick clay, sulphur
Sea	Salt (NaCl), magnesium, bromine

brass and bronze, for coinage. The 'Seven Metals of Antiquity' were now well known. **Zinc** and **arsenic** were used to make brass and other copper alloys but they were not known in the elemental state and were either naturally part of the smelted ore or added in non-metallic forms such as the zinc carbonate smithsonite. Other metals were used in their mineral form as **pigments**, e.g. cobalt and nickel minerals used to colour ceramics, enamels and glass.

The **Seven Metals of Antiquity**: copper, gold, iron, lead, mercury, silver, tin.

1.3 The Industrial Revolution

It was only when Abraham Darby leased the small **charcoal blast furnace** for smelting iron ore at Coalbrookdale in the U.K. in 1708 that the production and widespread use of metals really expanded. Iron making in the region had already led to a shortage of charcoal and Darby experimented successfully with **coke** for smelting iron ore. Before long, Darby was using iron from his coke fired furnace to mass produce common domestic articles — cooking pots, kettles, smoothing irons — as well as iron for machinery manufacture. He had taken the first and most difficult step towards freeing iron smelting from the tyrannical grip of **charcoal shortages**. This technological triumph led to the period of economic growth in Britain known as the **Industrial Revolution**. This change, from a traditional agriculturally based economy to one of the mechanized production of manufactured goods in large scale enterprises, has created the developed countries of the world and is more and more rapidly transforming the economies of nearly all the other countries of the world.

Even in the sixteenth and seventeenth centuries there was something like an industrial revolution evolving in Great Britain. By 1700 there were already a number of flourishing industries based on the mining industry, e.g. alum, bricks, copper, brass, nail making, ordnance, pottery; all these and the developments of the eighteenth century in general found the supply of indigenous, and usually local, earth resources sufficient for their needs.

Of course **prospecting** for mineral resources has always been essential to man's activities and much local exploration took place which was then vastly intensified by the Industrial Revolution. Later, as cheaper imports became available, e.g. tin from Malaya in the nineteenth century, whose import into Britain devastated the local, higher cost, underground producers of Cornwall and Devon (south-west England, Fig. 1.1), the interest in mineral prospecting became worldwide. One event which played a great part in initiating this worldwide interest was the discovery of gold in California in 1849. This led to further **gold rushes** in the U.S.A., Australia, Africa and other parts of the world. Rich deposits of other metals were

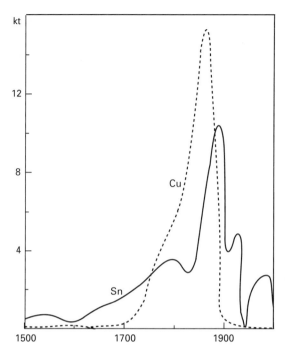

Fig. 1.1 Graphs for copper and tin production in the south-west of England showing the catastrophic effect of imports of much cheaper metals in the later part of the nineteenth century.

found by these gold seekers and soon mines and smelters in the U.S.A. were producing much cheaper metal products than in the older economies of western Europe which mostly worked much leaner ores.

1.4 The present situation

Through and since the Industrial Revolution technological discoveries have led to today's diverse, **widespread and massive use of earth resources** that can only be supplied by large, organized, worldwide mining and extractive industries prepared to sink vast amounts of capital into what are often **risk prone projects**. Some countries are better blessed with resources that can be exploited by these industries than others. As a result, wars have been fought over mineral resources and this lamentable result of mineral exploitation may well spread to disputes over **water supplies** in the Middle East in coming years (see Section 22.1).

A domestic mineral industry, particularly a source or sources of energy, is thus of inestimable value to a country because economic progress the world over depends fundamentally on minerals and metals. The Industrial Revolution in England owed its inception to the proximity of coal, iron ore, limestone and clay. Since then, oil and natural gas have joined coal as important energy providers and **iron**, mainly in the form of steel, has become even more **the dominant metal**. Local supplies of bulky, low value

industrial minerals, which cannot be economically transported over long distances, are essential because these, unlike metals or ores, cannot be imported other than in exceptional circumstances. Raw or refined products (crude oil, copper, gold, liquid natural gas etc.) can be invaluable foreign currency earners for developing countries.

1.5 Conclusions

The exploitation of our earth resources has had a long and fascinating history from the search for stone tool-making materials to the sophisticated mining and extractive industries of today, without whose development civilization as we know it could never have come about. Only a few centuries ago the Seven Metals of Antiquity provided most of our metallic needs, oil and coal were hardly used and the only remotely large scale exploitation was of industrial minerals for building and ceramics. Now we make use of most elements in the periodic table, both metals and non-metals, produce prodigious amounts of energy from coal, oil, natural gas and nuclear fuels (with geothermal heat under development) and mine vast volumes of industrial minerals. To enable us to maintain these feats we must strive to understand how, why and where minerals deposits were formed and the effect such exploitation might have on the natural environment and global economic development in coming decades.

It is regrettably true that in our desire for more possessions and a higher standard of living we have spoilt large areas of the earth with mines, quarries, superhighways and lines of pylons. Our cities have spread ever outwards and the addition of large buildings has turned them into 'concrete jungles'. Forests have been destroyed for ever. As a result, the world's consumption of raw materials has grown and grown. The present world population of over 5000 million will be 7000 million by 2000 A.D. Unfortunately, *more* people require *more* food, *more* water, *more* raw materials and *more* land. The earth's resources will not last for ever. Coal will last at present rates of extraction for centuries but other resources, such as oil, will last for much shorter times unless we use them more sparingly.

1.6 Further reading

Kennedy P. (1993) *Preparing for the Twenty-first Century*. HarperCollins, London. Contains a useful discussion, pp. 3–47, of population growth and its global impact. The rest of the book is also rewarding and stimulating reading for those concerned about the exploitation of natural resources and our planet's future.

Raistrick A. (1972) *Industrial Archaeology*. Eyre Methuen & Paladin, London. Although published twenty-four years ago, this classic volume remains a wonderful introduction to the history, technology and development of the Industrial Revolution in the U.K. Later reprints are available.

Wolfe J.A. (1984) *Mineral Resources: A World Review*. Chapman and Hall, New York. A chatty and stimulating book, the writer speaks his mind on many controversial points. Certain sections now need some revision but the book still provides most valuable reading.

2: Some Aspects of Mineral Economics

2.1 Introduction

A large economically mineable mineral deposit, e.g. 50 Mt underlying an area of 2 km², is minute in comparison with the earth's crust and in most countries the easily discovered deposits cropping out at the surface have nearly all been found. The deposits we now search for are largely concealed and require sophisticated exploration methods to find them. We usually refer to the target material in these deposits as ore, unless we use a more specific term such as coal, gas, oil or water. So **what is ore?** And what sort of ore is nowadays financially viable to look for and exploit? To answer these questions we must have some knowledge of mineral economics.

The main topics covered in this chapter are:
- nature of ore, industrial minerals and bulk materials,
- world production of metals and industrial minerals and their relative importance,
- commodity prices,
- metal and mineral prices,
- factors governing the economic exploitation of mineral deposits,
- ore reserves and mineral resources.

2.2 Mineral economics

2.2.1 Ore

The word **ore** formerly, and strictly, meant material from which a metal or metals could be won at a profit. An economically mineable ore deposit is normally called an **orebody**. There has, however, been a slow extension in the meanings of these words to include industrial minerals and bulk materials. **Industrial minerals** are any rock, mineral or other naturally occurring substance of economic value, exclusive of metallic ores, mineral fuels and gemstones. They are, therefore, minerals where either **the mineral itself**, e.g. baryte, industrial diamond or **the oxide** or **some other compound derived from any mineral** (but not the elemental metal) has an industrial application (end use). They include rocks such as granite, sand, gravel, limestone that are used for constructional purposes (in fragmental form these are often referred to as **aggregates** or **bulk materials**) as well as more valuable minerals with specific chemical or physical properties like **fluorite, kaolinite, perlite** and **phosphates**. Confusingly, some oxides and other compounds derived from 'metallic ores', e.g. Al_2O_3 from bauxite, normally considered as aluminium ore, TiO_2 from ilmenite (titanium ore), are also important raw materials for industrial mineral end uses and, as indicated above, are then classified with the industrial minerals.

All the above and many other pertinent considerations mean that no simple definition of ore can now be put forward. But there is no space in this book for the necessary discussion of what is at the moment providing a very vigorous debate in the literature, so we must content ourselves with a simple statement such as **ore** is a solid, naturally occurring mineral aggregate usable as mined or from which one or more valuable constituents may be economically recovered (see Sections 2.3.2 & 3).

Gangue is the unwanted material, minerals or rock, with which ore minerals are usually intergrown. Mines commonly possess processing plants in which the raw ore is milled before the separation of the ore from the gangue minerals to provide **ore concentrates** and **tailings** which are made up of the gangue.

2.2.2 The relative importance of ore and industrial minerals

Metals always seem to be the **focus of attention** for various reasons, such as their use in warfare, rapid and cyclical changes in price, occasional occurrence in very rich deposits (gold bonanzas, for example), with the result that **the great importance of industrial minerals** to our civilization is overlooked. Today industrial minerals permeate every segment of our society. They occur as components in durable and non-durable consumer goods and in many industrial activities and products, from the construction of buildings to the manufacture of ceramic tables or sanitary ware, the use of industrial minerals is obvious but often unappreciated. With numerous other goods, ranging from books to pharmaceuticals, the consumer is frequently unaware that industrial minerals play an essential role.

In more developed countries such as the U.K. and U.S.A., **industrial mineral production is far more important than metal production** from both the tonnage and financial viewpoints, as in fact it is on a worldwide basis (Table 2.1).

Graphs of the world production of the traditionally important minerals (Figs 2.1–2.3) show interesting trends. The world's appetite for the major metals appeared to be almost insatiable after W.W. 2 and post-war production increased with great rapidity; however, in the mid-seventies an abrupt slackening in demand occurred, triggered by the coeval oil crisis but clearly continuing up to the present day. These curves suggest that consumption of major metals is following **a wave pattern** in which the various crests may not be far off in time. Indeed, lead is clearly over the crest as also are tin and tungsten, and zinc appears to be joining them. Various factors are probably at work here: **recycling**, more economical use of metals and **substitution** by ceramics and plastics—industrial minerals are much used as a filler in plastics.

Table 2.1 World production of some mineral commodities in 1994; metals are in italics.

Rank	Commodity	Tonnage (Mt)	Rank	Commodity	Tonnage (Mt)
1	Aggregates	7500	23	Feldspar	6.5
2	Coal	4451	24	*Lead*	5.1
3	Crude oil	3117	25	Borates	4.73
4	Portland cement	1370	26	Baryte	4.5
5	*Iron ore*	975	27	Fuller's earth	4.2
6	Clay	200	28	*Titania*	3.9
7	Salt (NaCl)	184	29	Fluorspar	3.6
8	Silica sand	111	30	Asbestos	2.6
9	Phosphate rock	106	31	Perlite	1.8
10	Gypsum	96	32	Diatomite	1.5
11	Sulphur	50	33	Graphite	1.0
12	Kaolin	22.8	34	*Zirconium minerals*	0.92
13	Potash	22.6	35	*Nickel*	0.88
14	*Manganese ore*	22.1	36	Nepheline-syenite	0.84
15	*Aluminium*	19.1	37	Sillimanite minerals	0.61
16	*Chromium ores*	10.0	38	Vermiculite	0.50
17	*Copper*	9.4	39	*Magnesium*	0.27
18	Trona (Na_2CO_3)	9.1	40	Mica	0.23
19	Magnesite	8.9	41	Wollastonite	0.19
20	Bentonite	8.2	42	*Tin*	0.18
21	Talc	8.0	43	Strontium minerals	0.11
22	*Zinc*	6.7			

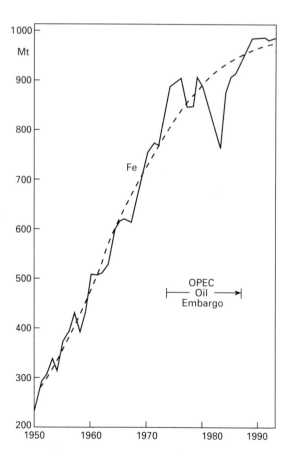

Fig. 2.1 World production of iron ore from 1950 to 1994. General trend superimposed.

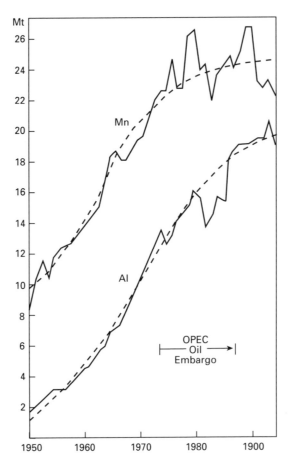

Fig. 2.2 World production of manganese ore and primary aluminium from 1950 to 1994. General trends superimposed.

Production of plastics rose by a staggering 1529% between 1960 and 1985 and a significant fraction of the demand behind this is attributable to metal substitution. In Table 2.2 the increases in production of selected metals and industrial minerals provide a striking contrast and one that explains why for some years now many large **metal mining companies** have been **moving into industrial mineral production**. Are we soon to pass from the Iron Age into a ceramic–plastic age? Readers are urged to monitor this *tentative* prophecy by keeping these graphs up to date using data from the same source (in this case the British Geological Survey's World Mineral Statistics). A factor of small but growing importance is the **strong demand for non-ferrous metals** in the non-O.E.C.D. countries; this has grown by over 6% p.a. during the present decade, compared with less than 1% in the O.E.C.D. countries, and it may increase sufficiently in the coming decade to influence present trends in demand for these metals. This demand too should be monitored. Finally, although the increase in demand for the major, high tonnage production metals is decreasing at the

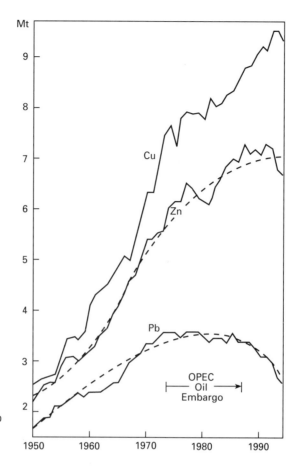

Fig. 2.3 World production of copper, zinc and lead from 1950 to 1994. General trends for zinc and lead superimposed.

present time, the future is bright for certain minor, low tonnage metals such as P.G.M., R.E.E., tantalum and titanium.

2.2.3 Commodity prices—the market mechanism

Most mineral trading takes place within the market economies of the world and the prices of minerals or mineral products are governed by the factors of supply and demand. Modern transport leads to many commodities having **a world market**; a price change in one part of the world affects the price in the rest of the world. Such commodities include wheat, cotton, rubber, gold, silver and base metals. These commodities have a wide demand, are capable of being transported and the costs of transport are small compared with the value of the commodity. The market for diamonds is worldwide but that for bricks is local.

Over the last few centuries **formal organized markets** have developed. Base metals are traded on the London Metal Exchange (L.M.E.), gold and silver on the London Bullion Market. Similar markets exist in many other

Metal/mineral	Increase
Aluminium	163
Cobalt	−16
Copper	47
Diatomite	15
Feldspar	158
Gold	38
Graphite	138
Gypsum	84
Iron ore	27
Kaolin	62
Lead	−24
Mica	21
Molybdenum	−25
Nickel	33
Perlite	140
Phosphate	27
P.G.M.	98
Potash	22
Sillimanite minerals	89
Silver	43
Sulphur	61
Talc	69
Tin	−4
Vanadium	71
Vermiculite	66
Zinc	24

Table 2.2 Increases (%) in world production of some metals and industrial minerals, 1970–1994; metals are in italics. (Recycled metal production is not included.)

countries, e.g. the New York Commodity Exchange (Comex). Because these markets are composed of specialist buyers and sellers and are in constant communication with each other, prices are sensitive to any change in worldwide supply and demand. The prices of some metals on Comex and the L.M.E. are quoted daily by many newspapers whilst more comprehensive guides to current metal and mineral prices can be found in the *Engineering and Mining Journal*, *Industrial Minerals*, the *Mining Journal* and other technical journals.

FORCES DETERMINING PRICES

(a) *Demand and supply.* Demand may change over a short period of time for a number of reasons. Where one commodity substitutes to a significant extent for another and the price of the latter falls, then the substituting commodity becomes relatively expensive and less of it is bought. Copper and aluminium are affected to a degree in this way. A change in technology may increase the demand for a metal, e.g. the use of titanium in jet engines, or decrease it, e.g. tin—development of thinner layers of tin on tinplate and substitution (see Table 2.2). The expectation of future price rises or short-ages will induce buyers to increase their orders to have more of a commodity in stock.

Supply refers to how much of a commodity will be offered for sale at a given price over a set period of time. This quantity depends on the price of the commodity and the conditions of supply. High prices stimulate supply and investment by suppliers to increase their output. A fall in prices has the opposite effect and some mines may be closed completely or put on a care-and-maintenance basis in the hope of better times in the future. **Conditions of supply** may change fairly quickly through: (1) changes due to abnormal circumstances such as natural disasters, war, other political events, fire, strikes at the mines of big suppliers; (2) improved technique in exploitation; (3) discovery and exploitation of large new orebodies.

(b) *Government action*. Governments can act to stabilize or change prices. Stabilization may be attempted by building up a **stockpile**, although the mere building up of a substantial stockpile increases demand and may push up the price! With a substantial stockpile in being, sales from the stockpile can be used to prevent prices rising significantly and purchases for the stockpile may be used to prevent or moderate price falls. As commodity markets are worldwide, it is in most cases impossible for one country acting on its own to control prices. Groups of countries have attempted to exercise control over tin (I.T.C.) and copper (C.I.P.E.C.) in this way but with little success and, at times, signal failure.

Governmental action that will increase consumption of platinum and rhodium is the adoption of new regulations on car exhausts by the E.U. countries. The worldwide effort to diminish harmful exhaust emissions resulted in a record industrial purchase of 1.7 million troy ounces of platinum in 1993. Comparable actions by governments stimulated by environmental lobbies will no doubt occur in the coming years.

(c) *Recycling* is already having a significant effect on some product prices. Economic and particularly environmental considerations will lead to increased recycling of materials in the immediate future. Recycling will **prolong resource life** and **reduce mining wastes and smelter effluents**. Partial immunity from price rises, shortages of primary materials or actions by cartels will follow. A direct economic and environmental bonus is that **energy requirements** for recycled materials are usually **much lower** than for treating ores — in the case of aluminium, 80% less electricity is needed. In the U.S.A., the use of ferrous scrap as a percentage of total iron consumption rose from 35% to 42% over the period 1977–87 and aluminium from 26% to 37%; but copper has remained mainly in the range 40–45% and zinc, 24–29%. Of course the potential for recycling some materials is much greater than for others. Unlike metals, the **potential for recycling industrial minerals is much lower** and is confined to a few commodities such as bromine, fluor-compounds, industrial diamonds, iodine and feldspar and silica in the form of glass; so prices will be less affected by this factor.

(d) *Substitution and new technology* may both lead to a diminution in demand. We have already seen great changes such as the development of longer lasting car batteries that use less lead, substitution of copper and plastic for lead water pipes and a change to lead-free petrol; all factors that have contributed to a downturn in the demand for lead (see Fig. 2.3). But a factor affecting all metals was the O.P.E.C. shock in 1973 (see Figs 2.1–2.3) which led to huge increases in the prices of oil and other fuels, pushed demand towards materials having a low sensitivity to high energy costs and favoured the use of lighter and less expensive substitutes for metals.

METAL AND MINERAL PRICES

(a) *Metals.* Metal prices are **erratic and hard to predict** (see Figs 2.5 & 2.6). In the short run, prices fluctuate in response to unforeseen news affecting supply and demand, e.g. strikes at large mines or smelters, unexpected increases in warehouse stocks. This makes it difficult to determine regular behaviour patterns for some metals. Over the intermediate term (several decades) the prices clearly respond to rises and falls in **world business activity** which is some help in attempts at forecasting price trends (Figs 2.5 & 2.6). Despite an upturn in price for many metals during the last few years, the general outlook is not promising for most of the traditional metals, in particular iron, manganese (Fig. 2.4), lead (Fig. 2.6), tin and tungsten. Some of the reasons for this prognostication have been discussed above. It is the minor metals such as titanium, tantalum and others that are likely to have a brighter future.

Gold has had a different history since W.W.2. From 1934 to 1972, the price of gold remained at $U.S.35 per troy ounce. In 1971 President Nixon removed the fixed link between the dollar and gold and left market demand to determine the daily price. The following decade saw gold soar to a record price of $U.S.850 an ounce, a figure inconceivable at the beginning of the 1970s; it then fell back to a price little higher in real terms than that of the 1930s (Fig. 2.7). (Using the U.S. Consumer Price Index, the equivalent of $35 in 1935 would have been $289 in 1987, so that at $400 oz^{-1} the gold price in that year would be up by only 38% in real terms on the 1935 price.) Citizens of many countries were again permitted to hold gold either as bars or coinage and many have invested in the metal. Unfortunately for those attempting to predict future price changes, **demand** for

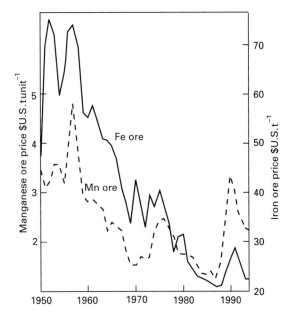

Fig. 2.4 Annual average prices of iron and manganese ores for 1950 to 1994. The iron ore price is for Brazilian ore (64.5% Fe) f.o.b. North Sea ports expressed in constant 1985 U.S. dollars. The manganese ore price is for Indian ore (46–48% Mn) c.i.f. U.S. ports expressed in constant 1985 U.S. dollars.

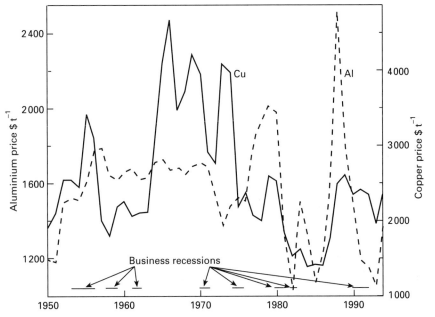

Fig. 2.5 Annual average prices of primary aluminium and top grade copper on the London Metal Exchange for 1950 to 1994, both expressed in constant 1985 U.S. dollars.

this metal is not **determined** so much by industrial demand as **by fashion and sentiment**—two notoriously variable and unpredictable factors! The main destinations of gold at the present day are carat jewellery and bars for investment purposes.

The **rise in the price of gold since 1971** has led to a great increase in prospecting and the discovery of many large deposits. This trend is

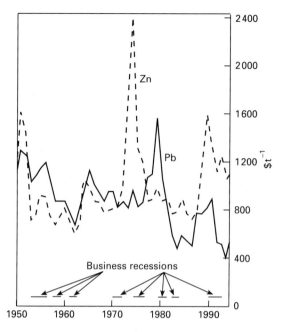

Fig. 2.6 Annual average prices of pig lead and special high grade zinc in New York for 1950 to 1994, expressed in constant 1985 U.S. dollars.

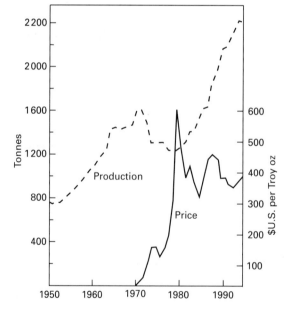

Fig. 2.7 World production of gold from 1950 to 1994 and the actual annual average price in U.S. dollars per troy ounce (i.e. no correction for inflation).

continuing at an increasing rate and gold production since reaching a low point in 1979 has been increasing rapidly (Fig. 2.7)—**will fashion and sentiment absorb such an annually sustained increase in supply?**

(b) *Industrial minerals.* **Most industrial minerals can be traded internationally;** exceptions are the low value commodities such as sand, gravel and crushed stone which have a low unit value and are mainly produced for local markets. However, minor deviations from this statement are beginning to appear, such as crushed granite being shipped from Scotland to the U.S.A., sand from Western Australia to Japan and filtration sand and water from the U.K. to Saudi Arabia! Lower middle unit value minerals from cement to salt can be moved over intermediate to long distances provided they are **shipped in bulk by low cost transport**. Nearly all industrial minerals of higher unit value are internationally tradeable, even when shipped in small lots.

Minerals with a low unit value will increase greatly in cost to the consumer with increasing distance to the place of use. Consequently, low unit value commodities are normally of little or no value unless available close to a market. Exceptions to this rule may arise in special circumstances, e.g. the south-eastern sector of England (including London) where demand for aggregates cannot now be met from local resources. Considerable additional supplies now have to be brought in by rail and road over distances in excess of 150 km. For high unit value minerals like industrial diamonds, sheet mica and graphite, **location is largely irrelevant**.

Like metals, industrial minerals respond to changes in the intensity of business activities but, as a group, to nothing like the extent shown by metals and their **prices are generally much more stable**. Already discovered world reserves of most industrial minerals are adequate to meet the expected demand up to at least 2000 A.D. and so no significant increases in real long-term prices are expected. Exceptions to this are likely to be sulphur, baryte, talc and pyrophyllite. **Growth rates** are expected to **rise steadily**, rates exceeding 4% p.a. are forecast for nine industrial minerals and 2–4% for twenty-nine others. These figures may well prove to be conservative estimates. Unlike metals, the **recycling potential** of industrial minerals is, with some exceptions, **low** and competing substitutional materials are frequently more expensive or less efficient (e.g. calcite for kaolinite as a cheaper paper filler).

2.3 Important factors in the economic recovery of minerals

2.3.1 Principal steps in the exploration for, and exploitation of, an orebody

These may be briefly summarized as follows:
1 *mineral exploration:* to discover an orebody;
2 *feasibility study:* to prove its commercial viability;
3 *mine development:* establishment of the entire infrastructure;
4 *mining:* extraction of ore from the ground;
5 *mineral processing (ore dressing):* milling of the ore, separation of ore

minerals from gangue, separation of the ore minerals into concentrates, e.g. copper concentrate; separation and refinement of industrial mineral products;

6 *smelting:* recovering metals from the mineral concentrates;

7 *refining:* purifying the metal;

8 *marketing:* shipping the product (or metal concentrate if not smelted and refined at the mine) to the buyer, e.g. custom smelter, manufacturer.

2.3.2 Some important factors in the evaluation of a potential orebody

'Now a miner, before he begins to mine the veins, must consider seven things, namely: the situation, the conditions, the water, the roads, the climate, the right of ownership and the neighbours.' [Georgius Agricola in *De Re Metallica* (1556)]

(a) *Ore grade.* The concentration of a metal in an orebody is called its **grade**, usually expressed as a percentage or in parts per million (ppm). The process of determining these concentrations is called **assaying**. Various economic and sometimes political considerations will determine the lowest grade of ore that can be produced from an orebody; this is termed **the cut-off grade**. In order to delineate the boundaries of an orebody in which the level of mineralization gradually decreases to a background value, many samples will have to be collected and assayed. The boundaries thus established are called **assay limits**. Being entirely economically determined, **they may not be marked by any particular geological feature**. If the price received for the product increases, then it may be possible to lower the value of the cut-off grade and thus increase the tonnage of the ore reserves. This will have the effect of lowering the overall grade of the orebody, but for the same daily production, it will increase the life of the mine. Grades vary from orebody to orebody and, clearly, **the lower the grade, the greater the tonnage of ore required** to provide an economic deposit. The general tendency in metalliferous mining during this century has been to mine ores of lower and lower grade. This has led to the development of more large scale operations with outputs of 40 kt or more of ore per day being not unusual. It will also be necessary to estimate, if possible by comparison with similar orebodies, what the **head grade** will be. This is the grade of the ore as delivered to the mill (mineral dressing plant). Often the head grade is lower than the measured ore grade because of mining dilution — the inadvertent or unavoidable incorporation of barren wall rock into the ore during mining.

Transmutation! Technological advances may transform waste into ore. For example in the 1970s the introduction of solvent extraction enabled Nchanga Consolidated Copper Mines in Zambia to treat 9 Mt of tailings to produce 80 kt of copper.

The **grade of an industrial mineral deposit** is not always as critical as that for a metal deposit. The important criteria for assessing the usefulness of non-metallic deposits include both **chemical and physical properties**, and many types of deposit are used *en masse*. This means that deposit homogeneity is important; patches with different properties must either be discarded or **blended to form a uniform product**. For example, in an aggregate to be used for roadstone the properties that matter are the aggregate crushing, impact and abrasion values (A.C.V., A.I.V. and A.A.V.), the

10% fines value, the polished stone value (P.S.V.), the size grading possible from the plant and the petrography of the pebbles. As another example, limestone has a wide variety of uses, depending on such properties as the **chemical purity** (for making soda ash or sea water magnesia), the **colour**, **grain-size distribution** and **brightness of a powder** (paper and other filler applications) or its **oil absorption** (putty manufacture). For a new industrial mineral deposit to be worked at a profit, it is essential firstly that the properties of the material either before or after processing match the specification for intended use, and secondly that there are **adequate reserves** to meet the expected demand. From many deposits a number of products with different properties can be made; **a variety of different markets** may therefore be required to achieve the most economical exploitation of the deposit.

(b) *By-products.* In some ores several metals are present and the sale of one may help finance the mining of another. For example, silver and cadmium can be by-products of the mining of lead–zinc ores and uranium is an important by-product of many South African gold ores. Among industrial minerals the recovery of by-product baryte and lead from fluorspar operations can be cited.

(c) *Commodity prices.* The price of the product to be marketed is a vital factor and this subject has been discussed above (see 'Metals and mineral prices' in Section 2.2.3).

(d) *Mineralogical form.* The properties of a mineral govern the ease with which existing technology can extract and refine certain metals and this may affect the **cut-off grade**. Thus, nickel is recovered far more readily from sulphide than from silicate ores, and sulphide ores can be worked down to about 0.5% whereas silicate ores must assay about 1.5% in order to be economic.

Industrial mineral deposits present different problems. For example, for a silica sand deposit to be utilized for high quality glass making, the Fe_2O_3 content should be less than 0.035%. Some brown-looking sands with much more Fe_2O_3 can be upgraded if most of the iron is present as a coating on the grains, which can be removed either by scrubbing or by acid-leaching. If the iron is present as inclusions within the quartz grains then upgrading may be impossible.

(e) *Grain size and shape.* The **recovery** is the percentage of the total metal or industrial mineral contained in the ore that is recovered in the concentrate. A recovery of 90% means that 90% of the metal in the ore is recovered in the concentrate and 10% is lost in the tailings. It might be thought that if one were to grind ores to a sufficiently fine grain size then complete separation of mineral phases might occur and make 100% recovery possible. With present technology this is not the case as **most mineral processing techniques fail in the ultra-fine size range**. As mentioned above, the grain size distribution is critical in the use of a number of different industrial rocks and minerals. Aggregate in concrete is used in specified size ranges depending on the end use. Each different mineral filler application (paper, rubber, plastics) requires different, carefully specified, often narrow, ranges.

(f) *Undesirable substances.* Deleterious substances may be present in both ore and gangue minerals. For example, tennantite ($Cu_{12}As_4S_{13}$) in copper ores can introduce **unwanted arsenic** and sometimes **mercury** into copper concentrates. These, like **phosphorus** in iron concentrates and **arsenic** in nickel concentrates, will lead to custom smelters imposing **financial penalties**.

(g) *Size and shape of deposits.* The size, shape and nature of ore deposits also affect the workable grade. Large, low grade deposits that occur at, or near, the surface can be worked by cheap open pit methods (Fig. 2.8) whilst thin tabular vein deposits will necessitate more expensive underground methods of extraction, although generally they can be worked in much smaller volumes so that a relatively small initial capital outlay is required. Although the initial capital outlay for larger deposits may be higher, open pitting, aided by the savings from bulk handling of large daily tonnage of ore (say >30 kt), has led to a trend towards the large scale mining of low grade orebodies. As can be seen from Fig. 2.8, a time comes during exploitation when the **waste-to-ore ratio** becomes too high for profitable working; for low grade ores this is around 2 : 1 and the mine must be aban-

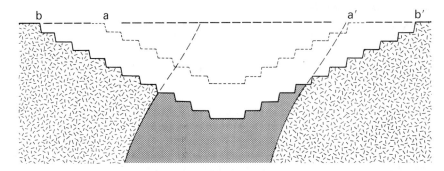

Fig. 2.8 Development of an open pit mine. During the early stages (a–a') more ore (black) is removed than waste rock; then, as the pit becomes deeper, the ratio of waste to ore mined becomes greater until at stage b–b' it is about 1.6 to 1. (After Barnes, 1988, *Ores and Minerals*, Open University Press, with permission.)

doned or converted into an underground operation. Many small mines start as small, cheaply worked open pits and then develop into underground operations (Fig. 2.9). Haulage always used to be by narrow gauge, electrically operated railways but now, if the orebody size permits, rubber-tyred equipment is used to produce larger tonnages more economically and shafts are then gentle spiral declines up which ore can be hauled out by diesel trucks (**trackless mining**).

(h) *Ore character*. A loose unconsolidated beach sand deposit can be mined cheaply by **dredging** and does not require crushing. Hard compact ore must be **drilled, blasted and crushed**. In hard-rock mining operations a related aspect is the strength of the country rocks. If these are badly sheared or fractured they will be weak and require roof supports in underground working, and in open pitting a gentler slope to the pit sides will be required, which in turn will affect the waste-to-ore ratio adversely.

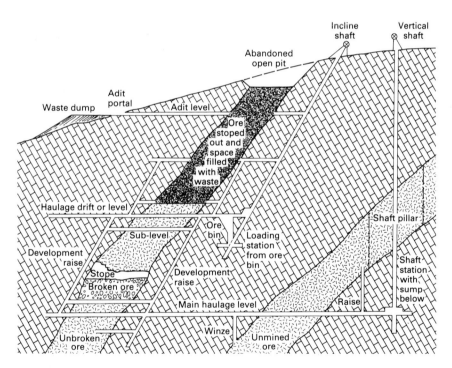

Fig. 2.9 Mining terminology. Ore was first mined at the outcrop from an open pit; then an adit was driven into the hillside to intersect and mine the ore at a lower level. An inclined shaft was sunk later to mine at even deeper levels and, eventually, a vertical shaft was sunk to serve operations to two orebodies more efficiently. Ore is mined by driving two haulage drifts at different levels and connecting them by raises which are then connected by sublevels. Ore is mined upwards from the lower sublevel to form a stope. Broken ore can be left in the stope to form a working platform and to support its walls (shrinkage stoping), or withdrawn and waste from the mill pumped in (cut-and-fill stoping). Ore between haulage and sublevel is left as supporting pillars until the level is abandoned. A shaft pillar is also left unmined. (After Barnes, 1988, *Ores and Minerals*, Open University Press, with permission.)

(i) *Cost of capital*. Big mining operations have now reached the stage, thanks to inflation, where they require enormous initial capital investments and sums in excess of $U.S.500 M are not uncommon. This means that the stage has been reached where few companies can afford to develop a mine with their own financial resources. They must **borrow the capital** from banks and elsewhere, capital which has to be **repaid with interest**. Thus the revenue from the mining operation must cover the running costs, the payment of taxes, royalties, the repayment of capital plus interest on it, and provide a **profit** to shareholders who have **risked their capital** to set up or invest in the company.

Capital outlays
The development of the Pipeline gold deposit in the western U.S.A. will cost $U.S.319 M and the Lihir gold deposit in P.N.G. $U.S.673 M!

(j) *Location*. Geographical factors may determine whether or not an orebody is economically viable. In a **remote location** there may be no electric power supply, roads, railways, houses, schools, hospitals, etc. All or some of these infrastructural elements will have to be built, the cost of transporting the mine product to its markets may be very high and wages will have to be high to attract skilled workers.

(k) *Environmental considerations*. These will be dealt with in Chapter 3.

(l) *Taxation*. Greedy governments may demand so much tax that mining companies cannot make a reasonable profit. On the other hand, some governments have encouraged mineral development with taxation incentives, such as a waiver on tax during the early years of a mining operation. This proved to be a great attraction to mining companies in the Irish Republic in the 1960s and brought considerable economic gains to that country.

(m) *Political factors*. Many large mining houses will not now invest in politically unstable countries. Fear of **nationalization** with perhaps very inadequate or even no compensation is perhaps the main factor. Nations with a history of nationalization generally have poorly developed mining industries. Possible political turmoil, civil strife and currency controls may all combine to increase greatly the financial risks of investing in certain countries.

Ore reserve types. In many countries these, or equivalent, words have nationally recognized definitions and legal connotations. The practising geologist must therefore know the local definitions thoroughly and make sure that he uses them correctly. See under Miskelly (1994) in Section 2.5.

2.3.3 Ore or mineral reserve classification

In delineating and working an orebody the mining geologist often has to classify his **ore reserves** into three classes: **proved, probable and possible**; frequently used synonyms are: **measured, indicated and inferred**. Proved ore has been sampled so thoroughly that we can be certain of its outline, tonnage and average grade, within certain limits. Elsewhere in the orebody, sampling from drilling and development workings may not have been so thorough, but there may be enough information to be reasonably sure of its tonnage and grade; this is probable ore. On the fringes of our exploratory workings we may have enough information to infer that ore extends for some way into only partially explored ground and that it may amount to a certain volume and grade of possible ore.

2.3.4 Mineral resources

These represent the total amount of a particular commodity (e.g. tin) and usually they are estimated for a nation as a whole and not for a company. They consist of ore **reserves**, known but **uneconomic deposits** and **hypothetical deposits** not yet discovered. The estimation of the undiscovered potential of a region can be made by comparison with well explored areas of similar geology. Theoretically, world resources of most metals are enormous. Taking copper as an example, there are large amounts of rock running 0.1–0.3% and enormous volumes containing about 0.01%. The total quantity of copper in such deposits probably exceeds that in proven reserves by a factor of from 10^3 to 10^4. Nevertheless, the enormous amount of such material does not at present imply a virtually endless resource of metals. As grades approach low values, a concentration (**the mineralogical limit**) is reached, below which an element no longer forms a distinct physically recoverable mineral phase.

2.3.5 Geochemical considerations

It is traditional in the mining industry to divide metals into groups with special names. These are as follows:

Table 2.3 Concentration factors.

	Average crustal abundance (%)	Average minimum exploitable grade (%)	Concentration factor
Aluminium	8	30	3.75
Iron	5	25	5
Copper	0.005	0.4	80
Nickel	0.007	0.5	71
Zinc	0.007	4	571
Manganese	0.09	35	389
Tin	0.0002	0.5	2500
Chromium	0.01	30	3000
Lead	0.001	4	4000
Gold	0.0000004	0.0001*	250

* 1 ppm.

1   *precious metals*—gold, silver, platinum group (P.G.M.);
2   *non-ferrous metals* — copper, lead, zinc, tin, aluminium (the first four being commonly known as *base metals*);
3   *iron and ferroalloy metals* — iron, manganese, nickel, chromium, molybdenum, tungsten, vanadium, cobalt;
4   *minor metals and related non-metals* — antimony, arsenic, beryllium, bismuth, cadmium, magnesium, mercury, R.E.E., selenium, tantalum, tellurium, titanium, zirconium, etc.;
5   *fissionable metals*—uranium, thorium (radium).

For the formation of an orebody the element or elements concerned must be enriched to a considerably higher level than their normal crustal abundance. The degree of enrichment is termed the **concentration factor** and typical values are shown in Table 2.3.

2.4 Economics of coal, natural gas, oil and water extraction

Rather than overburdening the reader with financial discussions long before the chapters on these commodities are reached, these will be dealt with in the relevant chapters towards the end of this book.

2.5 Further reading

Evans A.M. (1993) *Ore Geology and Industrial Minerals: An Introduction.* Blackwell Scientific Publications, Oxford. Much of the above chapter is based on Chapter 1 of this book which will provide the reader with references and an amplified discussion of many points.

Kesler S.E. (1994) *Mineral Resources, Economics and the Environment.* Macmillan College Publishing, New York. Chapters 1, 4, 6 and 13 of this excellent book form a good complement to much of this chapter.

Miskelly N. (1994) A Comparison of International Definitions for Reporting Mineral Resources and Reserves. *Minerals Industry International*, July, 28–36. This paper contains an invaluable comparison of the definitions of these terms for Australasia, Canada, R.S.A., U.K. and U.S.A. but is only recommended to those readers who wish to delve deeply into this complex area.

Wolfe (1984)—see Section 1.6.

3: The Environmental Impact of Mineral Exploitation

3.1 Introduction

New mines bring prosperity to the areas in which they are established but they are bound to have an **environmental impact**. The relatively new base metal mine at Neves-Corvo in southern Portugal raised that country's copper output by 93 000% and its tin production by 9 900%! The total labour force is about 900. When it is remembered that one mine job creates about three indirect jobs in the local community in service and construction industries, the impact clearly is considerable. Impacts of this and even much smaller size have led to conflicts over land use and opposition to the exploitation of mineral deposits by environmentalists, particularly in the more populous of the developed countries. The resolution of such conflicts may involve the payment of compensation and the eventual cost of rehabilitating mined out areas, or the abandonment of projects; 'whilst political risk has been cited as a barrier to investment in some countries, environmental risk is as much of a barrier, if not greater in others' (Select Committee of the British Parliament in 1982). Opposition by environmentalists to exploration and mining was partially responsible for the abandonment of a major copper mining project in the U.K. in 1973.

In its report in 1987, *Our Common Future*, the United Nations World Commission on Environment and Development, headed by Mrs Brundtland, Norway's Prime Minister, pointed out that the world manufactures seven times more goods today than it did in 1950. The Commission proposed '**sustainable development**', a marriage of economy and ecology, as the only practical solution, i.e. growth without damage to the environment. In 1989 James Stevenson of Rio Tinto Corporation (the world's largest mining company) admitted that sustainable growth is an awkward concept for the extractive industry—'How does mining fit in? How can you regard a copper mine as a sustainable development?'—remembering that **all mines have a finite life**, some of twenty years or even less.

As David Munro, a leading environmentalist, has written, **sustainable development** and **sustainability** have become buzz words for everyone concerned with environment and development. These terms 'have been used to characterize almost any path to the kind of just, comfortable and

> This chapter discusses the following topics:
> ■ concepts of sustainability and sustainable development,
> ■ environmental effects of mining operations,
> ■ examples of past failures to safeguard the environment,
> ■ modern efforts to deal with these and to establish preventive measures for the future,
> ■ environmental impact statements.

secure future to which everyone aspires' and they have come to be 'misunderstood and misused with increasing frequency'. 'Do they still have any meaning beyond rhetoric?' Munro thinks they do: 'sustainability must be the main criterion for judging development.' But here we must return to Stevenson's predicament—orebodies **are finite and non-renewable** and if exploited cannot be replaced. The best the miner can do is to show **environmental responsibility** and return the mined area as nearly as possible to its pre-mining form or provide some other use for the desolation that might have resulted from his work, such as turning an opencast pit into a sailing lake and waste heaps into material for the construction industry. This approach can be backed up by using every method we can to cut down the demand for non-renewable resources by using them more efficiently, developing renewable substitutes, recycling and so on, but the most important immediate objective having probably the greatest impact would be **population stabilization** followed by a decrease in world population. However, these policies are the responsibility of governments, not mining companies! A final silent comment on the non-applicability of the concept of sustainable development to mineral resource exploitation comes from a recently published book on the subject which has no reference to mines, mining, minerals, coal or oil in its comprehensive index!

In the next section we will look at the major environmental impacts of mineral extraction and discuss how these can be mitigated or even eliminated.

3.2 Environmental effects of mining operations

Most **mines** have an on-site **mineral processing plant** and many metal mines have a nearby **smelter**. For a general assessment of the environmental impact of developing new mining operations we must consider the effects of all three. The term mining is here taken to include all extractive operations, e.g. **quarrying**. The major areas of concern are dealt with below.

3.2.1 Possible environmental consequences of mining and quarrying

(a) *Damage to land.* It has been estimated that the cumulative world use of land for mining between 1976 and 2000 will be about $37\,000\,km^2$; that is about 0.2% of the land surface. More developed countries have a greater proportion of disturbed ground than the less developed. The degree of reclamation of this ground is now accelerating rapidly and good use being made of many old holes for the disposal of old mine, domestic and other wastes. Other mined over areas have been turned into nature reserves and recreational parks. Future mines may be less likely to produce sites for waste disposal as most are now **backfilled** (the exception being those for bulk materials). This is a very necessary operation as each year an estimated 27 000 Mt of non-fuel minerals and overburden are taken from the earth's crust.

(b) *Release of toxic substances.* Metals are not only important for the uses we make of them, they are also an integral part of our make-up and that of other living organisms. However, while some metallic elements are essential components of living organisms, either **deficiencies** or **excesses** of these can be very **damaging to life**. Excesses in the natural environment can arise when it is penetrated by mine waters which may issue from the mine itself or from waste heaps. Some metals, e.g. cadmium, mercury and metalloids, e.g. antimony, arsenic, which are very common in small amounts in many polymetallic sulphide ores and are indeed often recovered as by-products, are highly toxic, even in small quantities, particularly in a soluble form which can be absorbed by living organisms. The same applies to **lead**, but mercifully it is fairly unreactive unless ingested and fortunately most lead minerals that form in nature are very insoluble in groundwater. **Cyanide** has long been used for gold extraction in mineral processing plants and in the world's largest goldfield, the Witwatersrand Basin, R.S.A., there is major contamination of surface waters with Co, Mn, Ni, Pb and Zn as a result of the cyanidation process and oxidation by acid mine waters. Cyanide itself is not a problem as it breaks down under the influence of ultraviolet light in near surface layers. Nevertheless, in developed countries legislation now requires the establishment of **cyanide neutralization plants** at all industrial undertakings using this chemical.

(c) *Acid mine drainage.* Acid waters generated by present or past mining result from the **oxidation**, in the presence of air, water and bacteria, of sulphide minerals, particularly pyrite. They may therefore develop in coalfields as well as orefields. **Sulphuric acid** and iron oxides are generated. The acid attacks other minerals, producing solutions which may carry toxic elements, e.g. cadmium, arsenic, into the local environment. **Acid water generation** may occur during the **exploration**, **operation** and **closure stages of a mine**. These waters may issue from three main sources: the mine's dewatering system, tailings disposal facilities and waste heaps (Fig. 3.1). These discharges may only produce minor effects such as local discoloration of soils and streams with precipitated iron oxides, or lead to extensive areal pollution of whole river systems and farmland. In some mining fields this problem is at its worst *after* **mine closures** have taken place. This is due to the **water table rebound** that occurs after pumping equipment is removed (Fig. 3.2) and this has become an urgent problem in

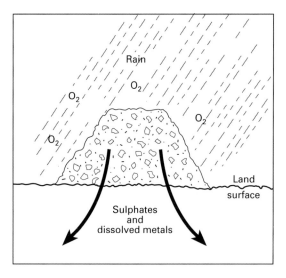

Fig. 3.1 Generation of polluting acid water by oxidizing rain water percolating through a waste heap containing sulphides.

Fig. 3.2 Generation of acid water during mine dewatering followed by closure. During mining, drainage through the adit from strata above adit level and water pumped into it from lower levels produces an acid flow into the neighbouring river. After mine closure, water table rebound produces an initial, large outflush of concentrated acid water with high metal content through the adit and/or surrounding surface strata into the surface drainage, followed by a continuing lower volume, but still highly polluting, outflow.

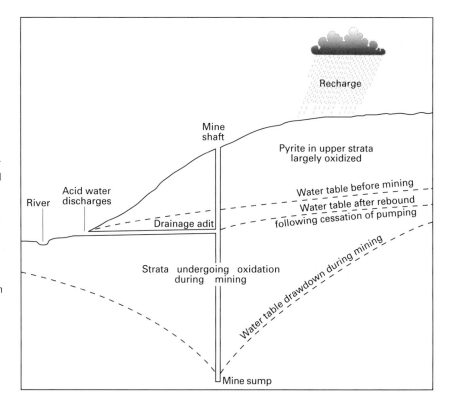

British coalfields, which were and are dominantly underground mines working **high sulphur coals**, as mine closures accelerated over the last decade. In 1950 the U.K. had about 1 800 working coal mines; now the figure is about 30! The frightening lesson from these coalfields is that

Keep pollutants at bay—wall them in . . .

At the world's biggest man-made hole, the Bingham Canyon Copper–Gold Mine in Utah (4 km wide and 0.8 km deep), to prevent acid water flowing into local aquifers Kennecott Utah Copper have built a chain of impermeable cut-off walls deep into bedrock to dam back the flow of subsurface water. Behind these walls are the repositories for the dumping of waste from clean up operations. Water from the repositories together with that flowing from waste dumps is diverted to processing plants to make it suitable for reuse in the mine and to recover dissolved copper.

or put up a fence!

AEA Technology, part of the British Atomic Energy Authority, and the Dutch environmental company Geokinetics, have developed an electrokinetic fence of alternate positive and negative electrodes drilled into the ground at 5 m intervals. These attract and concentrate groundwater pollutants around them, which are then pumped to the surface for the water to be cleansed and returned to the ground. The reclaimed heavy metals will help to pay for the running costs. The fence is now on trial and all being well will be on sale to industry before the end of 1996.

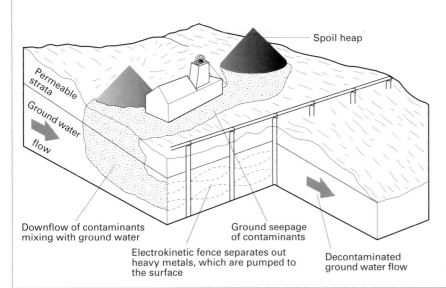

Fig. B.3.1 For explanation see text above.

Labels in figure:
- Spoil heap
- Permeable strata
- Ground water flow
- Downflow of contaminants mixing with ground water
- Ground seepage of contaminants
- Electrokinetic fence separates out heavy metals, which are pumped to the surface
- Decontaminated ground water flow

Brown and bloody rivers

The British National Rivers Authority has developed a compact, cheap process to stop rivers running orange through iron pollution from abandoned coal mines. The mine effluent is treated with sodium hydroxide to precipitate the iron oxides. A pilot plant on the Little Don River in Yorkshire has lowered the iron in the water from 30–70 to 5–15 mg l^{-1}.

Brick makers are interested in buying the precipitate for use as a pigment.

although the waters from working coal mines are but moderately saline, well buffered and typically contain only $2\,mg\,l^{-1}$ total dissolved iron and therefore depositing little or no ochre, after closure the water table rebound may produce **very acid, polluting waters** with a **high iron content**. Planning for closure must now take place before a mine goes into operation if the environment is to be safeguarded!

(d) *Health and safety of workers.* We are concerned here with the **exposure** of workers in mines, processing plants and smelters **to toxic materials** derived from the ore, e.g. cadmium, lead and mercury, as well as chemical reagents used particularly in the ore processing. **In uranium mines and plants radiation exposure must be minimal**. To this end these mines require **a high level of ventilation** to remove mineral dust and radon gas.

(e) *Dust.* Dust control is important in any mine where silica dust may be generated to prevent miners contracting **pneumoconiosis** and related lung ailments. Dust must also be kept to a minimum in all mines and quarries to protect local inhabitants.

(f) *Noise.* Noise is now the most widespread occupational hazard and both workers and those near industrial sites must be adequately protected from **harmful** and distracting **sound levels**.

(g) *Waste dumps.* Over the past two hundred years or more these have despoiled many of our mining fields and present practice is to mitigate this nuisance as far as possible; for example, by backfilling, using waste to create new landforms to screen the mining operation and cut down noise emission and processing waste for use in the construction industry. In many developed countries spoil heaps have and are being partially or wholly removed or landscaped.

(h) *Smelters.* One of the most polluting emissions from these, and from coal-fired electricity generating stations, has been SO_2 which, with NO_x and CO_2, gives rise to acid rain. Unpolluted rainfall has a pH ~5.7 but in the eastern U.S.A. and western Europe it may get as low as 2.9 and lakes in such areas have suffered severe declines in fish stocks. Older mining areas with a number of smelters may be surrounded by **barren land** where the vegetation was killed and the topsoil eroded. The dead zone around the copper–nickel smelters of Sudbury, Ontario extended over $100\,km^2$, but is now improving as emissions have been cut by 50%. Every effort is now made to cut down or eradicate these gaseous emissions by using state of the art smelters in which metal pouring and other processes are completely automated and enclosed to prevent the escape of contaminating gases and

Nature's revenge

Britian has now killed many complete coalfields but their corpses are bleeding red and brown iron oxides. One of a plethora of examples comes from the South Wales Coalfield. Here the Lower Ynysarwed Colliery worked the No. 2 Rhondda Seam up to 1938. The same seam was later worked by the Blaenant Colliery which lies in the Dulais Valley with a mountain between it and Ynysarwed. A 90 m safety barrier of unworked coal was left between the workings of the two mines. Blaenant was closed and pumping discontinued in 1991. Any subsequent polluting mine water discharge was expected to occur in the Dulais Valley near Blaenant. But Nature decided otherwise. A strong discharge (~1.3 Ml d^{-1}) appeared from an old adit at Lower Ynysarwed in April 1993. By the end of the year it had doubled and is now 3–3.5 Ml d^{-1}. The coal barrier had been breached by the build up of ground water and a highly acidic discharge (pH ~4.2) resulted. Total dissolved iron concentrations are 300–400 mg l^{-1}. The acid waters have flowed into the Neath Canal and wiped out virtually all aquatic life over a 12 km stretch. The canal bed is heavily coated with ochre. Emergency works to prevent pollution of the River Neath and to protect a major industrial abstraction from the Neath Canal had to be rapidly implemented. The polluted water has been diverted from the canal into the Neath estuary and clean water from the river pumped into the canal.

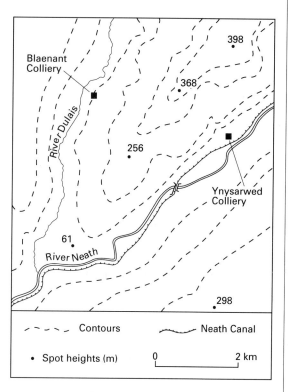

Fig. B.3.2a Map showing the locations of the places mentioned in the text. Regional location shown in Fig. B.3.2b (see next Box).

particulates carrying As, Cd, Pb etc. However, the situation is not straightforward. Some areas with high SO$_2$ emissions, like China, do not suffer widespread acid rainfall whereas some Pacific islands lacking SO$_2$ emissions experience acid rain. Nevertheless, the more developed countries now have stringent regulations concerning SO$_2$ and other emissions.

(i) *Legislation and cost.* Legal means of enforcing anti-pollution measures are very necessary, although it must be pointed out that many international mining companies are now adhering to the strictest self-regulation in countries where such legislation is minor or non-existent. **Legislation** has been brought in at an accelerating rate in developed countries with regions of high population such as Europe and the U.S.A. (see Fig. B.3.3).

Dramatic sequel to closing a metal mine

At Wheal Jane in Cornwall (wheal is the Cornish, Celtic word for mine so to write Wheal Jane Mine would be to commit a tautology) mining proceeded intermittently from the early 1700s until March 1991, when closure became inevitable because of the low tin price. The ore was complex and refractory: veins of very fine-grained tin in a *sulphur-rich* matrix. As tin mines go it was a fair-sized one producing Sn, Cu, Zn, W and Ag and it was one of the wettest mines per tonne of ore in the world. Water pressures as high as 1.38 MPa were encountered during drilling and 1 215 Ml water were pumped to the surface each month at a cost of over £1M p.a.—a complicating factor was that Wheal Jane is hydrologically linked to the neighbouring Mount Wellington Mine. The two mines had ore reserves exceeding 10 Mt and were worked together from 1979 to 1991 producing 360 000 t p.a. In June 1991 pumping stopped and by November the water table was close to the surface and minor leakage into the nearby Carnon River had been detected. This river discharges 3 km downstream into an estuary that is a popular sailing area *and* a shell fishery. The mine water is very acidic (pH 2–3) with high concentrations of As, Cd, Cu, Fe, Pb and Zn. The National Rivers Authority therefore decided to control the water level by pumping water from the shafts into tanks where lime would be used to raise the pH and

Fig. B.3.2b Location of mines.

precipitate toxic metals. All went well until 4 January 1992 when bad storms stopped the pumping. Within nine days water pressure increased sufficiently to burst a concrete plug sealing an adit and 320 Ml of polluted water poured into the estuary discolouring it and depositing poisonous metals; e.g. Cd levels in the river reached 600 mg l^{-1} relative to a U.K. water quality standard of 1 mg l^{-1}. There were also significant quantities of Cu, Fe and Zn. A great public outcry led to a resumption of pumping while a solution to the problem was sought. Under U.K. law, as the mine was by then abandoned, the former mine owners were not liable and the cost to date of £8M has been borne by the U.K. government.

This has come about through democratic government in the interests of everybody including minority groups. **Compromises or co-operation** with the extractive industries is **still rare**, preservationist opposition being very much the norm.

It must be realized that by piling regulation upon regulation and tax upon tax on one side of the scales, something has to go on the other pan to keep the economic balance, and this is price increase, i.e. **inflation** and, if a mining or quarrying operation is prevented from opening, **unemployment**. The materials required may have to be purchased from abroad, using foreign exchange, and **dug out of another country's environment**.

In European open pit mining and quarrying **environmental protection costs** may now amount to 50% of mining production costs. Lignite in Germany is a good example. Here the large environmental overheads

Proliferation of legislation

The sudden growth in the number of laws governing waste disposal in Germany is a good example of the wave of new laws that industrial planners and operators have to cope with. In some instances this vast expansion in legislation has rendered some operations bankrupt, their labour force workless.

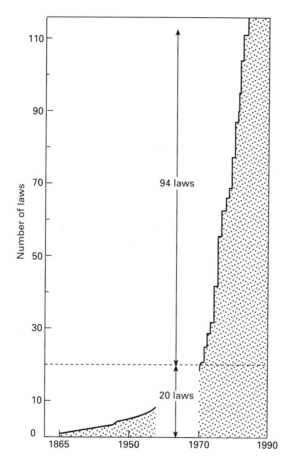

Fig. B.3.3 Expansion of German waste disposal laws.

usually include resettlement of local people, overburden disposal, water-related issues during operation and finally rehabilitation. In 1991 alone the German open pit mining and quarrying industry had to invest more than $U.S.100M in environmental protection — two thirds for clean air and the rest in equal parts for noise reduction, waste disposal and water protection.

(j) *Industrial minerals.* Industrial mineral operations have the same general environmental impacts on land and ground water disturbance as metalliferous or coal mining, although the impact is generally less marked since the mines are usually smaller and shallower, and normally less waste is produced because in most cases ore grades are higher than in metal mining. **Pollution hazards** owing to heavy metals or acid waters are **low** or non-existent and **atmospheric pollution**, caused by the burning of coal or the smelting of metallic ores, is **much less** serious or absent. The excavations created by industrial mineral operations are often close to conurbations, in which case these holes in the ground may be of great value as **landfill sites** for city waste. A British brick company recently sold such a site for £30M!

(k) *Environmental impact statements.* In many countries it is now mandatory that a company proposing to apply for planning permission to commence a mineral operation prepares such a statement. This will cover every aspect from effects on vegetation, climate, air quality, noise, ground and surface water to the proposed methods of **ground reclamation** at the termination of the operation. In some countries a bond must be deposited to assure that reclamation does take place. These statements must include records of the condition of the environment in the potential mining area when planning permission is applied for. Companies now collect such data during the exploration stage, including surface descriptions and photographs, geochemical analyses showing background levels of metals and acidity and details of the flora and fauna.

From the planning and regulatory authorities' point of view these reports present the most effective means of minimizing detrimental effects from the outset, but they may also be of great benefit to the developer because: (i) they will help to obtain planning permission in the shortest possible time; and (ii) they often reveal aspects of the operation requiring attention at the outset and thus avoid expensive modifications in the future.

(l) *Bugs and* in situ *mining.* Many sulphide deposits, e.g. porphyry coppers, are overlain by oxidized ores. Such ores can be mined, if necessary, by fracturing them by blasting and then **pumping acid solutions** through the rock **to dissolve metals** such as copper and uranium. The metal-bearing solutions are pumped to the surface and the metals recovered. Very low grade, small and otherwise economically non-viable deposits can be exploited and the process can be used at considerable depths. At Santa Cruz, Arizona the U.S. Bureau of Mines in cooperation with some mining companies is carrying out just such a project on an orebody estimated to contain 4.5 Mt of 1.5% Cu.

In 1947 it was recognized that bacteria in acid solutions played a key role in mineral oxidation and plans are afoot to drill wells into sulphide ore, fracture it and then inoculate it with bacteria to promote oxidation before extracting the metals in solution. Not only do these processes promise **much less ground disturbance and waste production** in the future but they are also **much less energy hungry**. The present drawback is that they are much slower than present technology in processing ores. Research is in progress to develop bacteria capable of removing sulphur from coal.

(m) *The outlook.* Measures such as those just described, together with recycling and substitution (see Section 2.2.3), and new materials technology will play their parts in reducing the impact of mineral exploitation on the environment, but in the immediate future we have to look to an **increasing sense of responsibility** by all those involved in any way with the industry, be they developers or regulators. There are many hopeful signs that this is taking place; for example, in 1992 nineteen major mining corporations from five continents joined together to form the International Council on Metals and the Environment whose remit is 'to promote the **development, implementation and harmonization of sound environmental and health policies and practices which will ensure the safe production, use, recycling and disposal of metals'**.

3.3 Further reading

Cambridge M. (1995) Use of Passive Systems for the Treatment and Remediation of Mine Outflows and Seepages. *Minerals Industry International*, May, 35–42.

Craig J.R., Vaughan D.J. and Skinner B.J. (1996) *Resources of the Earth: Origin, Use and Environmental Impact.* Prentice-Hall, Upper Saddle River. Chapter 4 is an excellent overview of the environmental effects that can result from resource exploitation.

Gaunt R. and Bliss N. (1993) Bauxite Mining Rehabilitation at Trombetas in the Amazon Basin. *Minerals Industry International*, March, 21–6.

Kesler S.E. (1994) *Mineral Resources, Economics and the Environment.* Macmillan College Publishing, New York. Chapter 3.

Miller G.T. Jr (1995) *Living in the Environment.* International Thomson Publishing, Belmont, Calif. A very readable and thought provoking volume which everyone who thinks he is a responsible citizen should read and chew over.

Papers by Ricks and others in *Minerals Industry International* for January 1995. These deal in particular with reclamation plans in environmental impact statements.

Pollock S.H.A. and Henry J.J. (1996) Mineral Extraction and United Kingdom Policies for Sustainable Development. *Minerals Industry International*, January, 13–6. A succinct discussion of the appositeness of the concepts of sustainability and sustainable development as applied to the U.K. mineral industry.

4: The Nature and Morphology of the Principal Types of Ore Deposit

4.1 Introduction

In this chapter you will find:
■ descriptions of the principal types of orebodies with particular reference to metalliferous ores,
■ details of their geological occurrence, shape and nature,
■ mention of the more important materials for which they are mined,
■ definitions of the important terms: syngenetic, epigenetic and supergene.

Some ore deposits are simply suitably composed and situated bodies of igneous, sedimentary or metamorphic rock and this applies in particular to certain industrial mineral deposits, e.g. limestone. Others, like some iron orebodies, are part of a stratigraphical succession which formed at the same time as the enclosing sediments and are therefore termed **syngenetic** deposits. Yet others, shaped like igneous dykes, cut across the enclosing rock, are clearly later in time and so are termed **epigenetic**.

It is possible to **classify orebodies** in the same way as we divide up igneous intrusions according to whether they are **discordant** or **concordant** with the lithological banding (often bedding) in the enclosing rocks. Considering discordant orebodies first, this large class can be subdivided into those orebodies which have an approximately regular shape and those which are thoroughly irregular in their outlines.

4.2 Discordant orebodies

4.2.1 Regularly shaped bodies

(a) *Tabular orebodies*. These bodies, called **veins**, are extensive in two dimensions, but have a restricted development in their third dimension. Veins are often inclined, and in such cases, as with faults, we can speak of the **hanging wall** and the **footwall** (Fig. 4.1). Veins frequently **pinch and swell** out as they are followed up or down a stratigraphical sequence. This pinch-and-swell structure can create difficulties during both exploration and mining often because only the swells are workable. **Veins** are usually developed **in fracture systems** and therefore show regularities in their orientation (Fig. 4.2).

(b) *Tubular orebodies*. These bodies are relatively short in two dimensions but extensive in the third. When vertical or subvertical they are called **pipes** or **chimneys**, when horizontal or subhorizontal, **mantos**. In eastern Australia, along a 2400 km belt from Queensland to New South Wales,

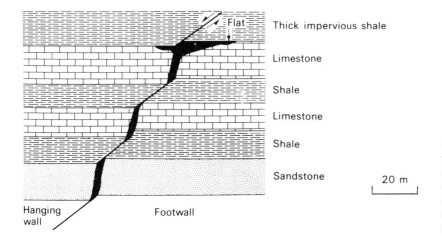

Fig. 4.1 Vein occupying a normal fault and exhibiting pinch-and-swell structure, giving rise to well mineralized sections in the dilatant zones. The development of a flat beneath impervious cover is also shown.

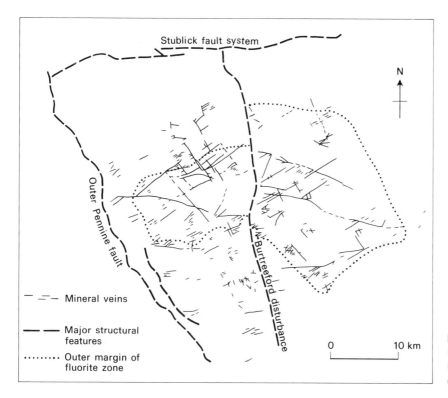

Fig. 4.2 Vein system of the Alston Block of the Northern Pennine Orefield, England. Note the three dominant vein directions.

there are hundreds of pipes in and close to granite intrusions. Most have quartz fillings and some are mineralized with bismuth, molybdenum, tungsten and tin; an example is shown in Fig. 4.3.

Mantos and pipes may **branch and anastomose**; an example of a branching manto is given in Fig. 4.4. Mantos and pipes are often found in association, the pipes frequently acting as feeders to the mantos. Some-

Fig. 4.3 Diagram of the Vulcan Pipe, Herberton, Queensland. The average grade was 4.5% tin.

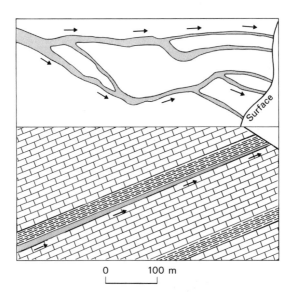

Fig. 4.4 Plan and section of part of the Hidden Treasure Manto, Ophir Mining District, Utah.

times mantos pass upwards from bed to bed by way of pipe connections, often branching as they go, an example being the Providencia Mine in Mexico where a single pipe at depth feeds into twenty mantos nearer the surface. In some tubular deposits formed by the subhorizontal flow of mineralizing fluid, ore grade mineralization may be discontinuous, thus creating pod-shaped orebodies.

4.2.2 Irregularly shaped bodies

(a) *Disseminated deposits.* In these deposits, ore minerals are peppered throughout the body of the host rock in the same way as accessory minerals are disseminated through an igneous rock; in fact, **they often *are* accessory minerals**. A good example is that of **diamonds in kimberlites**. In other deposits the disseminations may be wholly or mainly **along close-spaced veinlets** cutting the host rock and forming an interlacing network called a **stockwork** (see Fig. 15.1) or the economic minerals may be disseminated through the host rock and along veinlets (see Fig. 15.2). Whatever the mode of occurrence, mineralization of this type generally fades gradually outwards into subeconomic mineralization and the **boundaries** of the orebody **are assay limits**. They are, therefore, often irregular in form and may cut across geological boundaries. Stockworks occur as large bodies most commonly in acid to intermediate plutonic igneous intrusions, but they may cut across the contact into the country rocks, and a few are wholly or mainly in the country rocks. Disseminated deposits produce most of the world's copper and molybdenum and they are also of some importance in the production of tin, gold, silver, mercury and uranium.

(b) *Irregular replacement deposits.* Many ore deposits have been formed by the **replacement of pre-existing rocks** at low to medium temperatures (<400°C), e.g. magnesite deposits in carbonate-rich sediments and pyrophyllite orebodies in altered pyroclastics. Other replacement processes occurred at high temperatures, at contacts with medium-sized to large igneous intrusions forming skarn deposits. The orebodies are characterized by the development of calc-silicate minerals such as diopside, wollastonite, andradite garnet and actinolite. These deposits are **extremely**

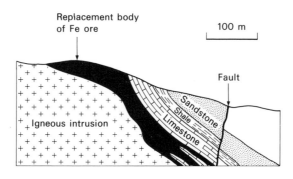

Fig. 4.5 Skarn deposit at Iron Springs, Utah.

irregular in shape (Fig. 4.5); tongues of ore may project along any available planar structure—bedding, joints, faults etc.—and the distribution within the contact aureole is often apparently capricious. Structural changes may cause abrupt termination of the orebodies. The principal materials produced from skarn deposits are: iron, copper, tungsten, graphite, zinc, lead, molybdenum, tin, uranium, garnet, talc and wollastonite.

4.3 Concordant orebodies

4.3.1 Sedimentary host rocks

Concordant orebodies in sediments are very important producers of many different metals, being **particularly important for base metals and iron**, and are of course concordant with the bedding. They may be an integral part of the stratigraphical sequence, as is the case with Phanerozoic iron-stones (syngenetic ores formed by sedimentary processes), or they may be epigenetic infillings of pore spaces or replacement orebodies. Usually these orebodies show a considerable development in two dimensions, i.e. *parallel* to the bedding and a limited development *perpendicular* to it (Figs 4.6 & 4.7) and for this reason such deposits are referred to as **stratiform**. This term must not be confused with **strata-bound**, which refers to any type or types of orebody, concordant or discordant, which are restricted to a particular part of the stratigraphical column. Thus the veins, pipes and flats of the Southern Pennine Orefield of England can be designated as strata-bound, as they are virtually restricted to the Carboniferous limestone of that region. A number of examples of concordant deposits that occur in different types of sedimentary rocks will be considered.

(a) *Limestone hosts.* Limestones are very common host rocks for base metal sulphide deposits. In a dominantly carbonate sequence, ore is often developed in a small number of preferred beds or at certain sedimentary interfaces. These are often zones in which the **permeability** has been **increased by dolomitization or fracturing**. When they form only a minor part of the stratigraphical succession, limestones, because of their solubility and reactivity, can become favourable horizons for mineralization. For example, the lead–zinc ores of Bingham, Utah, occurred in limestones, which make up 10% of a 2300 m succession mainly composed of quartzites. At Silvermines in Ireland, lead–zinc mineralization occurred (mining ceased in 1982) as syngenetic stratiform orebodies in a limestone sequence. The larger stratiform orebody occurred in massive, partly brecciated pyrite at the base of a thick sequence of dolomite breccias. The maximum thickness of the ore was 30 m, and, at the base, there was usually an abrupt change to massive pyrite from a footwall of nodular micrite, shale biomicrite and other limestones. The upper contact was always sharp. Pyrite and marcasite made up 75% of the ore, sphalerite formed 20% and galena, 4%.

(b) *Argillaceous hosts.* Shales, mudstones, argillites and slates are important host rocks for concordant orebodies, which are often remarkably continuous and extensive. In Germany, the Kupferschiefer of the Upper Permian is a prime example. This is **a copper-bearing shale** a metre or so thick which, at Mansfeld, occurred in orebodies that had plan dimensions of 8, 16, 36 and 130 km². Mineralization occurs at exactly the same horizon in Poland where it is being worked extensively, and across the North Sea in north-eastern England where it is subeconomic. The **world's largest, single lead–zinc orebody** occurs at Sullivan, British Columbia. The host rocks are late Precambrian argillites. Above the main orebody (Fig. 4.6) there are a number of other mineralized horizons with concordant mineralization. This deposit appears to be syngenetic, and the lead, zinc and other metal sulphides form an integral part of the rocks in which they occur. They are affected by sedimentary deformation, such as slumping, pull-apart structures, load casting etc., in a manner identical to that in which poorly consolidated sand and mud respond. The orebody occurs in a single generally conformable zone between 60 and 90 m thick and runs 6.6% lead and 5.9% zinc. Other metals recovered are silver, tin, cadmium, antimony, bismuth, copper and gold. Before mining commenced in 1900, the orebody contained at least 155 Mt of ore. The footwall rocks consist of

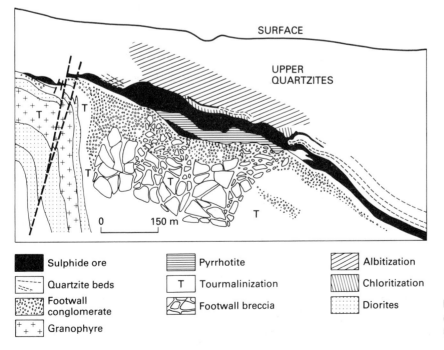

Sulphide ore

Quartzite beds

Footwall conglomerate

Granophyre

Pyrrhotite

T Tourmalinization

Footwall breccia

Albitization

Chloritization

Diorites

Fig. 4.6 Cross section through the ore zone, Sullivan Mine, British Columbia.

graded impure quartzites and argillites and, in places, conglomerate. The hanging wall rocks are more thickly bedded and arenaceous. **The ore zone is a mineralized argillite** in which the principal sulphide–oxide minerals are pyrrhotite, sphalerite, galena, pyrite and magnetite, with minor chalcopyrite, arsenopyrite and cassiterite. Beneath the central part of the orebody there are **extensive zones of brecciation** and tourmalinization, which extend downwards for at least 100 m. In places, the matrix of the breccias is heavily mineralized with pyrrhotite and occasionally with galena, sphalerite, chalcopyrite and arsenopyrite. This zone may have been a **channelway** up which solutions moved to debouch on to the sea floor, to precipitate the ore minerals among the accumulating sediment. Other good examples of concordant deposits in argillaceous rocks, or slightly metamorphosed equivalents, are the lead–zinc deposits of Mount Isa, Queensland, many of the Zambian Copperbelt deposits and the copper shales of the White Pine Mine, Michigan.

(c) *Arenaceous hosts.* Not all the Zambian Copperbelt deposits occur in shales and metashales. Some orebodies occur in altered feldspathic sandstones (Fig. 4.7). The Mufulira copper deposit occurs in Proterozoic rocks on the eastern side of an anticline and lies just above the unconformity with an older, strongly metamorphosed Precambrian basement. The gross ore reserves in 1974 stood at 282 Mt assaying 3.47% copper and the largest orebody stretched for 5.8 km along the strike and for several kilometres down dip. Chalcopyrite is the principal sulphide mineral, sometimes being accompanied by significant amounts of bornite. Fluviatile and aeolian arenites form the footwall rocks. The ore zone consists of feldspathic sandstones which, in places, contain carbon-rich lenses with much sericite. The basal portion is coarse-grained and characterized by festoon cross-bedding in which bornite is concentrated along the cross-bedding

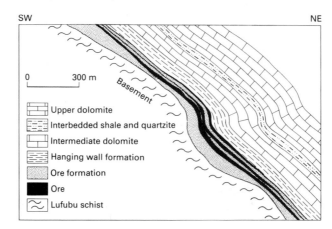

Fig. 4.7 Cross section through the Mufilira Orebodies, Zambia.

SW   NE

0   300 m

Basement

- Upper dolomite
- Interbedded shale and quartzite
- Intermediate dolomite
- Hanging wall formation
- Ore formation
- Ore
- Lufubu schist

together with well rounded, obviously detrital zircon; whilst in other parts of the orebody, concentrations of sulphides occur in the hollows of ripple-marks and in desiccation cracks. These features suggest that **some of the sulphides are detrital in origin**. Mineralization ends abruptly at the hanging wall, suggesting a regression and at this sharp cut-off the facies changes from an arenaceous one to dolomites and shallow-water muds.

Many mechanical accumulations of high density minerals, such as magnetite, ilmenite, rutile and zircon, occur in arenaceous hosts, usually taking the form of layers rich in **heavy minerals** in Pleistocene and Holocene sands. As the sands are usually unlithified, the deposits are easily worked and no costly crushing of the ore is required. These ore-bodies belong to the group called **placer deposits—beach sand placers** are a good example (Fig. 4.8). Beach placers supply much of the world's tita-nium, zirconium, thorium, cerium and yttrium. They occur along present day beaches or ancient beaches where longshore drift is well developed and frequent storms occur. **Economic grades can be very low** and sands running as little as 0.6% heavy minerals are worked along Australia's eastern coast. The deposits usually show a topographical control, the shapes of bays and the position of headlands often being very important; thus in exploring for buried orebodies **a reconstruction of the palaeogeo-graphy is invaluable**.

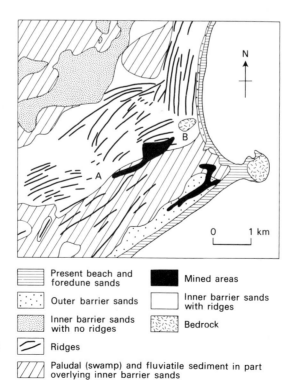

⊟ Present beach and foredune sands		■ Mined areas	
⋮ Outer barrier sands		Inner barrier sands with ridges	
Inner barrier sands with no ridges		Bedrock	
⟋ Ridges			
⫽ Paludal (swamp) and fluviatile sediment in part overlying inner barrier sands			

Fig. 4.8 Geology and mining areas of the beach sand deposits of Crowdy Head, New South Wales.

(d) *Rudaceous hosts.* Alluvial gravels and conglomerates also form important recent and ancient placer deposits. **Alluvial gold deposits** are often marked by '**white runs**' of vein quartz pebbles, as in the White Channels of the Yukon, the White Bars of California and the White Leads of Australia. Such deposits form one of the few types of economic placer deposits in fully lithified rocks, and indeed the majority of the world's gold has been won from Precambrian deposits of this type in South Africa. Figure 4.9 shows the distribution of the gold orebodies in the East Rand Basin where the **vein quartz pebble conglomerates** occur in quartzites of the Witwatersrand Supergroup. Their fan-shaped distribution strongly suggests that they occupy distributary channels. **Uranium** is recovered as a by-product of the working of the Witwatersrand goldfields.

(e) *Chemical sediments.* Sedimentary iron, manganese, evaporite and phosphorite formations occur scattered throughout the stratigraphical column, forming very extensive beds conformable with the stratigraphy. They are described in Chapters 19 and 21.

4.3.2 Igneous host rocks

(a) *Volcanic hosts.* There are two principal types of deposit to be found in

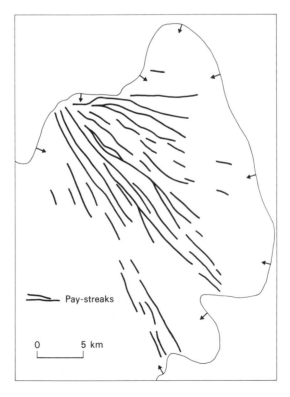

Pay-streaks

0   5 km

Fig. 4.9 Distribution of paystreaks (gold orebodies) in the Main Leader Reef of the East Rand Basin of the Witwatersrand Goldfield of South Africa. The arrows indicate the direction of dip at the outcrop or suboutcrop. For the location of this ore district see Fig. 19.9.

orthomagmatic deposit is the nickel–copper sulphide orebody formed by the sinking of an immiscible sulphide liquid to the bottom of a magma chamber containing ultrabasic or basic magma. These are known to the learned as **liquation deposits** and they may be formed at the base of lava flows as well as in plutonic intrusions. The **sulphide usually accumulates in hollows** in the base of the igneous body and generally forms sheets or irregular lenses conformable with the overlying silicate rock. From the base upwards, massive sulphide gives way through disseminated sulphides in a silicate gangue to lightly mineralized and then barren rock (see Figs 13.2 & 13.6).

4.3.3 Metamorphic host rocks

Apart from some deposits of metamorphic origin, such as the irregular replacement deposits already described and deposits generated in contact metamorphic aureoles, e.g. of wollastonite, andalusite, garnet and graphite, metamorphic rocks are mainly important for the metamorphosed equivalents of deposits that originated in sedimentary and igneous rocks and which have been discussed above.

4.3.4 Residual deposits

Protore—mineral material in which an initial but uneconomic concentration of metals has occurred that may by further natural processes be upgraded to the level of ore.

These are deposits formed by the **removal of non-ore material from protore**. For example, the leaching of silica and alkalis from a nepheline-syenite may leave behind a surface capping of hydrous aluminium oxides (**bauxite**). Some residual bauxites occur at the present surface, others have been buried under younger sediments to which they form conformable basal beds. The weathering of feldspathic rocks (granites, arkoses etc.) can produce important **kaolin deposits** which, in the Cornish granites of England, form funnel or trough-shaped bodies extending downwards from the surface for as much as 230 m. Other examples of residual deposits include some **laterites** sufficiently high in iron to be worked and **nickeliferous laterites** formed by the weathering of peridotites (see Figs 20.1 & 20.2).

4.3.5 Supergene enrichment

This is a process which may affect most orebodies to some degree. After a deposit has been formed, uplift and erosion may bring it within reach of circulating ground waters, which may leach some of the metals out of that section of the orebody above the water table. These dissolved metals may be redeposited in that part of the orebody lying beneath the water table volcanic rocks, **vesicular filling deposits** and **volcanic-associated massive sulphide deposits**. The first deposit type is not very important but the second type is a widespread and important producer of base metals, often with silver and gold as by-products. Volcanic-associated massive sulphide deposits often consist of over 90% iron sulphide usually as pyrite, although pyrrhotite is well developed in some deposits. They are **generally stratiform bodies**, lenticular to sheet-like (Fig. 4.10), developed at the interfaces between volcanic units or at volcanic–sedimentary interfaces. With increasing magnetite content, **these ores grade to massive oxide ores** of magnetite and/or hematite, such as at Savage River in Tasmania, Fosdalen in Norway and Kiruna in Sweden. Massive sulphide deposits **commonly occur in groups** and in any one area they are found **at one or a restricted number of horizons** within the succession. These horizons may represent changes in composition of the volcanic rocks, a change from volcanism to sedimentation, or simply a pause in volcanism. There is a close association with volcaniclastic rocks and many orebodies overlie the explosive products of **rhyolite domes**. These ore deposits are usually underlain by a **stockwork** that may itself be ore grade and which appears to have been the feeder channel up which mineralizing fluids penetrated to form the overlying massive sulphide deposit.

(b) *Plutonic hosts.* Many plutonic igneous intrusions possess **rhythmic layering** and this is particularly well developed in some basic intrusions. Usually the layering takes the form of alternating bands of mafic and felsic minerals, but sometimes minerals of economic interest, such as **chromite**, **magnetite** and **ilmenite**, may form discrete mineable seams within such layered complexes (see Fig. 12.2). These seams are naturally stratiform and may extend over many kilometres, as is the case with the chromite seams in the Bushveld Complex of South Africa (see Fig. 12.1). Another form of

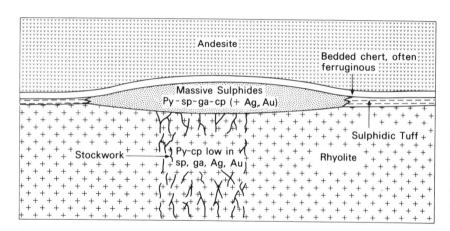

Fig. 4.10 Schematic cross section through an idealized volcanic-associated massive sulphide deposit showing the underlying feeder stockwork and typical mineralogy. Py, pyrite; sp, sphalerite; ga, galena; cp, chalcopyrite.

and this can lead to a considerable enrichment in metal values. Supergene processes are discussed in Section 20.3.

4.4 Further reading

Few authors have written on orebody morphology as a separate subject. A somewhat fuller discussion of the subject is to be found in Chapter 2 of Evans (1993) — see Section 2.5.

5: Textures and Structures of Ore and Gangue Minerals. Fluid Inclusions. Wall Rock Alteration

5.1 Introduction

This chapter describes:
■ the ways in which ore minerals may crystallize and grow,
■ the fluid inclusions they may contain, the knowledge these give us and the use we can make of them,
■ the principal types of wall rock alteration and their importance in mineral exploration and ore genetic theory.

The study of textures can tell us much about the genesis and subsequent history of orebodies. The textures of orebodies vary according to whether their constituent minerals were formed by **deposition in an open space** from a silicate or aqueous solution, or by **replacement of pre-existing rock** or ore minerals. Subsequent metamorphism may drastically alter primary textures. The manner in which ore minerals are intergrown is of great importance to the explorationist and the minerals processor as **complex intergrowths** may entail the installation of **expensive separating plant** in a mine's mill. The study of fluid inclusions and wall rock alteration is also of great importance in elucidating the origin of mineralization and in prospecting for it.

5.2 Open space filling

5.2.1 Precipitation from silicate melts

Critical factors in this situation are the time of crystallization and the presence or absence of simultaneously crystallizing silicates. **Oxide ore minerals**, such as chromite, **often crystallize out early** and thus may form good euhedral crystals, although these may be subsequently modified in various ways. Chromites deposited with interstitial silicate liquid **may suffer corrosion** and partial **resorption** to produce atoll textures (Fig. 5.1) and **rounded grains**, whereas those developed in monomineralic bands (see Fig. 12.2) may, during the cooling of a large parent intrusion, suffer **auto-annealing** and develop foam texture (the texture of polyhedral grain aggregates totally filling the space in which they occur). When oxide and silicate minerals crystallize simultaneously, **anhedral to subhedral granular textures** similar to those of granitic rocks develop, owing to mutual interference during the growth of the grains of all the minerals.

Sulphides, because of their lower melting points, crystallize after associated silicates and, if they have not segregated from the silicates, they will be present either as **rounded grain aggregates** representing frozen glob-

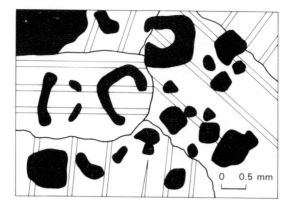

Fig. 5.1 Chromite grains in anorthosite, Bushveld Complex, R.S.A. The chromites are euhedral crystals which have undergone partial resorption, producing rounded grains of various shapes, including atoll texture.

ules of immiscible sulphide liquid, or as **anhedral grains** or **grain aggregates** which have crystallized interstitially to the silicates and whose **shapes are governed by those of the enclosing silicate grains.**

5.2.2 Precipitation from aqueous solutions

Open spaces, such as dilatant zones along faults, solution channels in areas of karst topography etc., may be permeated by mineralizing solutions. If the prevailing physicochemical conditions induce precipitation, then crystals will form. These will grow as the result of **spontaneous nucleation** within the solution, or, more commonly, **by nucleation on the enclosing surface**. This leads to the precipitation and **outward growth** of the first formed minerals **on vein walls**. If the solutions change in composition, there may be a change in mineralogy and crusts of minerals of different composition may give the vein filling a banded appearance (Fig. 5.2) called **crustiform banding**. Its development in some veins demonstrates that mineralizing solutions may change in composition with time and shows us the order in which the minerals were precipitated, this order being called the **paragenetic sequence**.

Open space deposition also occurs at the surface at sediment–water or rock–water interfaces during, for example, the formation of volcanic-associated massive sulphide deposits. Under such situations **rapid flocculation** of material occurs and a common primary texture that results is **colloform banding**. This is a very fine scale banding involving one or more sulphides, very like the banding in agate.

The textures of sedimentary iron and manganese ores are discussed in Chapter 19. If the hydrothermal fluid flows through the interstitial pores in a sedimentary rock and precipitates new minerals, we have a process similar to cementation which may be accompanied by granular replacement of rock minerals by ore.

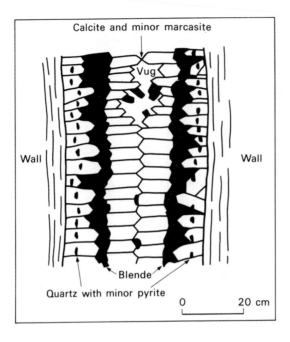

Calcite and minor marcasite

Vug

Wall

Wall

Blende

Quartz with minor pyrite

0   20 cm

Fig. 5.2 Section across a vein showing crustiform banding.

5.3 Replacement

The precipitation of new minerals at the expense of pre-existing ones is termed replacement. Replacement has been an important process in the formation of many of ore deposits, particularly the skarn class. This process involves not only the minerals of the country rocks, but also the ore and gangue minerals. Nearly all ores, including those developed in open spaces, show some evidence of the occurrence of replacement processes.

5.4 Fluid inclusions

The growth of crystals is never perfect and as a result samples of the fluid in which the crystals grew may be trapped in tiny cavities usually <100 μm in size; these are called fluid inclusions. Their study has proved to be useful in deciphering the formational history of many rock-types and particularly valuable in developing our understanding of ore genesis. Fluid inclusions are divided into various types.

(a) *Primary inclusions,* formed during the growth of crystals, provide us with samples of the ore-forming fluid. They also yield crucial **geothermometric data** and tell us something about the physical state of the fluid, e.g. whether it was boiling at the time of entrapment. They are common in most rocks and mineral deposits. The vast bulk of fluid inclusion work has of course been carried out on **transparent minerals**, such as quartz, fluorite and sphalerite. The principal matter in most fluid inclusions is **water**. Second in abundance is **carbon dioxide**. The commonest inclusions in ore deposits fall into four groups. Type I, moderate salinity inclusions, are generally two phase, consisting principally of water and a small bubble of water vapour, which forms 10–40% of the inclusions (Fig. 5.3). The presence of the bubble indicates trapping at an elevated temperature with formation of the bubble on cooling. Heating on a microscope stage causes **rehomogenization** to one liquid phase and the homogenization temperature indicates the **temperature of growth** of that part of the containing crystal, provided the necessary pressure correction can be made. If a pressure correction cannot be made, then the rehomogenization temperature can be assumed to be a minimum only, and not the actual temperature of trapping. Sodium, potassium, calcium and chlorine occur in solution and salinities range from 0 to 23 wt % NaCl equivalent. In some of these inclusions small amounts of **daughter salts** have been precipitated during cooling, among them carbonates and anhydrite.

Type II, gas-rich inclusions, generally contain more than 60% water vapour. Again, they are dominantly aqueous but CO_2 may be present in small amounts. They often appear to represent **trapped steam**. The simultaneous presence of gas-rich and gas-poor aqueous inclusions is good evidence that the fluids were **boiling** at the time of trapping. Type III, halite-bearing inclusions, have salinities ranging up to more than 50%. They contain well formed, cubic halite crystals and generally several other daughter minerals, particularly sylvite and anhydrite. Clearly the more numerous and varied the daughter minerals, the more complex was the ore fluid. Type IV, CO_2-rich inclusions, have CO_2/H_2O ratios ranging from 3 to over 30 mol %. They grade into type II inclusions and indeed there is a

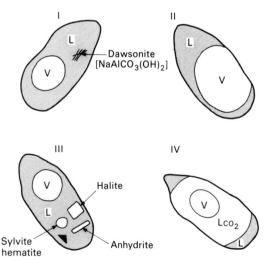

Fig. 5.3 Sketches of four important types of fluid inclusions. L, aqueous fluid; V, vapour; L_{CO_2}, liquid CO_2. For explanation see text.

general gradation in many situations, e.g. porphyry copper deposits, between the common types of fluid inclusion.

Perhaps one of the most surprising results of fluid inclusion studies is the evidence of the common occurrence of exceedingly **strong brines** in nature, brines more concentrated than any now found at the surface. They are compelling evidence that many **ore-forming fluids** are hot, saline aqueous solutions. They form a link between laboratory and field studies, and it should be noted that there is strong experimental and thermodynamic evidence which shows that chloride in hydrothermal solutions is a potent solvent for metals through the formation of **metal–chloride complex ions**, and indeed inclusions that carry more than 1 wt % of precipitated sulphides are known.

(b) *Secondary inclusions* are those formed by any process after the crystallization of the host mineral.

5.5 Wall rock alteration

Frequently alongside veins or around irregularly shaped orebodies of hydrothermal origin we find alteration of the country rocks. This may take the form of **colour, textural, mineralogical** or **chemical changes, or any combination of these**. Alteration is not always present, but when it is, it may vary from minor colour changes to extensive mineralogical transformations and complete recrystallization. Generally speaking, the higher the temperature of deposition of the ore minerals, the more intense is the alteration, but it is not necessarily more widespread. This alteration, which shows a spatial and usually a close temporal relationship to ore deposits, is called **wall rock alteration** or **hydrothermal alteration**.

The areal extent of the alteration can vary considerably, sometimes being limited to a few centimetres on either side of a vein, at other times forming a thick **halo around an orebody** and then, since it **widens the drilling target**, it may be of considerable exploration value. The associated orebodies in certain deposits (e.g. porphyry coppers, Chapter 15) may show a special spatial relationship to the zoning of their hydrothermal alteration, knowledge of which may be invaluable in probing for the deposit with a diamond drill.

There are two main divisions of wall rock alteration: hypogene and supergene. **Hypogene alteration** is caused by ascending hydrothermal solutions, and **supergene alteration** by descending meteoric water reacting with previously mineralized ground. In this chapter we will be concerned mainly with hypogene alteration. **Studies** of this alteration are **important because they**: (i) contribute to our knowledge of the nature and evolution of ore-forming solutions; (ii) are often valuable in exploration;

Exploration triumph
Back in the 1960s half the large San Manuel porphyry copper deposit appeared to have been faulted away. Drilling on the other side of the fault failed to find it until re-examination of the drill cores revealed that the outer alteration zones of a porphyry copper body had been penetrated. Deepening of the holes found the other half.

Drilling targets
In the Athabasca Basin uranium field, Saskatchewan, drilling targets are enlarged by a factor of 10 to 20 times by the wall rock alteration.

and (iii) produce minerals such as phyllosilicates that can be used to obtain radiometric dates on the wall rock alteration and, by inference, on the associated mineralization.

5.5.1 Types of wall rock alteration

There are many types of wall rock alteration. Only the principal types will be noted here.

(a) *Advanced argillic alteration.* This alteration is characterized by dickite, kaolinite [both $Al_2Si_2O_5(OH)_4$], **pyrophyllite** [$Al_2Si_4O_{10}(OH)_2$] and quartz. Sericite is usually present. This is one of the more intense forms of alteration, often present as an inner zone adjoining many base metal vein or pipe deposits and in telescoped shallow precious metal deposits.

(b) *Sericitization.* In orefields the world over this is **one of the commonest types** of alteration in aluminium-rich wall rocks such as slates, granites etc. The dominant minerals are **sericite** (muscovite) and **quartz**; pyrite often accompanies them. The reader must not assume that during this and other wall rock alteration processes the wall rocks necessarily become solid sericite or clay minerals, as the case may be. What we see is the appearance in significant amounts, or an increase in quantity, of the mineral or minerals concerned.

(c) *Intermediate argillic alteration.* Formerly known as argillic alteration and given its present name to avoid any confusion with advanced argillic alteration. The principal minerals are now kaolin- and **montmorillonite-group** minerals occurring mainly as alteration products of plagioclase.

(d) *Propylitic alteration.* This is a complex alteration generally characterized by chlorite, epidote, albite and carbonate (often with calcite, dolomite or ankerite). With the intense development of one of the main propylitic minerals we have what are sometimes considered as subdivisions of propylitization: *chloritization, albitization* and *carbonatization.*

(e) *Chloritization.* Chlorite may be present alone or with quartz or tourmaline in very simple assemblages; however, other propylitic minerals are usually present, and anhydrite also may be in evidence.

(f) *Carbonatization.* Dolomitization is a common accompaniment of low to medium temperature ore deposition in limestones, and **dolomite** is probably the commonest of the carbonates formed by hydrothermal activity. Dolomitization is associated most commonly with low temperature, lead–zinc deposits of 'Mississippi Valley-type'. These can have wall rocks of pure dolomite; usually this rock is coarser and lighter in colour than the surrounding limestone. Other carbonates may be developed in silicate rocks, especially where iron is available and ankerite may then be common, particularly in the calcium–iron environment of carbonatized basic igneous rocks and volcaniclastics. This is particularly the case with many Precambrian and Phanerozoic vein gold deposits.

(g) *Potassic alteration*. **Secondary potash feldspar** and/or **biotite** are the essential minerals of this alteration. Clay minerals are absent but minor chlorite may be present. Anhydrite is often important especially in porphyry copper deposits but near the surface has been dissolved away by weathering.

(h) *Silicification*. This involves an increase in the proportion of quartz or cryptocrystalline silica (i.e. cherty or opaline silica) in the altered rock. The silicification of carbonate rocks leading to the development of skarn is dealt with in Chapter 14.

5.6 Further reading

For textures and structures of ore and gangue minerals and a summary coverage of fluid inclusions, see Craig J.R. and Vaughan D.J. (1994) *Ore Microscopy and Ore Petrography*. Wiley, New York.

For in-depth discussions of fluid inclusions and their use in mineral exploitation see Roedder E. (1984) *Fluid Inclusions. Reviews in Mineralogy*, Vol. 12, Mineralogy Society of America, Resten, Virginia or Shepherd T.J., Rankin A.H. and Alderton D.H.M. (1985) *A Practical Guide to Fluid Inclusion Studies*. Blackie, Glasgow.

Wall rock alteration is discussed at greater length and additional references can be found in Evans A.M. (1993) *Ore Geology and Industrial Minerals: An Introduction*. Blackwell Scientific Publications, Oxford.

6: Some Major Theories of Ore Genesis

6.1 Introduction

Some indication has been given in Chapter 4 of the very varied nature of ore deposits and their occurrence. This variety of form has given rise over the last hundred years or more to an equally great variety of hypotheses of ore genesis. This chapter will be concerned only with current major theories of ore genesis and these will be divided for the sake of convenience into internal and surface processes. The reader should be warned, however, that very often several processes contribute to the formation of an orebody. Thus, where we have rising hot aqueous solutions forming an epigenetic stockwork deposit just below the surface and passing on upwards through it to form a contiguous syngenetic deposit under, say, marine conditions, even the above simple classification is in difficulties. This is the reason why ore geologists, besides producing a plethora of ore genetic theories, have also created a plethora of orebody classifications! A summary of the principal theories of ore genesis is given in Table 6.1.

In this chapter you will learn about:
- the major theories of ore genesis; these are listed in Table 6.1 and will not be repeated here;
- hydraulic fracturing—a pathway generator for ascending hydrothermal solutions.

6.2 Origin due to internal processes

6.2.1 Magmatic crystallization

This covers the ordinary processes of crystallization that provide us with volcanic and plutonic rocks. Some of these, such as granites and basalts, we may exploit as **bulk materials**; others may be important for their possession of one or more economically important minerals, e.g. **diamonds** in kimberlites, **feldspar** in pegmatites.

6.2.2 Magmatic segregation

The terms magmatic segregation deposit or **orthomagmatic deposit** are used for those ore deposits that have crystallized direct from a magma. Those formed by **fractional crystallization** are usually found in plutonic igneous rocks; those produced by **liquation** (separation into immiscible liquids) may be found associated with both plutonic and volcanic rocks.

Table 6.1 Simple classification of the theories of mineral deposit genesis.

Theory	Nature of process	Typical deposits
Origin due to internal processes		
Magmatic crystallization	Precipitation of ore minerals as major or minor constituents of igneous rocks in the form of disseminated grains or segregations	Diamonds disseminated in kimberlites, R.E.E. minerals in carbonatites. Lithium-tin-caesium pegmatites of Bikita, Zimbabwe. Uranium pegmatites of Bancroft, Canada and Rössing, Namibia. Bulk material deposits of granite, basalt, dunite, nepheline-syenite
Magmatic segregation	Separation of ore minerals by fractional crystallization and related processes during magmatic differentiation	Chromite layers in the Great Dyke of Zimbabwe and the Bushveld Complex, R.S.A.
	Liquation, liquid immiscibility. Settling out from magmas of sulphide, sulphide–oxide or oxide melts that accumulated beneath the silicates or were injected into wall rocks or, in rare cases, erupted on the surface	Copper–nickel orebodies of Sudbury, Canada; Pechenga, R.F. and the Yilgarn Block, Western Australia. Allard Lake titanium deposits, Quebec, Canada
Hydrothermal	Deposition from hot aqueous solutions, which may have had a magmatic, metamorphic, surface or other source	Tin–tungsten–copper veins and stockworks of Cornwall, U.K. Molybdenum stockworks of Climax, U.S.A. Porphyry copper deposits of Panguna, P.N.G. and Bingham, U.S.A. Fluorspar veins of Derbyshire, U.K.
Lateral secretion	Diffusion of ore- and gangue-forming materials from the country rocks into faults and other structures	Yellowknife gold deposits, Canada. Mother Lode gold deposit, U.S.A.
Metamorphic	Contact and regional metamorphism producing industrial mineral deposits	Andalusite deposits, Transvaal, R.S.A. Garnet deposits, N.Y., U.S.A.
	Pyrometasomatic (skarn) deposits formed by replacement of wall rocks adjacent to an intrusion	Copper deposits of Mackay, U.S.A. and Craigmont, Canada. Magnetite bodies of Iron Springs, U.S.A. Talc deposits, Luzenac, France
	Initial or further concentration of ore elements by metamorphic processes, e.g. granitization, alteration processes	Some gold veins, and disseminated nickel deposits in ultramafic bodies
Origin due to surface processes		
Mechanical accumulation	Concentration of heavy, durable minerals into placer deposits	Rutile–zircon sands of New South Wales, Australia and Trail Ridge, U.S.A. Tin placers of Malaysia. Gold placers of the Yukon, Canada. Industrial sands and gravels. Kaolin deposits, Georgia, U.S.A. Bauxites of Guyana
Sedimentary precipitates	Precipitation of particular elements in suitable sedimentary environments, with or without the intervention of biological organisms	Banded iron formations of the Precambrian shields. Manganese deposits of Chiaturi, R.F. Zechstein evaporite deposits of Europe. Floridan phosphate deposits, U.S.A.
Residual processes	Leaching from rocks of soluble elements leaving concentrations of insoluble elements in the remaining material	Nickel laterites of New Caledonia. Bauxites of Hungary, France, Jamaica and Arkansas, U.S.A. Kaolin deposits, Nigeria
Secondary or supergene	Leaching of valuable elements from the upper parts of mineral deposits and their precipitation at depth to produce higher concentrations	Many silver and gold bonanzas. The upper parts of a number of porphyry copper deposits
Volcanic exhalative (= sedimentary exhalative)	Exhalations of hydrothermal solutions at the surface, usually under marine conditions and generally producing stratiform orebodies	Base metal deposits of Meggen, Germany; Sullivan, Canada; Mt Isa, Australia; Rio Tinto, Spain; Kuroko deposits of Japan; black smoker deposits of modern oceans. Mercury of Almaden, Spain. Solfatara deposits (kaolin + alunite), Sicily

Magmatic segregation deposits may consist of layers within or beneath the rock mass (**chromite layers, subjacent copper–nickel sulphide ores**).

(a) *Fractional crystallization*. This includes any process by which early formed crystals are prevented from equilibrating with the melt from which they grew. During a period of monomineralic crystallization, these crystals may sink to the bottom of the magma chamber to form a single mineral layer. These precipitates are termed **cumulates** and they usually alternate with other minerals to form **rhythmic layering**. Both **chromite** and **ilmenite** may be concentrated in this way but whilst chromite accumulations are nearly all in **ultrabasic rocks**, ilmenite concentrations show an association with **anorthosites** or anorthositic gabbros. These striking rock associations are strong evidence for the magmatic origin of the minerals.

(b) *Liquation*. A different form of segregation results from liquid immiscibility. In exactly the same way that oil and water will not mix but form immiscible globules of one within the other, so in a mixed sulphide–silicate magma the two liquids will tend to segregate. **Sulphide droplets separate out** and coalesce to form globules which, being denser than the magma, **sink** through it to accumulate at the base of the intrusion or lava flow. Iron sulphide is the principal constituent of these droplets, which are associated with basic and ultrabasic rocks because sulphur and iron are both more abundant in these rocks than in acid or intermediate rocks. Chalcophile elements, such as **copper** and **nickel**, also **enter** ('partition into' is the pundits' phrase) **these droplets** and sometimes the **platinum group metals**. A basic or ultrabasic magma is generated by partial melting in the mantle and it may acquire its sulphur at this time, or later by assimilation in the crust. For significant sulphide segregation to occur, the magma must be **sulphide saturated**. If immiscible sulphides form and the proportion of sulphide formed is large, much of the available Ni and Cu in the magma are removed. The accumulation of Fe–Ni–Cu sulphide droplets beneath the silicate fraction can produce **massive sulphide orebodies**. These are overlain by a zone with subordinate silicates enclosed in a network of sulphides — **net-textured ore**, sometimes called disseminated ore. This zone is, in turn, overlain by one of weak mineralization which grades up into overlying peridotite, gabbro or komatiite, depending on the nature of the associated silicate fraction (see Fig. 13.2).

6.2.3 Hydrothermal processes

Hot aqueous solutions have played a part in the formation of many different types of mineral and ore deposit; for example, veins, stockworks of various types, volcanic-exhalative deposits and others. Such fluids are usually called **hydrothermal solutions** and many lines of evidence attest to their important role as mineralizers. The evidence from wall rock alteration and fluid inclusions has been discussed in Chapter 5. Homogenization of fluid inclusions in minerals from hydrothermal deposits and other geothermometers have shown that the **depositional range** for all types of deposit is approximately **50–650°C**. Analysis of the fluid has shown **water** to be **the common phase** and usually it has **salinities far higher than that of sea water**. Hydrothermal solutions are believed to be capable of carrying a wide variety of materials and of depositing these to form minerals as diverse as gold and muscovite, showing that the **physical chemistry** of such solutions is **complex** and very difficult to imitate in the laboratory. Our knowledge of their properties and behaviour is still somewhat hazy and there are many ideas about the origin of such solutions and the materials they carry. The **principal problems** are the **source** and **nature of the solutions**, the **sources of the metals and sulphur** in them and the **driving force** that moved the solutions through the crust, the **means of transport** of these substances by the solutions, and the **mechanisms of deposition**.

(a) *Sources of the solutions and their contents.* As explained in the section on fluid inclusions in Chapter 5, there is much evidence that saline hydrothermal solutions are, and have been, very active and widespread in the crust. In some present day geothermal systems, the circulation of hydrothermal solutions is under intensive study. **Whence the water of these solutions?** Data from water in mines, tunnels, drill holes, hot springs, fluid inclusions, minerals and rocks suggest that there are **five sources** of subsurface hydrothermal waters:

1 surface water, including groundwater, commonly referred to by geologists as meteoric water;

2 ocean (sea) water;

3 formation and deeply penetrating meteoric water;

4 metamorphic water;

5 magmatic water.

Measurements of the relative abundances of **oxygen and hydrogen isotopes** give us **information on the source of water** (Fig. 6.1), but there are problems of interpretation of the data that we obtain. Both formation and metamorphic water (produced by dehydration of minerals during metamorphism) may once have been meteoric water, but subsurface **rock–water reactions may change the isotopic compositions** and, if these reactions are incomplete, then a range of isotopic compositions will result. Another mechanism that may produce intermediate isotopic compositions is the **mixing of waters**, e.g. magmatic and meteoric.

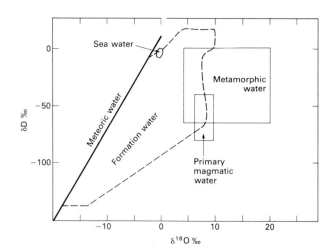

Fig. 6.1 Fields of isotopic composition of sea water, formation water, metamorphic water and magmatic water.

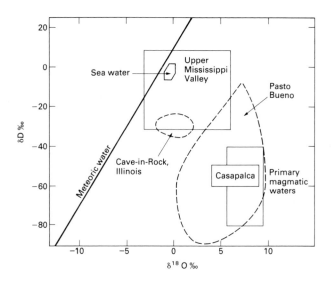

Fig. 6.2 Hydrogen *versus* oxygen isotopic plot showing observed values of hydrothermal fluids involved in the main stages of mineralization in four mineralized areas. For further discussion see text.

Examples of some hydrogen–oxygen isotopic studies are shown in Fig. 6.2 and in examining these please refer to Fig. 6.1 From the values obtained, it may be inferred that the main-stage mineralizing fluids at the Ag–Pb–Zn–Cu Casapalca Mine, Peru were of magmatic origin, whereas for the Pasto Buena tungsten-base metal deposit of northern Peru, the data suggest mixing of some meteoric (or metamorphic or formation) water with magmatic water during mineralization. By contrast, fluids in the Cave-in-Rock Fluorspar District, Illinois (the mines also produce zinc and lead) appear to have been dominantly meteoric-recharged formation water, and in the Upper Mississippi Valley Pb–Zn Field, the main-stage sulphide mineralization period has been ascribed to the action of somewhat modified meteoric water. Present evidence thus appears to show that **similar deposits can be formed from detectably different types of water** and, additionally, that **waters of at least two parentages have played an important role in the formation of some orebodies**.

Let us now turn our thoughts to the source of the dissolved contents. Firstly, what are they? Secondly, where do we look for this information? The answer to the second question is fluid inclusions (which provide most of our data), modern geothermal systems, hot springs and waters encountered during deep drilling operations in oilfields — **oilfield brines**. Data from these sources indicate that the **major constituents are sodium, potassium, calcium** and **chlorine**. Other elements and radicals, such as Mg, B, SO_4^{2-}, Fe, Zn, Cu, are usually present in amounts less than 1 000 ppm. Little can be said about the source of any of these except **helium, lead** and **strontium**.

The few **helium isotopic studies** carried out indicate a mantle (magmatic) origin. More numerous studies of **lead isotopes** in hydrothermal deposits suggest that lead may be derived at least in part from **underlying sediments, metamorphic basement rocks, associated igneous intrusions** or **direct from the mantle**. But the other metallic elements in a mineral deposit may not necessarily have come from the same source as the lead. Less work has been done on **strontium isotopes**, which tend to be concentrated in wall rock alteration zones. Results for the Cyprus copper deposits agree with the postulate of heated, recycled, Cretaceous sea water as the mineralizer; but for some molybdenum mineralization near Central City, Colorado, **a magmatic source** is indicated.

Our present knowledge indicates that most rocks can act as a source of **geochemically scarce elements**, which **can be leached out** under suitable conditions by hydrothermal solutions. For example, laboratory experiments have shown that heavy metals present in trace quantities will fractionate from greywacke into sea water or natural brine at 350°C. Because of the spatial relationship that exists between many hydrothermal deposits and igneous rocks, a strong school of thought holds that **consolidating magmas** are the **source of many**, if not all, **hydrothermal solutions**. The solutions are considered to be low temperature residual fluids left over after pegmatite crystallization, and containing the base metals and other incompatible elements that could not be accommodated in the crystal lattices of the silicate minerals precipitated by the freezing magma. This model derives not only the water, the metals and other elements from a hot body of igneous rock, but also **the heat to drive the mineralizing system**. The solutions are assumed to move upwards along fractures and other channelways to cooler parts of the crust where deposition of minerals occurs.

How much water do magmas contain? Water concentrations in felsic magmas appear to vary from 2.5 to 6.5 wt% with a median close to 3%. A rising monzogranitic magma with 3% water will begin to exsolve copious quantities at about 3.5 km depth (equivalent to a load pressure of 0.1 GPa). The same magma with 4% water will behave similarly at above 4.5 km depth. Below these depths, water remains in solution in melts because of the high containing pressure. As these magmas crystallize to produce a largely anhydrous mineral assemblage, an absolutely enormous volume of water can be given off by a cooling magma — **1 km³ of felsic magma with 3% water could exsolve approximately 100 Mt (10^{11} l).**

Now **water is only one of the volatile components of late stage magmatic fluids**; these also contain H_2S, HCl, HF, CO_2 and H_2. Of these, **H_2S and HCl** may be of particular importance and they would be expected to fractionate strongly into an exsolving aqueous phase. All base metals and many others that have been investigated so far can be extracted efficiently from a melt into an exsolving aqueous phase provided that sufficient water has exsolved. **With a 3 wt% water content of the melt, about 95% of the copper in a felsic magma would be extracted.** In December 1986, Mount Erebus, Antarctica discharged daily about 0.1 kg Au and 0.2 kg Cu — equivalent to 360 t Au in 10 000 years. This, and similar evidence from other volcanoes, demonstrates the ability of magmas to supply metal-bearing volatiles.

In many orefields, however, such as the Northern Pennine Orefield of England, there are **no** acid or intermediate **plutonic intrusions which might be the source of the ores**. Some workers have therefore postulated a more remote magmatic source, such as the **lower crust** or, more frequently, **magmatic processes in the mantle**, whilst an important body of opinion has favoured deposition from **formation solutions** — that is, water which was trapped in sediments during deposition and that has been driven up dip by the rise in temperature and pressure caused by deep burial. Such **burial** might occur in **sedimentary basins**, and solutions from this source are often called **basinal brines**. With a geothermal gradient of 1°C per 30 m, temperatures around 300°C would be reached at a depth of 9 km. Hot solutions from this source are believed to leach metals, but not necessarily sulphur, from the rocks through which they pass, ultimately precipitating them near the surface in shelf facies carbonates on the fringe of the basin, and far from any igneous intrusion (Fig. 6.3a). This, too, is a model favoured at present by many workers, particularly as an explanation for the genesis of **low temperature, carbonate-hosted lead–zinc–fluorite–baryte deposits** (Mississippi Valley-type). It has been suggested, however, that the available volumes of formation water are insufficient to carry the amount of metal that is present in such deposits and some workers favour a comparable flow of water, under a hydrostatic head, passing through sedimentary basins to produce the ore fluid (Fig. 6.3b).

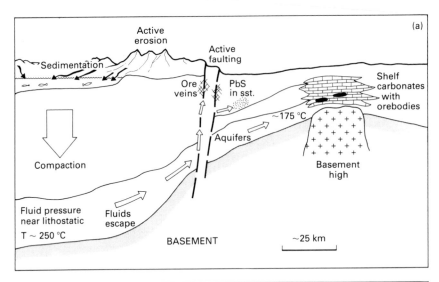

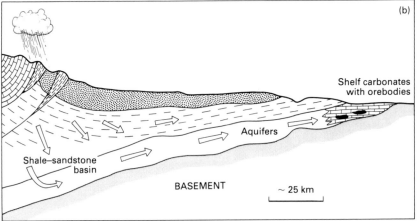

Fig. 6.3 Schemata for the formation of Mississippi Valley-type deposits. (a) Overpressured, hot pore fluids escape from a shale basin (perhaps aided by hydraulic fracturing) and move up aquifers to form deposits in cooler strata, filling fractures or forming other types of orebody. (b) Gravity-driven fluids flowing from a hydraulic head in a highland area flush through a basin driving out and replenishing the formation waters.

Geothermal systems could well be generators of orebodies. They form where a heat engine (usually magmatic) at depths of a few kilometres sets deep groundwaters in motion (Fig. 6.4a). These waters are usually meteoric in origin but, in some systems, formation or other saline waters (Salton Sea) may be present and magmatic water may be added by the heat engine. Dissolved constituents may be derived by the circulating waters from a magmatic body at depth, or from the country rocks which contain the system.

At Broadlands, New Zealand, an amorphous Sb–As–Hg–Tl sulphide precipitate enriched to ore grade in gold and silver has been formed, and in the nearby Rotokawa Geothermal System, base metal precipitates are accompanied by acanthite. Since the formation 6060 years ago of the large hydrothermal explosion crater in which Lake Rotokawa is now situated,

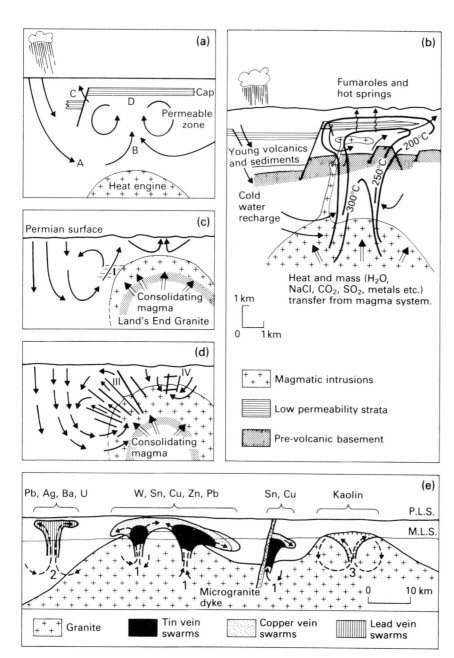

Fig. 6.4 (a) Schema showing some of the features of a geothermal system. (b) Schema showing the structure of a geothermal system like that of the Taupo Volcanic Zone, N.Z. (c and d) Schemata illustrating the evolution of some of the mineralization in a flank of the Land's End Granite; in detail these show: (c) initial emplacement of the pluton with the development of an H_2O-saturated carapace enclosing still-consolidating magma. It also shows formation of tin- and magnetite-bearing skarns (I) in aureole rocks, by aqueous solutions of a dominantly magmatic origin—at about 290–270 Ma ago. (d) Further crystallization of the pluton has taken place, and joints and fractures have formed in the crystallized carapace. With the formation of a water-rich phase that has separated from the H_2O-saturated melt, an extensive geothermal system has come into being. This has produced the main stage mineralization (III and IV) of tin- and copper-bearing quartz veins—at about 270 Ma (type II mineralization is that of pegmatites). (e) Schema of possible fossil geothermal systems associated with the granite batholith of south-west England, illustrating the different types and settings of mineralization in that region, and the district zoning developed there. 1, Emanative centres; 2, Cross course mineralization (late stage, post IV mineralization from basinal-type brines); 3, Kaolin deposits (weathering may have played a part in their formation).

about 370 Mt of gold may have been transported into the rocks beneath the crater.

The principal features of a geothermal system are shown in Fig. 6.4a, and Fig. 6.4b illustrates the structure of a geothermal system in a volcanic terrane like that of the Taupo Volcanic zone, New Zealand. Note that the hot waters are circulating through, reacting with and probably obtaining dissolved constituents from both the magmatic intrusion and the country rocks. In Figs 6.4c and d, geothermal systems are postulated to explain vein tin and copper mineralization in and adjacent to the Land's End Granite in south-west England. In Fig. 6.4e, we have a broader picture, with geothermal systems being invoked to explain some of the different types of mineralization in south-west England and the **zoning of metals** that is one of the well known features of this orefield.

(b) *Means of transport.* Sulphides and other minerals have such **low solubility** in pure water that it is now generally believed that the metals were transported in hydrothermal solutions as **complex ions**, i.e. the metals are joined to complexing groups (ligands). The most important are **HS⁻** or **H₂S, Cl⁻** and **OH⁻**. As a high concentration of H_2S and HS^- is required to keep these base complexes, e.g. $PbS(HS)^-$, stable, many workers favour the idea of metal transport in chloride complexes such as $AgCl_2^-$ and $PbCl^{3-}$. It is likely, however, that both, and other complexes, play a part in metal transport.

The **origin of the sulphur** at the site of deposition is also a problem; did it originate at this site or was it carried there with the metals in solution? Some mineralization situations, e.g. structurally deep sulphide veins in granites and quartzites, seem to demand that sulphur and the metals travelled together. Isotopic evidence, evidence from modern geothermal systems, ocean floor fissures venting hydrothermal solutions and evidence from some fluid inclusions, many of which carry daughter sulphides in amounts often implying quite high concentrations of metals and sulphur in the ore fluid, also favour this interpretation for many deposits. If the ore metals are transported as bisulphide complexes, then abundant sulphur will be available for the precipitation of sulphides at the site of deposition. On the other hand, the postulate that chloride complexes are the metal transporters, which is the most favoured hypothesis of those studying carbonate-hosted base metal deposits, creates difficulties as far as sulphur supply and metal transport are concerned in that chemical conditions dictate that at temperatures up to 150°C chloride solutions cannot carry enough Pb and H_2S at the same time to form orebodies. Three of the possible ways round this problem for low temperature carbonate-hosted deposits are as follows. The first way is to accept the hypothesis that sulphur is added from another solution to the ore fluid at the site of deposition—**the mixing model**. This sulphur could have been generated by the reduction of sulphate by organic matter at the site of deposition. A second way out of the problem, and one that allows us to keep **a single solution model** with metal and sulphur travelling together, is the hypothesis that organometallic complexes were the metal carriers, thus permitting metals and H$_2$S to travel together. A third possibility is the **co-transport of metals with sulphur as sulphate** which would then be reduced by reaction with organic compounds to precipitate metal sulphides.

(c) *Sulphide deposition.* This is another complex area of debate. We know that, when hydrothermal solutions rise into a zone saturated with ground water, **dilution will cause precipitation** of dissolved materials. It is also clear that an **increase in pH due to boiling**, in response to pressure decrease, **can induce precipitation. Cooling** too will reduce solubilities drastically. Of these three processes, boiling is by far the most efficient.

6.2.4 Lateral secretion

It has been accepted for many years that quartz lenses and veins in metamorphic rocks commonly result from the infilling of dilatational zones and open fractures by silica which has migrated out of the enclosing rocks, and that this silica may be accompanied by other constituents of the wall rocks including metallic components and sulphur. This **derivation of materials from the immediate neighbourhood** of the vein is called **lateral secretion**. Let us look at the probable behaviour of element levels in rocks adjacent to veins forming under different conditions. In Fig. 6.5a, we have a vein forming from an uprising hydrothermal solution supersaturated in silica. Some of this diffuses into the wall rocks and causes some silicification. The curve showing the level of silica decreases away from the source (i.e. the vein). In Fig. 6.5b, we have the opposite situation where silica is being

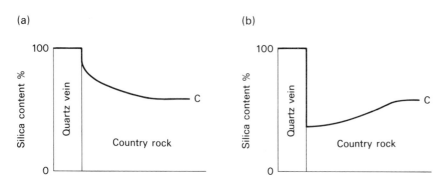

Fig. 6.5 Comparison of hypothetical profiles of silica. In (a) silica is added to the wall rocks from the hydrothermal solution, which is depositing quartz in the vein. In (b) silica is abstracted from the wall rocks and deposited as quartz in the vein. C indicates the normal level of silica in the country rocks.

supplied to the vein from the wall rocks. The curve now climbs as it leaves the vein, indicating a zone of silica depletion in the rocks next to the vein. Clearly, silica has been abstracted from the wall rocks and presumably has accumulated in the vein.

This is the situation in the wall rocks adjoining the quartz–carbonate–gold veins and lenses of the Yellowknife Goldfield of the Northwest Territories of Canada, where **more silica has been subtracted from the wall rocks than is present in the veins and lenses**. The evidence also points to the derivation of gold and other metals from the immediate wall rocks.

6.2.5 Metamorphic processes

(a) *Contact and regional metamorphism.* Isochemical metamorphism of many rocks can produce materials having an industrial use. An obvious example long used by mankind is **marble**, which may be produced by either contact or regional metamorphism of pure and impure limestones and dolomites. Another much used metamorphic rock is **slate**. **Other important industrial materials** of metamorphic origin are asbestos, corundum and emery, garnet, some gemstones, graphite, magnesite, pyrophyllite, sillimanite minerals, talc and wollastonite.

Allochemical metamorphism (**metasomatism**) may accompany contact or regional metamorphism. In the former case in particular, it may lead to the formation of **skarn deposits** carrying economic amounts of metals or industrial minerals. Their general morphology and nature have been summarized in Chapter 4. Their genesis will be further discussed in Chapter 14.

(b) *The role of other metamorphic processes in ore formation.* Some examples of lateral secretion are clearly the result of metamorphism. This subject has, however, already been covered above and will not be discussed further. In this section, we are concerned with those metamorphic changes that involve recrystallization and redistribution of materials by ionic diffusion in the solid state or through the medium of volatiles, especially water. Under such conditions, relatively **mobile ore constituents** may be **transported to sites of lower pressure**, such as shear zones, fractures or the crests of folds. In this way, the occurrence of quartz–chalcopyrite–pyrite veins in amphibolites and schists and many gold veins in greenstone belts may have come about. Recent studies of mass balance changes accompanying the development of foliation in metamorphic rocks have shown that regional metamorphic terranes are large hydrothermal systems analogous to the smaller scale systems in young oceanic crust. These systems have the capacity to leach a wide range of components, including ore-forming materials, from a very large volume of crust. For flow to take place

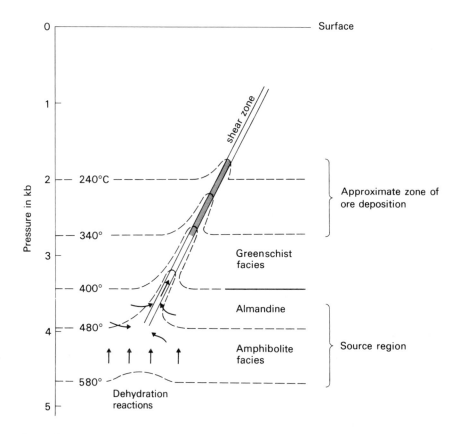

Fig. 6.6 Diagram of a shear zone where metamorphic water from a large volume of rock is rising to higher levels.

in such systems, **regional permeability** must be developed. Large shear zones can provide just such a conduit and such a mechanism is sketched in Fig. 6.6. This model, and variations of it, have been used to explain the formation of a number of Archaean gold deposits.

6.3 Origin due to surface processes

Processes involving mechanical and chemical sedimentation will not be dealt with here. The reader is referred to texts on sedimentology for the general principles involved. However, certain aspects will be touched upon in Chapter 19. Residual processes and supergene enrichment will be dealt with in Chapter 20. The remaining space in this chapter will be devoted to a consideration of **exhalative processes**. These, it should be noted, are a surface expression of the activity of hydrothermal solutions.

6.3.1 Volcanic-exhalative (sedimentary-exhalative) processes

We are concerned here with a group of deposits often referred to as **exhalites** and including massive sulphide ores. Some of the characteristics of these ores have been dealt with in Section 4.3.2(a). They frequently show a close spatial relationship to volcanic rocks, but this is not the case with all the deposits, e.g. Sullivan, Canada (see Fig. 4.6), which is sediment-hosted; this and similar examples are referred to commonly as **sedex** (sedimentary-exhalative) deposits. They are conformable and frequently banded. In the volcanic-associated types, the **principal constituent** is usually **pyrite** with varying amounts of copper, lead, zinc and baryte; precious metals together with other minerals may be present. Deposits of this type have been observed in the process of formation from hydrothermal vents (**black smokers**) at a large number of places along sea-floor spreading centres. These deposits are now often referred to as **volcanic-associated** (or volcanogenic) **massive sulphide deposits**. The ores with a volcanic affiliation show a progression of types. **Cyprus types** are associated with basic volcanics in ophiolites formed at oceanic or back-arc spreading ridges and are essentially cupriferous pyrite bodies. **Besshi-type** deposits occur in mafic volcanics in complex structural settings with thick greywacke sequences. They carry zinc as well as copper. **Kuroko deposits**, named after Miocene ores in Japan, are associated with felsic volcanics and carry copper, zinc, lead and often gold and silver. Large amounts of baryte, gypsum and quartz may be associated with them. There is today wide agreement that these deposits are **submarine-hydrothermal in origin**, but there is a divergence of opinion as to whether the **solutions responsible** for their formation are **magmatic** in origin **or** whether they represent **circulating sea water** as suggested in Fig. 6.7.

Black smokers were discovered in the late 1970s during ocean floor investigations using a submersible. They are plumes of hot, black, or sometimes white, hydrothermal fluid issuing from chimney-like vents that connect with fractures in the sea floor. The black smoke is so coloured by a high content of fine-grained metallic sulphide particles and the white by calcium and barium sulphates. The chimneys are generally less than 6 m high and are about 2 m across. They stand on **mounds of massive ore-grade sulphides** (Fig. 6.8) that **occur within the grabens and on the flanks**

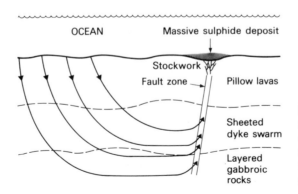

Fig. 6.7 Diagram showing how sea water circulation through oceanic crust might give rise to the formation of an exhalative volcanic-associated massive sulphide deposit.

of oceanic ridges. The mineralogy of the mounds is similar to that of massive sulphide deposits on land with high temperature copper–iron sulphides beneath lower temperature zinc- and iron-rich sulphides, baryte and amorphous silica. Gold and silver values have been found. After growing upwards at a rate of perhaps 8–30 cm a day, chimneys eventually become unstable and collapse, forming a mound of chimney debris mixed with anhydrite and sulphides upon which further chimney growth and collapse occur. Once a mound has developed, it grows both by accumulation of chimney debris on its upper surface **and** by precipitation of sulphides within the mound. The covering of chimney debris and sulphide and silica precipitation in the outer part of the mound decrease the permeability of the mound and form a crust that constrains fluid escape and leads to considerable circulation of high temperature solutions within the mound. The isotherms within the mound then rise, leading to the replacement of lower temperature mineral assemblages by higher temperature ones, thus producing **a similar zoning to that of volcanic-associated massive sulphide deposits found on land**. A model for their mode of origin is given in Fig. 6.9. The principal stages of development of this model are as follows.

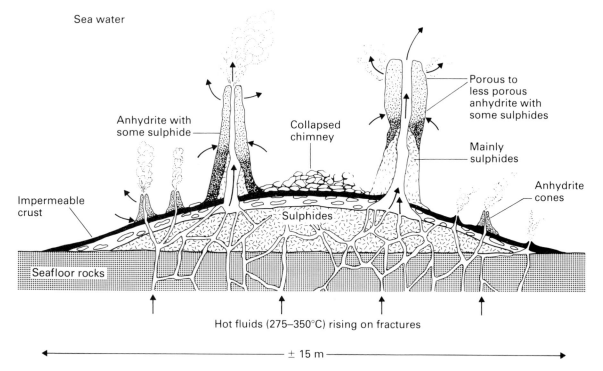

Fig. 6.8 Formation of chimneys and sulphide mounds on the sea floor. (After Barnes, 1988, *Ores and Minerals*, Open University Press, with permission.)

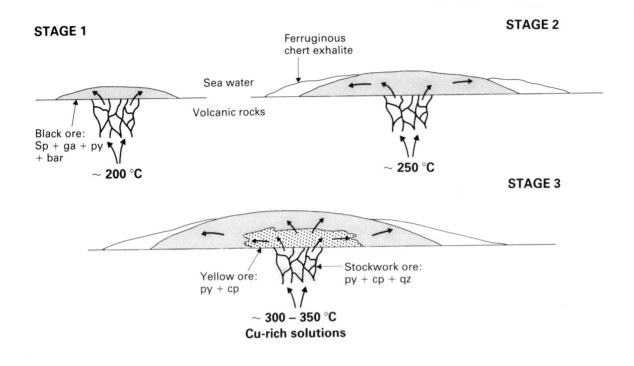

STAGE 1

Sea water

Volcanic rocks

Black ore:
Sp + ga + py
+ bar

~ 200 °C

STAGE 2

Ferruginous
chert exhalite

~ 250 °C

STAGE 3

Yellow ore:
py + cp

Stockwork ore:
py + cp + qz

~ 300 – 350 °C
Cu-rich solutions

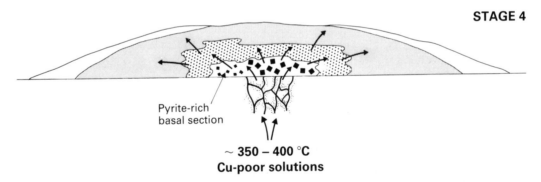

STAGE 4

Pyrite-rich
basal section

~ 350 – 400 °C
Cu-poor solutions

Fig. 6.9 Diagrams to illustrate the first four stages during the formation of volcanic-associated massive sulphide deposits as described in the text; bar = baryte, cp = chalcopyrite, ga = galena, py = pyrite, qz = quartz, and sp = sphalerite.

1   Precipitation of fine-grained sphalerite, galena, pyrite, tetrahedrite, baryte with minor chalcopyrite (black ore) by the mixing of relatively cool (~200°C) hydrothermal solutions with cold sea water.

2   Recrystallization and grain growth of these minerals at the base of the evolving mound by hotter (~250°C) solutions, together with deposition of more sphalerite etc.

3   Influx of hotter (~300°C) copper-rich solutions which replace the earlier deposited minerals with chalcopyrite in the lower part of the deposit (yellow ore). Redeposition of these replaced minerals at a higher level.

4   Still hotter, copper undersaturated solutions then dissolve some chalcopyrite to form pyrite-rich bases in the deposits.

5   Deposition of chert-hematite exhalites above and around the sulphide deposit. Similar exhalites will also have formed during previous stages. Silica is slow to precipitate; it needs silicate minerals to nucleate on and so, although much may be deposited in the underlying stockwork, the rest is mainly carried through the sulphide body to form exhalites above it.

6.4 Hydraulic fracturing

All the hydrothermal processes that have been described so far in this chapter require the development of sufficient rock permeability to allow the mineralizing solutions to flow from their source to the site of mineral deposition. This permeability was achieved by rock fracturing on scales from the microscopic to that of major crustal faults. Much of the fracturing is due to the solutions themselves, for it is now well known that fluids under high pressure can influence the mechanical behaviour of the crust dramatically and develop sites for precipitation of material held in solution. The fracturing of rocks by water under high pressure is known as **hydrofracturing or hydraulic fracturing**. The process is used artificially to increase the permeability in oil, gas and geothermal reservoirs.

The effect of pore pressure and the development of hydraulic fracturing. Fluids in the pore spaces of rocks support some of the load that would otherwise be supported by the rock matrix. A sufficiently high increase in pore pressure will cause hydraulic fracturing. Such large scale fracturing is associated with faults up which hydrothermal fluids have penetrated (Fig. 6.10).

Increases in pore pressure leading to hydraulic fracturing may well play a part in **increasing permeability** and **allowing formation water to move up dip in sedimentary basins**. One of the simplest situations in which high pore pressure regimes may develop is underneath low permeability zones, such as thick shale or evaporite sequences (Fig. 6.11). High pore pressure in such regions is often encountered during oil drilling in sedimentary basins.

The effects of hydraulic fracturing have been reported from a number of sedimentary basins where they have given rise to horizontal bands of breccia with shale fragments in carbonate and other diagenetic cements. In the Kupferschiefer of central Europe (see Chapter 16), i.e. the copper–silver-bearing ore shales of south-western Poland, crowds of horizontal and vertical gypsum, calcite and base metal veinlets are present. These appear to have been formed by hydraulic fracturing resulting from high pore pressures.

As might be predicted from the above discussions, hydraulic fracturing

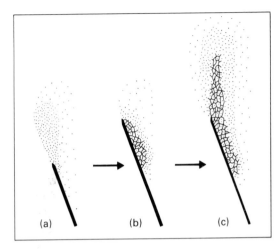

Fig. 6.10 Propagation of normal faults and accompanying development of breccia zones. (a) Hydrothermal solution saturates fault zone and permeates into the hanging wall. (b) Sudden fracturing extends fault upwards and a breccia zone forms in the hanging wall; with further similar extensions and brecciation a large breccia zone may form. (c) Vertical hydraulic fracturing may occur above the fault zone.

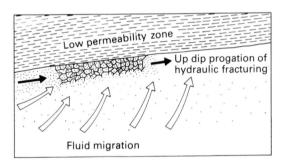

Fig. 6.11 Development of hydraulic fracturing in a sedimentary basin. Fluids accumulate beneath an impermeable zone, e.g. shale or evaporite sequence, pore pressure builds up and hydraulic fracturing is initiated and propagated up dip.

plays an important role in the **escape of the end-magmatic fluids from consolidating intrusions.** This leads to the formation of veins including sheeted vein systems, which may host economic Sn–W deposits, and other deposit types.

6.5 Further reading

Evans A.M. (1993) *Ore Geology and Industrial Minerals: An Introduction.* Blackwell Scientific Publications, Oxford. Chapter 4 of this book covers this subject in greater depth and will be especially valuable to the reader for the references therein. These can be used as an entry into more recent literature through the medium of the Science Citation Index.

7: Geothermometry, Geobarometry, Paragenetic Sequence, Zoning and Dating of Ore Deposits

7.1 Geothermometry and geobarometry

Ores are deposited at temperatures and pressures ranging from very high, at deep crustal levels, to atmospheric, at the surface. Some pegmatites and magmatic segregation deposits have formed at temperatures around 1000°C and under many kilometres of overlying rock, whilst placer deposits and sedimentary ores have formed under surface conditions. Most orebodies were deposited between these two extremes. Clearly, knowledge of the temperatures and pressures obtaining during the precipitation of the various minerals will be invaluable in assessing their probable mode of genesis and such knowledge also will be of great value in formulating exploration programmes. In this small volume it is only possible to touch briefly on a few of the methods that can be used.

7.1.1 Fluid inclusions

The nature of fluid inclusions and the principle of this method have been outlined in Section 5.4.

7.1.2 Inversion points

Some natural substances exist in various mineral forms (**polymorphs**) and some use of their **inversion temperatures** can be made in geothermometry. For example, β-quartz inverts to α-quartz with falling temperature at 573°C. We can determine whether quartz originally crystallized as β-quartz by etching with hydrofluoric acid and thus decide whether it was deposited above or below 573°C. Examples among ore minerals include:

177°C

acanthite (monoclinic) = argentite (cubic),

104°C

orthorhombic chalcocite = hexagonal chalcocite.

This chapter:
- gives a brief look at geothermometry and geobarometry, noting that we can now determine temperatures of hydrothermal ore fluids at their source as well as at their sites of mineral deposition,
- discusses paragenetic sequence and zoning in greater depth with particular reference to epigenetic and syngenetic hydrothermal deposits as well as sedex deposits,
- ends with the difficult task of dating epigenetic ore deposits.

7.1.3 Sulphide systems

Chemical variations in sulphide systems such as $ZnS-FeS$, $ZnS-FeS-FeS_2$, $FeAsS-FeAs_2$ have yielded some information on the temperatures and pressures at which sphalerite and arsenopyrite have crystallized.

A very interesting recent development is the **Ga/Ge-geothermometer** using sphalerite. This can be used to determine **temperatures in the source regions** of ore solutions and to estimate the degree of mixing of hot parental ore fluids with cool, near-surface waters. The first applications of this method appear to be very promising and have shown that for the vein mineralization of Bad Grund and Andreasberg in Germany, the ore fluid at its time of genesis had a temperature above 220°C, but must have mixed with waters of less than 130°C at the level of mineral deposition. However, at Ramsbeck there is no difference between the temperature deduced by Ga/Ge-geothermometry for the source region and the fluid inclusion homogenization temperatures, indicating that no significant fluid mixing occurred. For the sediment-hosted ore deposits, of various types investigated, source fluid temperatures of 180–270°C using this method were obtained and, on the assumption that these fluids were generated by sediment dewatering in sedimentary basins, this indicates a source depth of about 6–11 km, if geothermal gradients of 25–30°C km^{-1} are assumed.

Several empirical chemical geothermometers based on the Na, K and Ca contents of modern geothermal brines, which can be used to determine the temperatures in the fluid source region, have been developed. Work on specimens of fluorite, calcite and quartz from the Mississippi Valley-type deposits of the Alston Block of the Northern Pennine Orefield of England has given temperatures for the source region, considered to be a nearby sedimentary basin, of 150–250°C. These temperatures, as might be expected, are about 50°C higher than the temperatures of deposition of the fluorite. The source region temperatures are interestingly very close to those obtained on similar deposits using the Ga/Ge-geothermometer.

7.1.4 Sulphur isotopic ratios

Sulphur isotopic ratios in ore minerals can, in certain circumstances, be used to deduce the temperature of formation. Depositional temperatures obtained by fluid inclusion methods on transparent minerals, sulphur isotope methods on sulphides intergrown with them, and oxygen isotopic compositions of coexisting oxides, generally give results in close agreement, but the fluid inclusion method is by far the most useful, reliable and widely used.

7.2 Paragenetic sequence and zoning

The time sequence of deposition of minerals in a rock or mineral deposit is known as its paragenetic sequence. If the minerals show a spatial distribution, then this is known as zoning. The paragenetic sequence is determined from studying such structures in deposits as crustiform banding and from the microscopic observations of textures in polished sections.

7.2.1 Paragenetic sequence

Abundant evidence has been accumulated from worldwide studies of epigenetic hydrothermal deposits, indicating that there is a general order of deposition of minerals in these deposits. Exceptions and reversals are known but not in sufficient number to suggest that anything other than a common order of deposition is normally the case. **A simplified general paragenetic sequence** is as follows:

1 silicates;
2 magnetite, ilmenite, hematite;
3 cassiterite, wolframite, molybdenite;
4 pyrrhotite, löllingite, arsenopyrite, pyrite, cobalt and nickel arsenides;
5 chalcopyrite, bornite, sphalerite;
6 galena, tetrahedrite, lead sulphosalts, tellurides, cinnabar.

Of course, not all these minerals are necessarily present in any one deposit and the above list has been drawn up from evidence from a great number of orebodies.

7.2.2 Zoning

Zones may be defined by changes in the mineralogy of ore or gangue minerals or both, by changes in the percentage of metals present, or by more subtle changes from place to place in an orebody or mineralized district of the ratios between certain elements or even the isotopic ratios within one element. Zoning was first described from epigenetic vein deposits but it is present also in other types of deposit. For example, syngenetic deposits may show zoning parallel to a former shoreline, as is the case with the iron ores of the Mesabi Range, Minnesota; alluvial deposits may show zoning along the course of a river leading from the source area; some exhalative syngenetic sulphide deposits show a marked zonation of their metals and skarn deposits often show a zoning running parallel to the igneous-sedimentary contact. In this discussion, attention will be focused on the zoning of epigenetic hydrothermal, exhalative syngenetic and sedimentary syngenetic sulphide deposits.

Epigenetic deposits are often classified according to their depth and temperature of formation: **hypothermal** deposits being deep-seated, high temperature deposits; **mesothermal** deposits those formed at low temperatures and medium depths; and **epithermal** deposits being formed near the surface.

One of the factors controlling paragenetic sequence and zoning is clearly the temperature **gradient** away from the emanative centre(s), i.e. frequently decreasing depth, but the **relative stabilities of the complex ions** to which metals are linked during transport is also very important.

(a) *Epigenetic hydrothermal zoning.* Zoning of this type can be divided into three intergradational classes: **regional, district and orebody zoning**. *Regional zoning* occurs on a very large scale, often corresponding to large sections of orogenic belts and their foreland. Some regional zoning of this type, e.g. in the Andes, appears to be related to the depth of the underlying Benioff Zone, which suggests a deep level origin for the metals as well as the associated magmas. *District zoning* is the zoning seen in individual ore-fields such as Cornwall, England (Figs 6.4 & 7.1) and Flat River, Missouri (Fig. 7.2). Zoning of this type is most clearly displayed where the mineralization is of considerable vertical extent and was formed at depth, where changes in the pressure and temperature gradients were very gradual. If deposition took place near to the surface, then steep temperature gradients may have caused superimposition of what would, at deeper levels, be distinct zones, thus giving rise to the effect known as **telescoping** *Orebody zoning* takes the form of changes in the mineralization within a single orebody. A good example occurs in the Emperor Gold Mine, Fiji, where vertical zoning of gold–silver tellurides in one of the main ore shoots gives rise to an increase in the Ag to Au ratio with depth.

(b) *Syngenetic hydrothermal zoning.* This is the zoning found in stratiform sulphide bodies principally of volcanic affiliation [see Sections 4.3.2(a) & 6.3.1]. These deposits are frequently underlain by stockwork deposits and many of them appear to have been formed from hydrothermal solutions

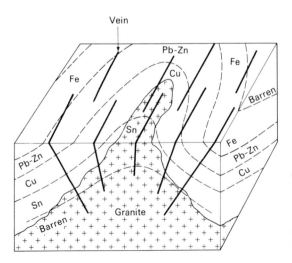

Fig. 7.1 Diagrammatic illustration of district zoning in the orefield of south-western England showing the relationship of the zonal boundaries to the granite–metasediment contact. N.B. The iron in the uppermost zone is in low temperature siderite and hematite.

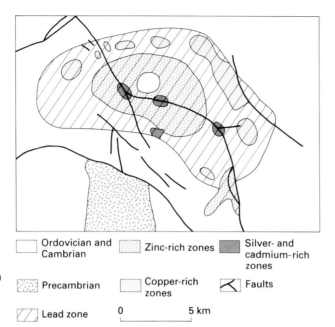

Fig. 7.2 District zoning in the Flat River area of the Old Lead Belt of south-east Missouri.

☐	Ordovician and Cambrian	▦	Zinc-rich zones	▨	Silver- and cadmium-rich zones
▨	Precambrian	☐	Copper-rich zones	◁	Faults
▨	Lead zone		0 5 km		

that reached the sea floor. The zonal sequence is not always clearly seen in this type of ore deposit, but where it has been established, it is identical to that of epigenetic hydrothermal ores, i.e. there is **a general sequence upwards and outwards** through the orebodies of Fe (in pyrite and/or magnetite) → (Sn) → Cu → Zn → Pb → Ag → Ba. As in vein deposits, the **zonal boundaries are normally gradational** with frequent **overlap** of zones. The cause and evolution of this zoning are discussed in Section 6.3.1.

(c) *Sedimentary syngenetic sulphide zoning.* This zoning is found in stratiform sulphide deposits of sedimentary affiliation usually of wide regional development such as the Permian Kupferschiefer of Germany and Poland and the Zambian Copperbelt deposits. Underlying stockwork feeder channelways are rarely known beneath these deposits and usually they have formed in **euxinic environments**. The zoning appears to show a relationship to the palaeogeography, and proceeding basinwards through a deposit it takes the form of **Cu + Ag → Pb → Zn**. In the Zambian Copperbelt, however, the zoning is principally one of copper minerals and pyrite, as lead, zinc and silver are virtually absent.

Euxinic environment—one of restricted circulation where stagnant or anaerobic (oxygen poor) conditions exist.

    Different patterns of *orebody zoning* are seen **in the lead–zinc sediment-hosted deposits** (see Section 16.2.5), as at Broken Hill, where the stack of orebodies shows a vertical zonation from a copper zone, in the lowest orebody, upwards to Pb–Zn orebodies, which themselves have an upward and outward decrease in Cu and Zn and an increase in Pb, Ag, Mn and F. A barium-rich horizon overlies these orebodies. This zoning strongly resembles that of volcanic-hosted massive sulphide deposits described in the foregoing section, and, like the zoning in these, it shows a spatial relationship to the feeder stockwork zone(s).

7.3 The dating of ore deposits

When orebodies form part of a stratigraphical succession, such as the Cambrian to Devonian Clinton ironstones of eastern North America, their age is not in dispute. Similarly, the ages of orthomagmatic deposits may be fixed almost as certainly if their parent pluton can be well dated. On the other hand, epigenetic deposits may be very difficult to date, especially as there is now abundant evidence that many of them may have resulted from **polyphase mineralization**, with epochs of mineralization being separated by intervals in excess of 100 Ma. There are three main lines of evidence that can be used: the **field data**, **radiometric** and **palaeomagnetic age determinations**. The field evidence may not be very exact. It may take the form of an unconformity cutting a vein and thus giving a minimum age for the mineralization. Similarly, if mineralization occurs in rocks reaching up to Cretaceous in age, we know that some of it is at least as young as the Cretaceous, but we cannot tell when the first mineralization occurred or how young some of it may be.

7.3.1 Radiometric dating

Sometimes a mineral clearly related to a mineralization episode will permit a direct dating, e.g. uraninite in a vein, but more commonly it is necessary to use wall rock alteration products such as micas, feldspar and clay minerals. In the latter case, the underlying **assumption** is that **the wall rock alteration is coeval** with some of or all **the mineralization**. It is also important to note that our radiometric clocks can be reset. Broken Hill, New South Wales, gives us a dramatic example of this phenomenon. The lead–zinc–silver orebodies of Broken Hill are generally held to be regionally metamorphosed syngenetic deposits. Rb–Sr and K–Ar ages of biotites (450–500 Ma), however, markedly post-date the 1700 Ma of the dominant metamorphism in the country rocks.

Porphyry coppers in many parts of the world have been dated using Rb–Sr and K–Ar methods. the results of this work have shown: (i) that it is usually the youngest intrusions in any particular area that carry mineralization; and (ii) that there is a close temporal relationship between porphyry copper mineralization and associated calc-alkaline magmatism.

Thus at Panguna, P.N.G., the mineralization dates at 3.4 ± 0.3 Ma, compared with intrusives 4–5 Ma old. At Ok Tedi, north-west Papua, alteration as young as 1.1–1.2 Ma has been found. This contrasts with the oldest mineralization yet found, a banded iron formation with associated stratabound copper sulphides at Isua, West Greenland, dated at 3760 ± 70 Ma.

7.3.2 Palaeomagnetic dating

The successful palaeomagnetic dating of ore deposits depends on a number of factors. Some of the most important are:

1 the development of magnetic minerals in a deposit or its wall rocks during one of the principal phases of mineralization;

2 a lack of complete oxidation or alteration, which may be accompanied by overprinting with a later period of magnetization;

3 the availability of an accurate polar wandering curve for the continent or plate in which the deposit occurs.

As is well known, magnetite and hematite are the two principal carriers of magnetization in rocks; this is true also for ore deposits. In general, magnetization carried wholly or in the main by magnetite is, with present palaeomagnetic techniques, measured and interpreted more easily. This mineral is, however, by no means common in epigenetic ore deposits, and hematite-mineralized specimens often have to be used. An important development in this method of dating was the perfecting of the **cryogenic magnetometer** in the early 1970s. This instrument is beginning to provide us with new data on mineralization ages, particularly for **Mississippi Valley-type deposits**, which present many problems to the geochronologist using radiometric dating methods. This type of mineralization in south-east Missouri has been dated as later Pennsylvanian to early Permian using a cryogenic magnetometer.

7.4 Further reading

The subjects of this chapter are all dealt with at greater length in Evans (1993) — see Section 6.5.

Part 2
Examples of the More Important Types of Ore Deposit

We were taken from the ore-bed and the mine,
*   We were melted in the furnace and the pit—*
We were cast and wrought and hammered to design,
*   We were cut and filed and tooled and gauged to fit.*
Some water, coal and oil is all we ask,
*   And a thousandth of an inch to give us play:*
And now, if you will set us to our task,
*   We will serve you four and twenty hours a day!*

It is easy! Give us dynamite and drills!
Watch the iron-shouldered rocks lie down and quake,
As the thirsty desert-level floods and fills,
And the valley we have dammed becomes a lake.

Rudyard Kipling (1865–1936), extracts from his poem
The Secret of the Machines

8: Mineral Deposit Classification

There are various ways in which mineral deposits can be classified. In this book we will look at them as deposit types largely based on an environmental–rock association classification. **Deposit types can be polymetallic** and a simple reading may not make it clear to the tyro which are the most important deposit types for the production of particular metals and to resolve this difficulty Table 8.1 has been drawn up. Of course it might seem desirable, and it is possible, to classify and discuss metallic deposits metal by metal but this necessitates much duplication and cross referencing.

The deposit types in succeeding chapters are dealt with in the following general order: moving outwards, as it were, from intramagmatic

Table 8.1 Deposit types and their products.

Deposit types	Important metal and mineral production
Kimberlites and lamproites	Diamonds
Carbonatites	Phosphates, Fe, Nb, Zr, R.E.Es, fluorite, baryte, strontianite, lime, vermiculite, one carbonatite produces Cu (Palabora)
Pegmatites	Be, Li, Sn, W, Rb, Cs, Nb, Ta, R.E.Es, U, feldspar, mica, quartz
Orthomagmatic	
fractional crystallization	Cr, P.G.M., Ti, Fe-V
liquation	Cu, Ni, P.G.M.
Skarn	Cu, Mo, Fe, W, Zn, Pb, Sn, Bi
Stockworks	Cu, Mo, Au, Ag, Sn, Rh
Stratiform	
sedimentary-affiliated	Cu, Co, Pb, Zn, Cd, Ag, baryte
volcanic-associated sulphide facies	Cu, Zn, Pb, Ag, Au, Sn, Hg, baryte
volcanic-associated oxide facies	Fe, Cu, U, Au, Ag, Sn, W
Vein association including pipes and mantos and replacement deposits	Au, Ag, Sn, U, Cu, W, Pb, Zn, Sb, As, 3e, fluorite, baryte
Strata-bound	Pb, Zn, fluorite, baryte, U, V, Ge
Sedimentary	
placers	
residual-alluvial	Apatite, Nb, Zr, Sn, Au, Pt
beach	Sn, diamonds, Au, Ti, Fe, R.E.Es, Ce, Y, Zr, quartz and garnet sands
fossil	Au, U
autochthonous	Fe, Mn, coal, evaporites, phosphates, sulphur
Residual	Al, Fe, Ni, Co, R.E.Es
Supergene enrichment	Fe, Mn, Cu, Ag, Au, Ge, Ga, U

deposits, where the wanted elements or minerals are part of an intrusion, to orthomagmatic deposits which are usually segregated from, but still within or adjacent to, their parent magma and on to hydrothermal deposits. These may be in or adjacent to igneous intrusives, which in many cases appear to be the source of the solutions. The highest temperature examples are skarn deposits which may be associated with plutonic disseminated and stockwork deposits. From this point the connexion with igneous activity is sometimes tenuous or non-existent. Among the stratiform deposits, there is little or no connexion as far as those with a sedimentary affiliation are concerned, but the volcanic-associated group are at least dependent on subjacent magma bodies as a heat engine to drive their circulation, as are many vein deposits. Some strata-bound deposits also show a spatial relationship to igneous bodies which are often too old to have been the source of the mineralizing fluid, but which may have served as structural guides to their circulation and provided some radiogenically generated heat.

Next, we proceed through sedimentary and residual deposits to industrial minerals. Some of these, e.g. diamonds and titania, are included in the above chapters and more could be added but, for the sake of convenience, most industrial minerals are considered together in Chapter 21 for the reasons given in that chapter's introduction. Finally, we have water, energy resources, metamorphism and the global and temporal distribution of ore deposits.

9: Diamond Deposits in Kimberlites and Lamproites

9.1 Introduction

Long known as the hardest of naturally occurring minerals, diamonds are now known to be **among the oldest minerals in the earth**, capable of being picked up as exotic fragments by magmas generated at great depths and then surviving both this ordeal and subsequent violent volcanic activity at and near the earth's surface. Their units of weight are also ancient — carats, one carat = 0.2 g. The ultimate bedrock sources of diamonds are the igneous rocks kimberlite and lamproite and it is these occurrences that are dealt with in this chapter. Important amounts of diamond are recovered from beach and alluvial placer deposits, which are described in Chapter 19.

Not all kimberlites and lamproites contain diamond and, in those that do, it is present only in minute concentrations. For example, in the famous Kimberley Mine, R.S.A., 24 Mt of kimberlite yielded only 3 t of diamond, or one part in eight million. From the revenue point of view, it is not necessarily the highest grade mines that provide the greatest return — **what matters is the percentage of gem quality diamonds**. Each mining region and even each mine has a different percentage, and some may be 90% or more while others may rely on only an occasional gemstone to boost their income. Total world production of natural diamonds is about 107 Mct p.a., of which about half are of industrial grade. In addition, over 250 Mct p.a. of synthetic industrial diamonds are manufactured. The leading world producers are shown in Table 9.1 and world bedrock diamond fields and the general occurrence of kimberlites in Fig. 9.1. Most diamonds are marketed by the **Central Selling Organization** (C.S.O.) which is controlled by De Beers, itself a subsidiary of the great Anglo-American Corp. of the R.S.A. The C.S.O. policy is to maintain a stable world diamond price. This **cartel** has been remarkably successful and long lived. It has undoubtedly been of benefit to both buyers and producers, especially those producers in developing countries. However, Argyle Diamonds, the Australian producer, announced in June 1996 that they will in future sell all their production direct to the world market.

In this chapter you will learn about:
- a very rare mineral,
- its unusual and restricted occurrence,
- its very deep genesis.

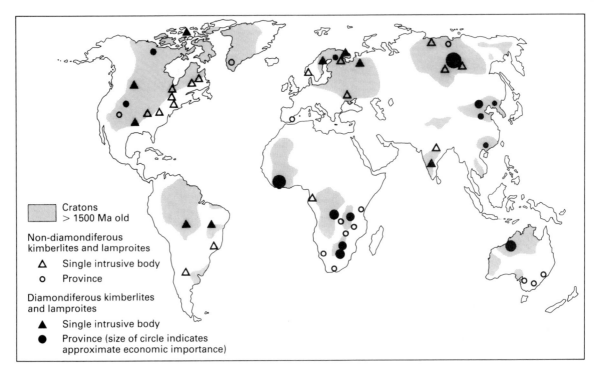

Fig. 9.1 Distribution of diamondiferous and non-diamondiferous kimberlites and lamproites.

Country	Mct
Australia	43.8
Zaïre	18.0
Botswana	15.6
Russian Federation	11.5
R.S.A.	10.2
South America	3.0
Angola	1.4
Namibia	1.3
Ghana	0.6
C.A.R.	0.5
Sierra Leone	0.4
Others	1.2
Total	107.5

Table 9.1 World production, in Mct, of natural diamonds in 1994 (bedrock and placer deposits).

9.2 Morphology and nature of diamond pipes

Many near surface, diamond-bearing kimberlites and lamproites occur in pipe-like **diatremes** (often just called **pipes**), which are small, generally less than 1 km² in horizontal area and often grouped in clusters. Some diatremes are known to coalesce at depth with **dykes** of non-fragmental kimberlite (Fig. 9.2). These dykes are thin, usually less than 10 m thick but may be as much as 14 km long. Some recently formed diatremes terminate at the surface in **maars**; these volcanic craters may be filled with lacustrine sediments to a depth of over 300 m. The sediments, which may be diamondiferous with a heavier concentration of more and bigger diamonds near the crater rim shoreline, are sometimes affected by subsidence, perhaps of a cauldron nature. The kimberlite beneath the sediments may be relatively barren. Below the flared crater area is the vertical or near vertical pipe itself which typically has walls dipping inwards at about 82° and a fairly regular outline producing the classic, **carrot-shaped diatreme** (Fig. 9.3) that may exceed 2 km in depth.

At the surface, kimberlite may be weathered and oxidized to a hydrated **'yellow ground'**, which gives way at depth to fresher **'blue ground'** and **'hardebank'** (resistant kimberlite that often crops out and does not disintegrate easily upon exposure). In the upper levels of pipes, the kimberlite is usually in the form of so-called agglomerate (really a tuffisitic breccia with many rounded and embayed fragments in a finer grained matrix) and tuff. The rounded fragments are often **xenoliths of metamorphic rocks** from deeper crust, or **garnet–peridotite or eclogite from the upper mantle**. Their rounded nature is attributed to a gasfluidized origin. Magmatic kimberlite bodies are confined mainly to the root zones of pipes, where they are often in the form of intrusion breccias which grade up with gradational boundaries into the agglomerate, or to sills and dykes.

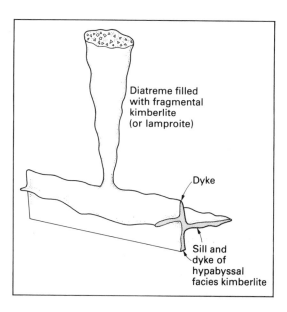

Fig. 9.2 Schematic drawing showing the relationship between an explosive kimberlite diatreme and deep-seated sills and a feeder dyke filled with non-fragmental kimberlite magma that solidified *in situ*.

Diatreme filled with fragmental kimberlite (or lamproite)

Dyke

Sill and dyke of hypabyssal facies kimberlite

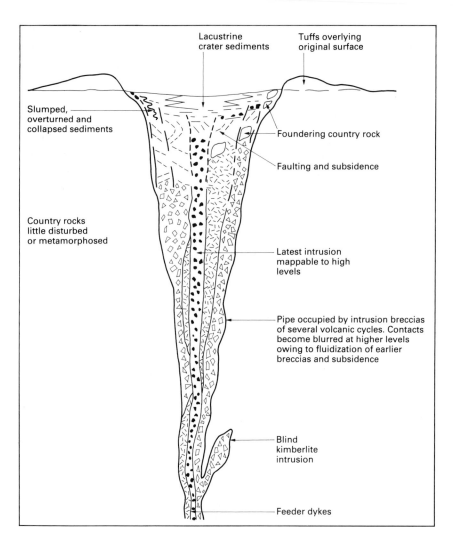

Lacustrine
crater sediments

Tuffs overlying
original surface

Slumped,
overturned and
collapsed sediments

Foundering country rock

Faulting and subsidence

Country rocks
little disturbed
or metamorphosed

Latest intrusion
mappable to high
levels

Pipe occupied by intrusion breccias
of several volcanic cycles. Contacts
become blurred at higher levels
owing to fluidization of earlier
breccias and subsidence

Blind
kimberlite
intrusion

Feeder dykes

Fig. 9.3 Schematic diagram of a kimberlite diatreme (pipe) and maar (volcanic crater below ground level and surrounded by a low tuff rim). The maar can be up to 2 km across.

9.3 Kimberlites and lamproites and their emplacement

Kimberlite may be defined as a potassic ultrabasic hybrid igneous rock containing large crystals (megacrysts) of olivine, enstatite, Cr-rich diopside, phlogopite, pyrope-almandine and Mg-rich ilmenite in a fine-grained matrix containing several of the following minerals as prominent constituents: olivine, phlogopite, calcite, serpentine, diopside, monticellite, apatite, spinel, perovskite and ilmenite. **Lamproites** are potash- and magnesia-rich lamprophyric rocks of volcanic or hypabyssal origin comprising mineral assemblages containing one or more of the following primary phenocrystal and/or groundmass phases: leucite, Ti-rich phlogopite, clinopyroxene, amphibole (typically Ti-rich, potassic richterite), olivine and sanidine. Accessories may include priderite, apatite, nepheline, spinel, perovskite, wadeite and ilmenite. Xenoliths and xenocrysts, including olivine, pyroxene, garnet and spinel of upper mantle origin, may be present and **diamond** as a rare accessory. **Fertile lamproites** appear to be the silica-saturated **orendites** and **madupites** which carry sanidine rather than leucite.

Kimberlites and lamproites are generally regarded as having been intruded upwards through a series of deep-seated tension fractures, often in **areas of regional doming** and **rifting**, in which the magmas started to consolidate as dykes. Then highly gas-charged magma broke through explosively to the surface at points of weakness, such as cross-cutting fractures, to form the explosion vent, which was filled with fluidized fragmented kimberlite or lamproite and xenoliths of country rock.

9.4 Some examples

(a) *The Argyle AK1 Lamproite in Western Australia.* This extremely productive deposit (Fig. 9.4) contrasts strongly with the kimberlite diatreme of Fig. 9.3. It is richer in diamond than any known kimberlite — proven

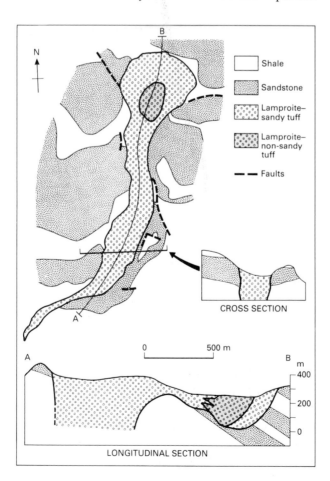

Fig. 9.4 Map and sections of the Argyle AK1 Lamproite.

reserves in 1990 were 94.2 Mt at 4 ct t^{-1} and probable reserves 5.9 Mt at 2.8 ct t^{-1}. However, the gem content is only 5%; cheap gem, 40%, and industrial diamonds, 55%. Ore processing exceeds 5 Mt p.a.

(b) *Kimberlites of Lesotho*. Among the many non-productive pipes of this small country there are a small number of very profitable but low grade pipes, for example the Letseng-La-Trae whose grade is only 0.309 ct t^{-1}. However, **13% of the output is of 10 ct and larger stones of great value**.

(c) *Russian Federation*. The Siberian kimberlite pipes (see Fig. 7.1) were discovered in 1954 and have been important producers over recent decades. Some 25% of production is of gem quality. Recent finds in the Kola Peninsula will increase Russian production and this field appears to extend into Finland.

9.5 Origin of diamonds

Graphite is the stable dimorph of carbon at surface temperatures and pressures; **diamond is metastable** and, if heated, will burn at 700°C to CO_2. Extremely high temperature and pressure are required to form diamond rather than graphite from pure carbon — of the order of 1 000°K and 3.5 GPa; equivalent, in areas of 60 km thick continental crust, to a depth of about 117 km. For many decades there has been a very active debate as to whether diamonds crystallized from the magmas which cooled to form the igneous rocks in which they are now found (phenocrysts), or whether they were picked up by these magmas as exotic fragments derived from the diamond stability field within the upper mantle (xenocrysts). The much greater abundance of diamonds in eclogite xenoliths than in the surrounding kimberlite suggests that they have been derived from disaggregated eclogite. Convincing radiometric dating evidence that **many, perhaps most, diamonds are xenocrysts** and often much older than the magmas that transported them to the surface has been accumulating since 1984. This and other evidence is leading workers to the opinion that diamonds grew stably within the upper mantle in eclogite and ultramafic rocks. The estimated conditions of equilibration for ultramafic suite minerals coexisting with diamonds suggest that this growth takes, or took, place in a layer between about 132 and 208 km in depth beneath continents and 121–197 km beneath oceans, at temperatures of 1 200–1 600°K, provided that carbon is (or was) present. Thus, any magma that samples a diamondiferous zone may bring diamonds to the surface if it moves swiftly enough to prevent the inversion of diamond to graphite. The **speed of ascent of such magmas** has been calculated to be **about 70 km h^{-1}**. Slow ascent could allow time for the **resorption** of diamonds by transporting magmas as the pressure decreased. That resorption is important appears from the fact that although some pipes produce good crystals (cubes, octahedra or cubo-octahedra), good crystals are rare in many kimberlites and only rounded forms are present. Such rounded forms are rare or absent in diamonds included in xenoliths and thus protected from corrosive magma.

9.6 Further reading

Craig J.R., Vaughan D.J. and Skinner B.J. (1996) *Resources of the Earth: Origin, Use and Environmental Impact.* Prentice-Hall, Upper Saddle River. Parts of Chapters 3 and 10.

Evans A.M. (1993) *Ore Geology and Industrial Minerals: An Introduction.* Blackwell Scientific Publications, Oxford. Chapter 7 provides references and further discussion.

10: The Carbonatite-alkaline Igneous Ore Environment

10.1 Carbonatites

In this chapter we look at:
■ some unusual igneous rocks and their crustal occurrence,
■ the common and not so common materials we can win from them.

Carbonatite complexes consist of intrusive magmatic carbonates and associated alkaline igneous rocks, mainly ranging in age from Proterozoic to Recent; **only a few Archaean examples are known**. They belong to alkaline igneous provinces and are **generally found as clusters in stable cratonic regions sometimes with major regional doming and rift faulting**, such as the East African Rift Valley and the St Lawrence River Graben **or at major fault intersections** as in the Kola Peninsula, R.F. There are, however, exceptions where carbonatite complexes are not associated directly with any alkalic rocks; moreover, not all alkalic rock provinces and complexes have associated carbonatites.

10.2 Economic aspects

The most important products of this environment are **phosphorus** (from apatite), **magnetite**, **niobium** (pyrochlore), **zirconia** and **rare earth elements** (monazite, bastnäsite). To date, only one carbonatite complex is a major producer of **copper** (Palabora, R.S.A.). Other economic minerals in carbonatites include **fluorite**, **baryte** and **strontianite**, and the carbonatites themselves are useful as a source of **lime** in areas devoid of good limestones. About twenty-two mines are now operating on nineteen different carbonatites in fourteen countries. Carbonatites and associated alkaline rocks carry an impressive resource of economic minerals and a growing number are under development. They supply most of the world's niobium (Brazil, Canada), R.E.E. (U.S.A., China) and vermiculite (R.S.A.). (The other important source of R.E.E. is monazite in beach placers.) **Carbonatites** are highly susceptible to weathering and therefore **may have residual, eluvial and alluvial placers associated with them**.

10.3 Some examples

(a) *The Mountain Pass occurrences, California.* The Mountain Pass Carbonatite of California occurs within the Rocky Mountain area, but it is a Precambrian intrusion (1 400 Ma) into Precambrian gneisses so that its present tectonic environment is a function of considerably later deformation, and this may be the reason why it is not now in an obviously alkaline igneous province. The deposits lie in a belt about 10 km long and 2.5 km wide. The metamorphic country rocks have been intruded by potash-rich igneous rocks and the R.E.E.-bearing carbonate rocks are related spatially, and probably genetically, to these granites, syenites and shonkinites. The rare earth elements are carried by **bastnäsite** and **parisite**, these minerals being **in carbonate veins** that are most abundant in and near the largest shonkinite-syenite body. Most of the 200 veins that have been mapped are less than 2 m thick. One mass of carbonate rock, however, is about 200 m in maximum width and about 730 m long and is a huge repository of R.E.E. Carbonate minerals make up about 60% of the veins and the large carbonate body; they are chiefly calcite, dolomite, ankerite and siderite. The other constituents are baryte, bastnäsite, parisite, quartz and variable small quantities of twenty-three other minerals. The **R.E.E. content** of much of the orebody is **5–15%**.

(b) *The Palabora Igneous Complex.* This Proterozoic Complex (*c.* 2 047 Ma) lies in the Archaean of the north-eastern Transvaal. It resulted from alkaline intrusive activity in which there were emplaced in successive stages pyroxenite, syenite and ultrabasic pegmatoids. The first intrusion was that of apatite-rich, phlogopite pyroxenite, kidney-shaped in outcrop (but forming a pipe in depth) and about 6 × 2.5 km (Fig. 10.1). Ultrabasic pegmatoids were then developed at three centres within the pyroxenite pipe. In the central one, phoskorite (magnetite-olivine-apatite rock) and banded carbonatite were emplaced to form the Loolekop carbonatite–phoskorite pipe, which is about 1.4 × 0.8 km. Fracturing of this pipe led to the intrusion of a dyke-like body of transgressive carbonatite and the development of a **stockwork of carbonatite veinlets**. The zone along which the main body of transgressive carbonatite was emplaced suffered repeated fracturing, and mineralizing fluids migrated along it depositing **copper sulphides** which healed the fine discontinuous fractures. These near vertical veinlets occur in parallel-trending zones up to 10 m wide, although individually the veinlets are usually less than 1 cm wide and do not continue for more than 1 m. Chalcopyrite with minor cubanite is the principal copper mineral in the core, especially in the transgressive carbonatite, which runs 1% Cu, while bornite is dominant in the lower grade banded carbonatite and phoskorite. Magnetite and **baddeleyite** are recovered from both phoskorite and carbonatite and **apatite** from the phoskorite. A variety of evidence suggests that the copper mineralization is, like these minerals, part of the carbonatite magmatism.

Diamond drilling has shown that the orebody continues beyond

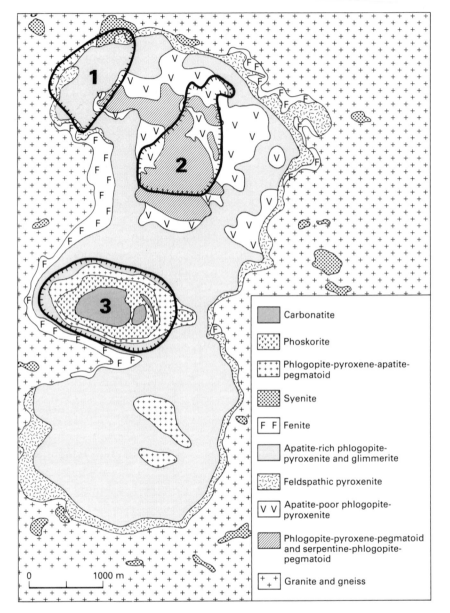

Fig. 10.1 Simplified geological map of the Palabora Alkaline Igneous Complex, Transvaal, R.S.A. The north-westerly trending dolerite dyke swarm has been totally omitted and some of the satellite intrusions of syenite. **1**, Open pit operation of Foskor which produces phosphate. **2**, The vermiculite open pit. **3**, The Palabora open pit at Loolekop, which produces copper, magnetite etc. Both **2** and **3** are operated by the Palabora Mining Co., who sell their apatite concentrate to Foskor. North is parallel to the vertical map margins.

Legend:

- Carbonatite
- Phoskorite
- Phlogopite-pyroxene-apatite-pegmatoid
- Syenite
- F F Fenite
- Apatite-rich phlogopite-pyroxenite and glimmerite
- Feldspathic pyroxenite
- V V Apatite-poor phlogopite-pyroxenite
- Phlogopite-pyroxene-pegmatoid and serpentine-phlogopite-pegmatoid
- + + Granite and gneiss

0   1000 m

1000 m below the surface. Ore reserves are about 300 Mt grading 0.69% Cu. This is the **world's second largest open pit mine** and the size of the operation can be judged from the fact that in the first twenty-four years of operation to 1989 the total of material mined was 1680 Mt, the ore milled 558 Mt and 2.7 Mt of copper was produced. With a **cut-off grade of 0.15% Cu** and an average mill grade of 0.48–0.54%, Palabora is **one of the most efficient low grade copper mines in the world.**

There are **three separate open pits at Palabora**. The carbonatite and phoskorite are worked for copper with by-product magnetite, apatite, gold, silver, P.G.M., baddeleyite (ZrO_2), uranium, nickel sulphate and sulphuric acid. About 2 km away in the same alkaline complex is the **Foskor Open Pit** in an apatite-rich pyroxenite that forms **the world's largest igneous phosphate deposit**. Reserves in the pit area alone, which only covers part of the pyroxenite, are 3 Gt of apatite *concentrates* (36.5% P_2O_5). In 1984, Foskor produced 2.6 Mt of phosphate concentrate, nearly 9 kt of baddeleyite concentrate and 43 kt of copper concentrate carrying 35% Cu. In 1990 Foskor increased its mill capacity to 3.7 Mt p.a. In a neighbouring pit (**2** on Fig. 10.1) **vermiculite**, a weathering product of phlogopite, is worked. This is the second largest vermiculite mine in the world; it started production in 1946 and in 1990 produced nearly 218 kt of concentrate (90% vermiculite). Evaluation of the adjoining P.P. & V. deposit in 1990 indicated the presence of 46 Mt of vermiculite, which makes it the world's largest orebody of this type.

(c) *The Kola Peninsula–Northern Karelia Alkaline Province.* Alkaline igneous complexes and their associated carbonatites show a broad spatial relationship to areas of hot spot activity, which may be accompanied by **doming and fracturing**, the Kola Peninsula (Fig. 10.2) being an excellent example. The Upper Palaeozoic alkaline igneous and carbonatite complexes of this region host a number of extremely large orebodies, of which the most important are Khibina, Kovdor and Sokli. The R.F. was the world's second largest producer of phosphate rock but now, due to the sharp decline in fertilizer use in the G.I.S., it is fourth after the U.S.A., Morocco and China. Much of the R.F. production comes from Khibina. This is a **ring complex** about 40 km across with inward dipping, **layered intrusions** of various alkaline rock types. One apatite-nepheline orebody forms an arcuate,

---

Vermiculite

Vermiculite sold at $U.S.130–200 a short ton (2 000 lb) in August 1996 depending on the particle size. Nearly all crude vermiculite is transported in the unexfoliated (i.e. unexpanded) form because of the far higher bulk density of the raw material. Exfoliation plants are sited close to final markets and are thus in most cases thousands of kilometres from the mines. Vermiculite is a form of mica which loses water and expands up to twenty times in size when it is heated. The light, porous product finds a wide range of uses—as an insulating material, particularly in the building industry, for fireproofing, for acoustic damping, as a carrier for fertilizers, and in the horticultural industry. It has extreme lightness, useful cation exchange properties and is free from health hazards. World production in 1994 was about 503 kt of which about 223 kt (44%) was in R.S.A. About 35% came from the U.S.A. and much of the rest from the Kovdor deposit, R.F. and Brazil.

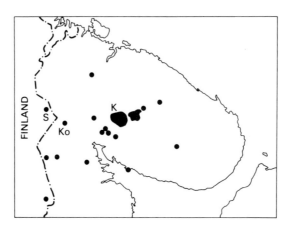

Fig. 10.2 Alkaline igneous and carbonatite complexes in the Kola Peninsula–Northern Karelia Alkaline Province. K, Khibina; Ko, Kovdor; S, Sokli.

irregular lens-shaped mass with a strike length of 11 km and a proved dip extension of 2 km. The thickness ranges from 10 to 200 m and averages 100 m, and at least 2.7 Gt of ore averaging 18% P_2O_5 are present. The apatite concentrates contain significant SrO and Re_2O_3 values. Annual production was 19 Mt in 1990 from four mines, by 1994 it was down to 7 Mt. Nepheline concentrates are also produced for the manufacture of alumina.

At Kovdor there is a late stage development of **apatite**–forsterite rocks and **magnetite** ores. The orebody is mined by open pit methods. Ore reserves are about 708 Mt grading 36% Fe and 6.6% P_2O_5. **Baddeleyite** forms a by-product and **vermiculite** is produced from a separate mining operation. In its mineralization and rock types this complex has affinities with Palabora. A drawback at Kovdor is the **high magnesium content** of the apatite concentrate, which presents processing **problems at fertilizer plants**.

Across the border in Finland is the **Sokli Complex**. With an areal extent of about 18 km², it is one of the world's largest carbonatites and it is remarkable for its partial cover of residual ferruginous phosphate rock, forming an apatite-francolite regolith of a type hitherto regarded as being of tropical origin. Proved reserves are over 50 Mt averaging 19% P_2O_5, with possible lower grade reserves of 50–100 Mt. The deposits vary from a few metres to 70 m in thickness. Pyrochlores with both U–Ta and Th enrichment occur in the carbonatites and apatite–magnetite mineralization in metaphoskorites which runs 11–22% Fe. Sokli is not at present being exploited but if phosphate prices rise it could well become a producer.

Not far away at **Siilinjarvi** is the only phosphate mine in western Europe. It produces about 0.5 Mt p.a. of apatite concentrates from a 10 km long carbonatite dyke dated at 2 600 Ma—probably the world's oldest carbonatite. Average ore is 10% apatite, 5% richterite, 15% calcite, 5% dolomite and 65% phlogopite. The by-products are well used: calcite and dolomite concentrates to make magnesium-rich agricultural lime, phlogopite concentrates go to paint and plastic industries and waste rock is crushed to provide aggregate for building and road construction.

As in a number of other parts of the world, **kimberlite occurrences accompany the carbonatites** and some of these have proved to be diamond bearing.

10.4 Further reading

Notholt A.J.G., Highley D.E. and Deans T. (1990) Economic Minerals in Carbonatites and Associated Alkaline Igneous Rocks. *Trans. Instn Min. Metall. (Sect. B: Appl. Earth Sci.)* **99**, B59–80.

11: The Pegmatitic Environment

11.1 Some general points

■ Many of our rarer metals are won from pegmatites which are often a mineral collector's paradise.
■ Their nature, occurrence, genesis and economic importance are considered in this chapter.
■ Rival deposit types producing some of these rare metals are noted.

Pegmatites are very coarse-grained igneous or metamorphic rocks, generally of granitic composition. Those of granulite and some amphibolite facies terranes are frequently indistinguishable mineralogically from the migmatitic leucosomes associated with them, but those developed at higher structural levels and often spatially related to intrusive, late tectonic granite plutons, are often marked by minerals with volatile components (OH, F, B) and a wide range of **accessory minerals containing rare lithophile elements**. These include Be, Li, Sn, W, Rb, Cs, Nb, Ta, R.E.E. and U, for which pegmatites are mined [rare element or (better) rare metal pegmatites]. Chemically, the *bulk* composition of most pegmatites is close to that of granite, but components such as Li_2O, Rb_2O, B_2O_3, F and rarely Cs_2O may range up to or just over 1%.

Pegmatite bodies vary greatly in size and shape. They range from pegmatitic schlieren and patches in parent granites, through thick dykes many kilometres long and wholly divorced in space from any possible parent intrusion, to pegmatitic granite plutons many kilometres squared in area. They form simple to complicated fracture-filling bodies in competent country rocks, or ellipsoidal, lenticular, turnip-shaped or amoeboid forms in incompetent hosts. Pegmatites are often classed as simple or complex. **Simple pegmatites** have simple mineralogy and no well developed internal zoning, while **complex pegmatites** may have a complex mineralogy with many rare minerals, such as pollucite and amblygonite, but their marked feature is the arrangement of their minerals in a zonal sequence from the contact inwards. An example of a complex, **zoned pegmatite** from Zimbabwe is given in Fig. 11.1. The crystals in complex pegmatites can be very large and at Bikita, for example, the spodumene crystals are commonly 3 m long. The Bikita Pegmatite was one of the world's largest Li–Cs–Be deposits; the main pegmatite was about 2 km long and 45–60 m thick, and the minerals of major economic importance were petalite, lepidolite, spodumene, pollucite, beryl, eucryptite and amblygonite. Cassiterite, tantalite and microlite were disseminated through quartz-rich zones in lepidolite-greisen, but these had been mined out by 1950. Mining ceased

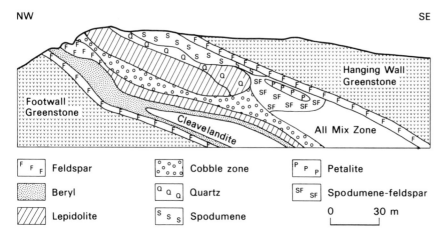

NW

SE

Footwall Greenstone

Hanging Wall Greenstone

Cleavelandite

All Mix Zone

F F F Feldspar

Beryl

Lepidolite

o o o
o o o Cobble zone

Q Q Q Quartz

S S S Spodumene

P P P Petalite

SF SF Spodumene-feldspar

0   30 m

Fig. 11.1 Section through the Bikita Pegmatite showing the generalized zonal structure and the important minerals of each zone. Cleavelandite is a lamellar variety of white albite.

in 1986 but the company will continue production for many years using stockpile material to produce petalite concentrates.

Pegmatites may occur singly or in swarms forming pegmatite fields, and these in turn may be strung out in a linear fashion to form **pegmatite belts**, one of the largest being the rare metal pegmatite belt of the Mongolian Altai, about 450 km long and 20–70 km broad, which contains over twenty pegmatite fields. Pegmatites in a particular field may show a **regional mineralogical–chemical zonation**. When a swarm of pegmatites is associated with a parent granite pluton, then the more highly fractionated pegmatites, enriched in rare metals and volatile components, are found at greater distances from the pluton. The degree of internal zoning also increases with an increase in rare metals and volatiles. The regional zoning may reflect the fact that melts enriched in Li, P, B and F have considerably **lower solidi** than H_2O-only saturated magma and can penetrate further from their source. This regional zoning is of course of great importance to the exploration geologist.

Most pegmatites, whether igneous or metamorphic in origin, have similar *bulk* compositions and these correspond closely to low temperature melts near the minima of Ab–An–Or–Q–H_2O systems. This is to be expected for melts developed by extreme magmatic differentiation or anatexis. Many rare metal pegmatites appear to be of magmatic origin; their unusual composition seems to be a consequence of retrograde boiling and the manner in which elements are partitioned between crystals, melt and volatile phases during the cooling of the magma. Bonding factors such as ionic size and charge largely prevent **many constituents, originally present in minor or trace concentrations** in the magmas, from being incorporated into precipitating crystals. They thus become more **highly concentrated in the residual melt** which, in the case of granites, is also enriched in water, since quartz and feldspars are anhydrous minerals.

The list of residual elements is long but includes Li, Be, B, C, P, F, Nb, Ta, Sn and W.

11.2 Some economic aspects

Pegmatitic deposits of spodumene, petalite, lepidolite and other Li minerals are exploited throughout the world for use in glass, ceramics, fluxes in aluminium reduction cells and the manufacture of numerous lithium compounds. These deposits often yield by-product Be, Rb, Cs, Nb, Ta and Sn. They may be internally zoned, or unzoned as are the highly productive lithium pegmatites of King's Mountain, North Carolina. The largest known pegmatitc lithium resources are in Zaïre in two laccoliths, each about 5 km long and 0.4 km wide. Reserves have been put at 300 Mt and Ta, Nb, Zr and Ti values have been reported. **Political instability**, however, is a major drawback as far as overseas investors are concerned and these projects are in abeyance. Pegmatites are also important as a source of **tantalum**, but it must be noted that the **largest reserves** — about 7.25 Gt of Ta — **are in slags** (running about 12% Ta_2O_5) produced during the smelting of tin ores in Thailand.

Lithium

The reader should note that lithium has already been produced from **brines** in Nevada, U.S.A., and a brine producer in Chile is now in production. In May 1989 the Japanese Industrial Research Institute announced the development of an efficient method of removing **lithium from sea water**. These new sources will present an economic challenge to the hard rock producers and the market price may suffer, as by 1984 lithium production had already exceeded demand.

Beryllium

The traditional source of beryllium (beryl) has now been overtaken by **bertrandite**, which occurs on a commercial scale in **hydrothermal deposits**. The main source of beryl is pegmatites, with the R.F. and Kazakhstan (900 t p.a.) being the world's largest producer, followed by Brazil with 850 t p.a. Hydrothermal bertrandite occurrences are typified by those at Spor Mountain, Utah where bertrandite is common in veins and disseminations in rhyolite intrusions and tuffs. The U.S.A. is now by far the biggest beryllium producer.

11.3 Uraniferous pegmatites

These pegmatites and pegmatitic granite have been exploited in a number of localities. Among the more important deposits are those of Bancroft, Ontario and the enormous Rössing Deposit in Namibia.

The Rössing Uranium Deposit, Namibia. This is **the world's largest uranium producer**. The operation is a **large tonnage**, **low grade** one — 15–16 Mt of ore p.a., grading 0.031% U_3O_8, being produced from an open pit and underground workings. The uranium mineralization occurs within a migmatitic zone, characterized by largely concordant relationships between uraniferous, pegmatitic granites and the country rocks. The pegmatitic granites have a very low colour index and are termed **alaskites**. About 55% of the uranium is in uraninite and **40% in secondary uranium minerals**. The arid climate of the Namib Desert played an important role in the formation of this deposit because, while giving rise to the enrichment of the primary mineralization with secondary minerals released by weathering, it prevented the leaching of this secondary mineralization by meteoric water.

11.4 Further reading

Čzerný P. (ed.) (1982) *Short Course in Granitic Pegmatites in Science and Industry.* Mineralogy Association of Canada, Winnipeg.

12: Orthomagmatic Deposits of Chromium, Platinum, Titanium and Iron Associated with Basic and Ultrabasic Rocks

Orthomagmatic ores of these metals are found almost exclusively in association with basic and ultrabasic plutonic igneous rocks—some platinum being found in nickel–copper deposits associated with extrusive komatiites (Chapter 13).

12.1 Chromium

Chromium is won only from **chromite** ($FeCr_2O_4$). This spinel mineral can show a considerable variation in composition, with magnesium substituting for the ferrous iron and aluminium, and/or ferric iron substituting for the chromium. It may also be so intimately intergrown with silicate minerals that these too act as an ore dilutant. Because of these variations there are **three grades of chromite ore**. The specifications for these are somewhat variable; typical figures are given in Table 12.1. **Metallurgical grade** ore was for many years marketed as ferrochrome for steel making, particularly stainless steel, but R.S.A. production is now moving towards value added steel products rather than raw ferrochrome. The non-metallurgical markets for chromite consume some 25% of the chromium ore mined and can be divided into **refractory**, **chemical** and **foundry** industries. The demand for refractory chromite has been reduced considerably by the demise of open-hearth steel furnaces, but some recent recovery has followed the use of chromite in steel refining ladles. Chemical grade material is enjoying a buoyant market with a firm demand for chromium chemicals for pigments, wood preservatives and leather tanning salts—three of the more important non-metallurgical uses. Sodium dichromate is also the raw material for chromium electroplating. As a foundry sand, chromite is suffering competition from zircon but does have some advantages over zircon sand.

Chromite ores are classified as lumpy or friable, depending on their cohesiveness. **Hard, lumpy ore** is required for the Perrin process, which produces low carbon ferrochrome (<0.03% C). **Fines** from the mining of lumpy ore and low grade ores are milled and concentrated to remove silicate gangue and, when present, magnetite is removed magnetically to improve the Cr/Fe ratio. The **fines** may be **pelletized** for smelting.

	Cr_2O_3	Cr/Fe	$Cr_2O_3 + Al_2O_3$	Fe	SiO_2
Metallurgical grade	>48%	>1.5	—	—	—
Refractory grade	>30%	Not critical	>57%	⪢10%	⪢5%
Chemical grade	>45%	—	—	—	⪢8%

Table 12.1 Chromite ore grades and specifications.

12.1.1 Occurrence

Three quarters of the world's chromium reserves are in the R.S.A. and 23% in Zimbabwe, most deposits outside these two countries being small. The only other countries with appreciable reserves are Kazakhstan, Albania, India and Turkey. **All economic deposits** of chromite **are in ultrabasic and anorthositic plutonic rocks.** There are two major types: stratiform and podiform (often referred to respectively as **Bushveld** and **Alpine types**). Each type yields about 50% of world chromite production, which was 10.0 Mt in 1994. Whilst podiform deposits belong to mobile belts, stratiform deposits (apart perhaps from the Fiskenaesset Complex of western Greenland) were intruded into stable Precambrian cratons. **Few environmental problems** arise from the mining and processing of chromite as in the Cr^{3+} state it is an essential nutrient.

12.1.2 Stratiform deposits

This type contains over 98% of the world's chromite resources. These deposits consist of layers (see Fig. 12.2) usually formed in the lower parts of stratified igneous complexes of either funnel-shaped intrusions (Bushveld, R.S.A., Great Dyke, Zimbabwe) or sill-like intrusions (Stillwater, U.S.A.; Kemi, Finland; Selukwe, Zimbabwe). The immediate country rocks of the complexes are ultrabasic differentiates of the parent gabbroic magma — dunites, peridotites and pyroxenites. The **layers** of these rocks **have great lateral extent, uniformity and consistent positions** within the complexes. The layers of massive chromite (chromitite) range from a few millimetres to over 1 m in thickness and can be traced laterally for as much as tens of kilometres. Orebodies may consist of a single layer or a number of closely spaced layers. **Chromite** in these deposits is **usually iron rich** but an outstanding exception is the Great Dyke with its high chromium ores.

(a) *Bushveld Complex.* This is **an enormous differentiated igneous complex** in the R.S.A. (Fig. 12.1), which is generally considered to have **resulted from the repeated intrusion of two main magma types** into partly overlapping conical intrusions — these eventually coalesced into

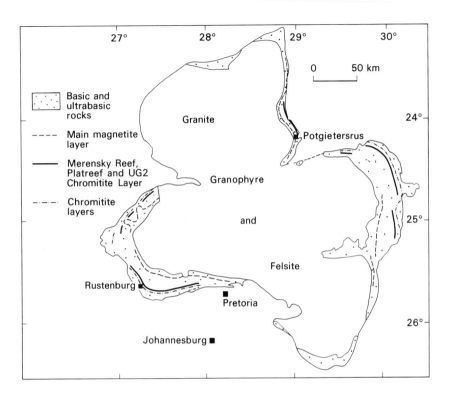

Fig. 12.1 Sketch map of the Bushveld Complex.

three larger magma chambers corresponding to the eastern, western and northern segments. The chromite occurs in the western and eastern outcrops of ultrabasic rocks, with layers a few centimetres to 2 m thick (Fig. 12.2). They make up an enormous tonnage. Assuming a maximum vertical mining depth of only 300 m gives a figure of 2.3×10^9 t, a figure that can be multiplied by 10 if lower grade deposits are included and the vertical mining depth is increased to 1 200 m. Potentially the largest orebodies are the LG3 and LG4 chromite layers present only in the western Bushveld. In these layers the **chromite grades 50% Cr_2O_3, Cr/Fe = 2.0**, the strike length is 63 km, the thickness 50 cm and the resources (300 m vertical depth) are 156×10^6 t. The **ore grades about 45% Cr_2O_3**. At present, however, the majority of mines exploit the 90–128 cm thick LG6 (Steelpoort) chromitite in the eastern Bushveld. The Bushveld chromite zone as a whole contains as many as 29 chromite layers or groups of layers. Above this zone is the **platinum-bearing Merensky Reef** (Figs 12.1 & 12.2), and near the top of the basic part of the complex, **vanadiferous magnetite layers occur**.

(b) *Great Dyke of Zimbabwe*. This early Proterozoic (2 470 Ma), mafic–ultramafic intrusion, 532 km long and 5–9.5 km broad, consists of **four narrow, layered complexes**. In cross section the mineral layers are synclinal (Fig. 12.3). Chromite layers occur along the entire length and individual

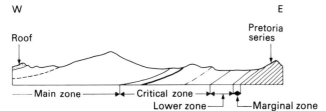

W   E

Roof   Pretoria series

Main zone   Critical zone

Lower zone   Marginal zone

Section showing major zones in the Bushveld Complex, north of Steelport. Length of section, 30.5 km. (After Hall, 1932)

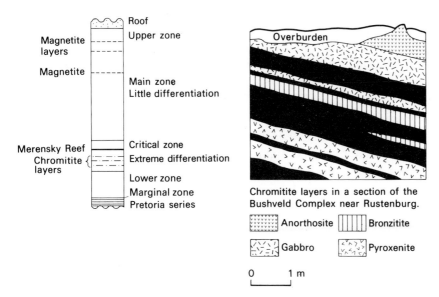

Roof
Upper zone

Magnetite layers

Magnetite

Main zone
Little differentiation

Merensky Reef
Chromitite layers

Critical zone
Extreme differentiation

Lower zone
Marginal zone
Pretoria series

Overburden

Chromitite layers in a section of the Bushveld Complex near Rustenburg.

| Anorthosite | Bronzitite |
| Gabbro | Pyroxenite |

0   1 m

Fig. 12.2 Sections showing the occurrence of econonic minerals in the Bushveld Complex.

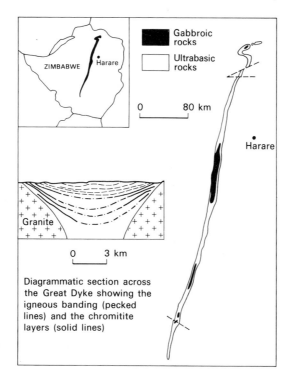

Gabbroic rocks

Ultrabasic rocks

ZIMBABWE   Harare

0   80 km

Harare

Granite

0   3 km

Diagrammatic section across the Great Dyke showing the igneous banding (pecked lines) and the chromitite layers (solid lines)

Fig. 12.3 Sketch diagrams illustrating the Great Dyke of Zimbabwe and the occurrence of chromitite layers in it.

layers extend across the entire width. The layers are in the range 5–100 cm and nearly all the **chromite is the high chromium variety**. Only layers 15 cm thick or more are mined. The Great Dyke has an estimated reserve of 10 000 Mt of chromite in as many as eleven persistent main seams.

12.1.3 Podiform deposits

The morphology of podiform chromite orebodies is irregular and unpredictable and they vary from sheet-like to pod-like in basic form, but can be very **variable in their shapes** (Fig. 12.4). Their **mass** ranges from **a few kilograms to several million tonnes**. Throughout most of the world, production comes from bodies containing 100 000 t or so and reserves greater than 1 Mt are most uncommon, except in the Urals where some very big bodies occur, as in the Kempirsai Ultramafic Massif, a large ophiolite complex in the southern Urals (Fig. 12.5). Most pod deposits are of high chromium type, but this deposit type is also the only source of **high aluminium chromite**. The chromite layers are in the range 1–40 cm, or above. **Podiform deposits occur in** irregular peridotite masses or peridotite–gabbro **complexes of the Alpine type** which are restricted mainly to orogenic zones, such as the Urals and Philippine island arc. The host intrusions are usually only a few tens of kilometres squared, or less, in area. Generally they are strongly elongated and lenticular in shape. Large numbers of these small intrusions occur in narrow zones (**serpentinite belts**) running parallel to regional thrust zones and the general trend of the orogen in which they occur. The intrusions are usually layered, but the layering does not often show the perfection nor the

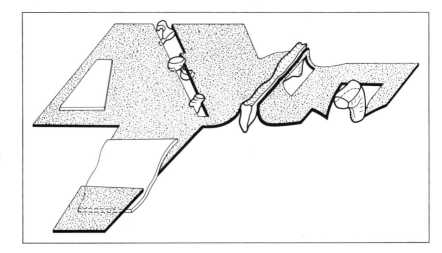

Fig. 12.4 Diagram showing the shapes of podiform chromite deposits in New Caledonia and their relationships to the plane of the foliation in the host peridotites. (The foliation is nearly always parallel to the compositional banding.) Note the disturbance of the foliation around the two deposit types on the right. The tabular deposits are of the order of 50–100 kt.

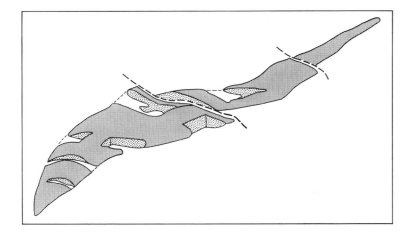

Fig. 12.5 The Molodezhoe Orebody, Southern Urals. Length 1.4 km, width 200–300 m, maximum thickness 140 m, ore reserves about 90 Mt.

continuity of the stratiform deposits. The largest known podiform deposits are those of the Urals, Albania, the Philippines and Turkey. Significant deposits occur in Cuba, New Caledonia, the former Yugoslavia and Greece.

12.1.4 Genesis of primary chromite deposits

In most, perhaps all, stratiform complexes, emplacement of the magma and crystallization took place in **a stable cratonic environment** and many delicate primary igneous and sedimentary features, in particular layering, are preserved. They were thus originally emplaced in the upper crust. The host rocks of podiform deposits, on the other hand, probably originally crystallized in the mantle and were then incorporated into **highly unstable tectonic environments** in the crust by movement up thrusts and reverse faults. They are usually part of the **ophiolite suite** and were probably developed originally at spreading ridges. Despite their tectonic deformation they often preserve layering and textures due to crystal settling, some identical with those in stratiform deposits, indicating a common origin. The transport of the peridotite and chromite from the upper mantle into the upper crust, where they now are, probably occurred by **plastic flowage** at high temperatures possibly over many kilometres; this fragmented the original layering and has produced many metamorphic features in the chromite and their host rocks. The peridotites have usually suffered a **high degree of serpentinization** in contrast to the stratiform deposits, in which it is relatively negligible. There is no doubt then that the chromite deposits of both stratiform and podiform deposits are **orthomagmatic**. Most workers now consider that these monominerallic cumulate layers are developed when a batch of new magma is injected into still crystallizing older magma resulting in a chemical interaction at the liquid interface.

12.2 Platinum group metals (P.G.M.)

These six elements—Pt, Pa, Rh, Ir, Ru, Os—always occur together in nature in the same restricted geological settings but platinum is by far the most abundant. Production reserve and grade figures are given in Tables 12.2 and 12.3.

12.2.1 Primary deposits

There are **two principal types of deposit** exemplified firstly by those associated with the basic intrusions of Noril'sk-Talnakh, R.F. and Sudbury, Ontario; and secondly by those of the basic–ultrabasic intrusions of the Bushveld Complex, R.S.A. and Stillwater, Montana. In the former type, platinum minerals occur in **liquation ores at the base of the intrusions** and in the latter, **platinum occurs in** the Merensky Reef and similar **layers** some 2000 m **above the base of the intrusions**. The platinum in these deposits occurs mainly as arsenides, sulphides and antimonides. A small production is won from platinum placers. The Sudbury copper–nickel ores will be covered in the next chapter. Attention will therefore be focused here on the Bushveld Complex.

Location	Production (kg)
R.S.A.	164 806
R.F.	58 000[e]
Canada	14 970
Japan	12 700
U.S.A.	8 400[e]
Colombia	1 130
Others	946
Total	~261 000

Table 12.2 Platinum and palladium production in 1994.

Deposit	Tonnage	Grade (ppm)
Bushveld Complex	60 388	8.27
Stillwater Complex	1 057	22.3
Great Dyke	7 892	4.7
Noril'sk-Talnakh	6 200	3.8
Sudbury	217	0.9

Table 12.3 Principal reserves of P.G.M. (in tonnes of metal) in major deposits and their grade.

P.G.M. smelters encounter **similar environmental emission problems to** those affecting **base metal smelters**. The principal toxic effect, platinosis, has to be guarded against only within smelters, refineries or chemical plants.

(a) *Bushveld Complex.* In this intrusion there are three very extensive deposits: the Merensky Reef, the UG2 Chromitite Layer (both in the western and eastern Bushveld) and the Platreef in the Potgietersrus area. The first-named occurs in the Merensky Zone—an exceedingly persistent igneous zone traced for over 200 km in the western Bushveld where it has been proved to contain workable ore for over 110 km (Fig. 12.1). This zone is also well developed in two other districts. It is 0.6–11 m thick and generally consists of dark-coloured norite.

The Merensky Reef is a thin sheet of coarsely crystalline pyroxenite with a pegmatitic habit that lies near the base of the Merensky Zone. It has a thickness of 0.3–0.6 m. Chromitite bands approximately 1 cm thick mark the top and bottom of the reef and are enriched with platinum metals relative to the reef. **Copper** (0.11%) and **nickel** (0.18%) are also recovered.

The UG2 Chromitite Layer lies 150–300 m below the Merensky Reef. It is 60–250 cm thick, carries 3.5–19 g t^{-1} platinum together with copper and nickel values and occurs within a bronzite cumulate. Reserves are about 5.42×10^9 t. **Chromium** is also produced.

The Platreef has been correlated with the Merensky Reef but it is much thicker, being up to 200 m with rich mineralization over thicknesses of 6–45 m and the mineralization is never far from the floor of the intrusion. Grades are very irregular, estimates ranging from 7 to 27 ppm total P.G.M. Ore reserves are about 4.08×10^9 t.

(b) *Other deposits.* The South African ores are essentially **platinum ores with by-product nickel and copper**. A horizon, the J-M Reef, similar to the Merensky Reef, has been traced within the Banded Zone of the Stillwater Complex, Montana over its entire exposed length of 39 km. Grades average about 22 g t^{-1} but Pt/Pa is about 0.33 : 1 compared with 3 : 1 in the Merensky Reef.

The genesis of this deposit type is still much debated and suggested hypotheses range from immiscibility to hydrothermal activity.

The second deposit type is certainly the product of immiscibility and carries much lower grade ores. At Sudbury, Ontario, the platinum metals are **by-products of copper–nickel ores** and the grades of the platinum metals are much lower, being about 0.6–0.8 g t^{-1}. Similar ores to those at Sudbury occur at Noril'sk in Siberia and the grade is said to average 3.8 g t^{-1}.

12.3 Titanium

Titanium metal and TiO_2 pigments are the two main products of titanium mines and on them large industries depend. The **pigment industry consumes over 90% of all titanium mined**. Only titanium-bearing oxide minerals with more than about 25% TiO_2 have present or potential economic value and all titaniferous silicate minerals are valueless. The most important oxides are **rutile** (>95% TiO_2) and **ilmenite** (about 52% TiO_2). The market price of a titanium mineral concentrate depends on its TiO_2 content **and** its suitability for a given industrial process. This means that different deposits obtain very different prices for their product. At present **shoreline placer deposits** supply **over half the annual production**, with nearly all the rest coming from primary magmatic sources. Titanium production was one of the few sectors of the minerals industry that suffered no downturn in the 1980s and prices are climbing. The development of substitutes for titanium pigments or metal seems remote and, as suggested in Chapter 2, titanium appears to be one of the few metals with a bright future. The market situation will change considerably over the coming decades as most currently known shoreline placers will be exhausted in about 30 years. New types of deposit are already being sought. **Titanium** is not toxic and its **mining should not present a chemical pollution threat** to the environment.

Future titanium deposit types could be eclogites carrying 6% or more TiO_2, rutile-bearing contact-metasomatic zones of alkalic anorthosites, perovskite-bearing pyroxenites, rutile by-products of porphyry copper deposits and placer deposits on continental shelves.

In this chapter we will consider **orthomagmatic ilmenite deposits**; these are always associated with anorthosite or anorthosite–gabbro complexes. At Allard Lake, Quebec, the ores grade 32–35% TiO_2 and occur in anorthosites. The **Lac Tio deposit** contains about 125 Mt ore and, at present, this single deposit supplies 19% of world production. It lies 40 km inland from Havre St Pierre on the north shore of the St Lawrence River in an uninhabited area and is reached by a company railway. The orebodies form **irregular lenses**, narrow **dykes**, large **sill-like masses** and various combinations of these forms. Some of these clearly cut the anorthosite and appear to be later in age.

The world's largest ilmenite orebody is at **Tellnes** in the anorthosite belt of southern Norway about 120 km south of Stavanger. The deposit is boat-shaped, elongated north-west, 2.3 km long, 400 m wide and about 350 m deep (Fig. 12.6). It **occurs in the base of a noritic anorthosite**. Proven reserves are 300 Mt of 18% TiO_2, 2% magnetite and 0.25% sulphides (pyrite and Cu–Ni sulphides). The ilmeno–norite intrudes the anorthosite with which it has sharp contacts. The ilmeno–norite intrusion's shape is apparently an original feature of intrusion. The annual production is 2.76 Mt of ore, and extraction of this by opencast methods means that 1.36 Mt of waste have to be removed.

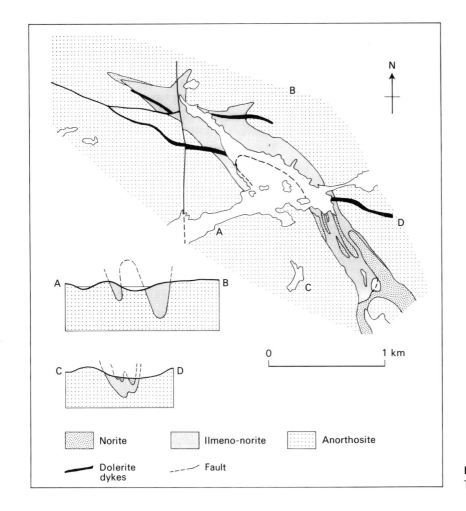

Fig. 12.6 Map and sections of the Tellnes titanium orebody.

Field relations, textures and the results of experimental petrology support the interpretation of these ores as igneous rocks formed from magmas generated by immiscibility within parent intrusions.

12.4 Iron and vanadium

Many small to medium-sized magnetite deposits occur in gabbroic intrusions but the really big tonnages occur in the stratiform lopoliths. Some of these deposits are worked for **vanadium** as well as for iron. This is the case with the vanadiferous magnetite in the Upper Zone of the Bushveld Complex (Fig. 12.1). This carries 0.3–2% V_2O_5. Values above 1.6% V_2O_5 are usually found in the Main Magnetite Layer and several thinner layers below it, but only the main layer (some 1.8 m thick) can be considered as

Vanadium from oil!
At the present time about 7% of vanadium production is from heavy crude oils rich in sulphur that may contain several hundred parts per million vanadium; such oils are exported from Venezuela and Mexico. Production from this source will probably increase in the future.

A minor amount of by-product vanadium is won from some uranium deposits (18.3).

ore. Based on a vertical limit of 30 m for opencast mining, the reserves are about 1 030 Mt. Much more iron ore than this is present, but it is spoilt by a titanium content of up to 19%. About 50% of total world vanadium production (~33 467 t in 1988) now comes from the Bushveld Complex. Similar occurrences are present in the R.F. (29%) and China (12%). Most vanadium production (about 85%) goes into steel manufacture, 10% into titanium alloys and 5% into chemicals. As **some vanadium compounds can be poisonous** care must be taken in their handling but **environmental problems** related to vanadium mining are **not significant**.

12.5 Further reading

Duke J.M. (1989) Magmatic Segregation Deposits of Chromite. In Roberts R.G. & Sheahan P.A. (eds) *Ore Deposit Models*, 133–43. Geological Association of Canada, Memorial University, Newfoundland.

Force E.R. (1991) *Geology of Titanium-mineral Deposits*. Spec. Pap. 259, Geological Society of America, Boulder, Colorado.

Naldrett A.J. (1989) *Magmatic Sulphide Deposits*. Oxford University Press, New York.

13: Orthomagmatic Copper–nickel–iron (-platinoid) Deposits Associated with Basic and Ultrabasic Rocks

13.1 Introduction

World **nickel** metal production is currently running at about 0.88 Mt p.a. It is a metal that **commands a high price**: $U.S. 6349 t^{-1} compared with $U.S. 2312 t^{-1} for copper in 1994. Nickel is produced from **two principal ore types**: nickeliferous laterites and nickel sulphide ores. We are concerned with the latter deposit type in this chapter. These deposits usually carry **copper**, often in economic amounts, and sometimes recoverable platinoids. Iron is produced in some cases from the pyrrhotite concentrates, which are a by-product of the dressing of these ores. Nickel sulphide deposits are not common and so **just a few countries** are **important for production of nickel sulphide ores**; Canada is pre-eminent, while the R.F. and Australia are the only other important producers. A small production comes from Zimbabwe, R.S.A., Botswana and Finland.

The mineralogy of these deposits is usually simple, consisting of pyrrhotite, pentlandite [$(Fe,Ni)_9S_8$], chalcopyrite and magnetite. Ore **grades** are somewhat **variable**. The lowest grade of a working deposit in western countries appears to be an Outukumpu (Finnish) mine working 0.2% Ni ore. This low grade can be compared with the very high grade sections of some Western Australian deposits that run about 12% Ni, and the Kambalda ores range from 8% to 22% Ni in 100% sulphide ore. Of course, the overall grade for Australian deposits is less than this because lower grade ore is mined with these high grades.

All nickel sulphide deposits are associated with basic or ultrabasic igneous rocks. There is both a spatial and a geochemical relationship in that deposits associated with gabbroic igneous rocks normally have a high Cu/Ni ratio (e.g. Sudbury, Ontario; Noril'sk, R.F.), and those associated with ultrabasic rocks a low Cu/Ni ratio (e.g. Thompson Belt, Manitoba; Western Australian deposits).

13.2 Ultrabasic and basic rocks with related nickel sulphide mineralization

There are a number of different types of basic and ultrabasic rocks; not all

Nickel—shortened form of *Kupfernickel*, literally copper demon, so called by German miners because it was mistakenly thought to contain copper, but now renowned as one of the metals of stainless steel and coinage—is won from lateritic and sulphide ores. We are concerned in this chapter with the latter and will note:
- their largely magmatic origin and association with basic and ultrabasic rocks,
- their mainly early and middle Precambrian age,
- the important komatiitic association in orogenic settings and the gabbroic association in cratonic regions,
- the importance of sulphur-rich magmas for their formation and the origin of the sulphur.

these have nickel sulphide deposits associated with them and therefore, in exploring for these deposits, **it is important to know which classes of basic and ultrabasic rocks are likely to have associated nickel sulphide ores**. Particular combinations of rock type and tectonic setting have proved to be especially productive. These are:

1 komatiitic and tholeiitic flows and intrusions in active orogenic settings e.g. Archaean greenstone belts (Kambalda, Agnew, Western Australia; Pechenga, R.F.);

2 intrusives emplaced in cratons which may be divided into:

    (a) noritic rocks intruded into an area that has suffered a catastrophic release of energy, e.g. an astrobleme (Sudbury, Canada); and

    (b) intrusions associated with flood basalts in intra-continental rift zones (Noril'sk-Talnakh, R.F.; Duluth, U.S.A.).

13.2.1 Bodies in an orogenic setting—the volcanic association

Two broad groups of ultrabasic and basic bodies can be seen in this setting: bodies coeval with plate margin volcanism and syntectonic intrusions. The first group occurs in the Archaean and Proterozoic greenstone belts and can be divided into the **tholeiitic and komatiitic suites**. The tholeiitic suite contains the picritic and anorthositic classes. The anorthositic class is important for titanium mineralization but so far no substantial nickel mineralization has been found in rocks of this class. The **picritic class is an important nickel ore carrier** and ultrabasic rocks in this class occur as basal accumulations in differentiated sills and lava flows, some having basal sulphide segregations — the Dundonald Sill of the Abitibi Greenstone Belt, Canada, is a good example. The tholeiitic activity in this and other areas was often contemporaneous with komatiitic volcanicity. The **komatiitic suite** is a **much more important carrier of nickel mineralization**. Komatiites are both extrusive and intrusive, and ultrabasic members are believed to have crystallized from liquid with up to 35 wt % MgO and carrying 20–30% of olivine phenocrysts in suspension. In some flows and near surface sills, quench textures (probably due to contact with sea water and consisting of platy and skeletal olivine and pyroxene growths) are present in the upper part. This is called **spinifex texture**. These flows clearly crystallized from magnesian-rich undifferentiated magma, extruded (in the case of 35 wt % MgO) at up to 1650°C. **Spinifex textures resemble those of silica-poor slags**, having a low viscosity and a high rate of internal diffusion — ideal conditions for the sinking of sulphide droplets to form accumulations at the flow bottom [see Section 6.2.2(b)]. The synorogenic intrusions are of little importance as carriers of nickel sulphide ores, but constitute a small resource for the future.

13.2.2 Bodies emplaced in a cratonic setting—the plutonic association

There are three main groups to be noted. The first is an important metal producer because it is that of the **large stratiform complexes**. In the last chapter, we noted the importance of the Bushveld Complex and its by-product nickel–copper won from the platinum-rich horizons. The **Sudbury intrusion** also belongs to this group and it **hosts the world's greatest known concentration of nickel ores**. The overall composition of this group is basic rather than ultrabasic but a lower ultrabasic zone is usually present. Sudbury is a notable exception to this rule but it possesses a sublayer rich in ultrabasic xenoliths probably derived from a hidden layered sequence.

The second group includes intrusions related to flood basalts and usually associated with the early stages of continental rifting. Very important here are the gabbroic intrusions of the **Noril'sk-Talnakh Nickel Field**. The Duluth Complex and the troctolite hosting the Great Lakes nickel deposit of Ontario are included in this group as they are both in a **rift setting** and are petrogenetically related to the Keweenawan flood basalts. Thirdly there is a group of medium-sized and small intrusions associated with rifted plate margins and ocean basins which we can divide into two subgroups. First are a number of intrusive and extrusive, mafic and ultra-mafic bodies (including komatiites) emplaced within Proterozoic sedi-mentary sequences that appear to be parts of **a rifting event along Archaean continental margins**, particularly the Circum-Superior Belt of Canada. This belt includes the **Thompson Belt, Manitoba and other belts carrying nickel deposits**. The second subgroup contains intrusions associ-ated with much more recent continental rifting, such as Skaergaard and Rhum, which carry nickel mineralization of only academic interest.

13.3 Relationship of nickel sulphide mineralization to classes of ultrabasic and basic rocks

Figure 13.1 indicates the relationship between known reserves plus past production of nickel in the main deposits of the world and the rock groups described above. **Apart from** the unique position of **Sudbury**, production from there being responsible for much of the Canadian section of group 2, **komatiitic magmatism [1(ii)] is clearly the most important**. Tholeiitic vol-canism [1(i)] is much less important. Deposits near the basal contacts of the Stillwater and Duluth Complexes (group 2) are low grade disseminated deposits which are unlikely to be producers in the near future. The Noril'sk-Talnakh field, like Sudbury, has many unusual features, which further emphasizes the importance of group 1(ii) as the best bet for further nickel exploration.

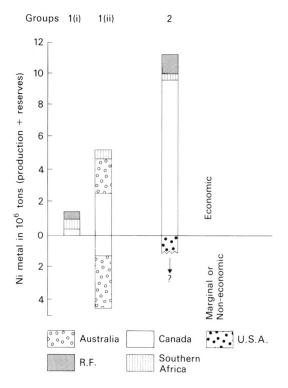

Fig. 13.1 Present reserves plus past production of sulphide nickel as a function of the type of host rock as described in Section 13.3.

13.4 Genesis of sulphur-rich magmas

For a rich concentration of magmatic sulphides it is necessary that: (i) the host magma is saturated in sulphur; and (ii) a reasonably high proportion of sulphide droplets can settle rapidly to form an orebody. Slow settling may give rise to a disseminated, uneconomic ore. **The production of a high proportion of immiscible sulphides is possible if the magma assimilates much sulphur from its country ricks**, e.g. Duluth, Noril'sk, **or if the magma carries excess amounts of mantle-derived sulphides**, e.g. some komatiites. We can often differentiate between these two sources (crustal and mantle) by examining variations in $^{34}S/^{32}S$. Published data for this ratio indicate that the **sulphur** in the Sudbury orebodies and the ore associated with komatiites at the Alexo Mine, Ontario and in many Western Australia orebodies was derived **from mantle sources**. At Noril'sk, on the other hand, the high ^{34}S values indicate **a crustal sedimentary source**. The Noril'sk Gabbro is Triassic (an unusual and perhaps unique situation as practically all economic nickel mineralization is early to middle Precambrian), and its magma is intruded through **gypsum beds**, where it is thought to have picked up ^{34}S-rich sulphur. A comparable example is the Water Hen intrusion of the Duluth Complex which also carries ^{34}S-rich sulphides. This is believed to have gained its sulphur by assimilating sulphur-rich sediments of the Virginia Formation. Clearly basic intrusions in sulphur-rich country rocks should be prospected carefully for nickel mineralization!

13.5 Nickel mineralization in time and depletion of sulphur in the mantle

The only large nickel sulphide deposits which are younger than 1700 Ma are Duluth (1115 Ma) and Noril'sk (Triassic). As we have seen, both host intrusions probably acquired most of their sulphur by assimilation of crustal rocks. This restriction of economically significant orthomagmatic nickel deposits to the Archaean and early Proterozoic may reflect **sulphur depletion of the upper mantle** caused by many cycles of plate tectonic or similar processes. On the other hand, the concentration of nickel sulphide mineralization early in the stratigraphical column may be related largely to the **concentration of komatiitic activity early in earth history**, no matter whether the necessary sulphur was acquired deep within the mantle or by assimilation at or near the surface.

13.6 Origin of the metals

There is no difficulty here. Ultrabasic and basic magmas are rich in iron and in trace amounts of copper, nickel and platinoid elements. These would be scavenged by the sulphur to form sulphide droplets.

13.7 Examples of nickel sulphide orefields

13.7.1 Volcanic association

Here we are concerned mainly with the komatiitic suite. In some of these, sulphides occur at or near the base of the flow or sill suggesting gravitational settling of a sulphide liquid. Typical sections through two deposits are given in Fig. 13.2. These have certain features in common:

1   massive ore at the base (the banding in the Lunnon orebody is probably the result of metamorphism);

2   a sharp contact between the massive ore and the overlying disseminated ore, which consists of net-textured sulphides in peridotite;

3   another sharp contact between the net-textured ore and the weak mineralization above it, which grades up into peridotite with a very low sulphur content.

   In nearly all examples of this deposit type, the orebodies are in former topographical depressions beneath the ultramafic lava flow.

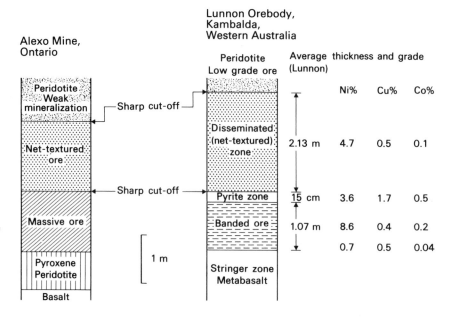

Fig. 13.2 Typical sections through two nickel sulphide orebodies associated with Archaean class 1(ii) ultrabasic bodies. The Alexo Mine is 40 km east-north-east of Timmins, Ontario.

The **Eastern Goldfields Province of Western Australia** is a typical Archaean region having a considerable development of greenstone belts (Fig. 13.3), with **nickel sulphide deposits of two main types**. The first consists of segregations of massive and disseminated ores at the base of small lens-like peridotitic to dunitic flows or subvolcanic sills at the bottom of thick sequences of komatiitic flows, e.g. Kambalda, Windarra, Scotia (Fig. 13.3.), which are termed volcanic-type deposits. The second type, dyke-like or sill deposits, occurs in largely concordant, but partially discordant, dunitic intrusions emplaced in narrow zones up to several hundred kilometres in length, e.g. Perseverance, Mount Keith. In the volcanic-type deposits most ores are at, or close to, the basal contact of the mineralized ultrabasic sequence (Fig. 13.4) and they are commonly associated with, or even confined to, embayments in the footwall. The **dyke-like deposits are much larger than the volcanic type**. For example, in the Perseverance deposit at Agnew, 33 Mt averaging 2.2% nickel with a cut-off grade of 1% were outlined and at Mount Keith 290 Mt grading 0.6%. They occur in long dunite dykes or sills, especially where these bulge out to thicknesses of several hundred metres; e.g. at Perseverance the host dyke thickens from a few metres to 700 m. The ores in these deposits are dominantly of disseminated type though some massive sulphides occur as at Perseverance. This deposit is now considered by some workers to be of volcanic-type in which case its large size contrasts strongly with the normal 1–5 Mt orebody size of this type.

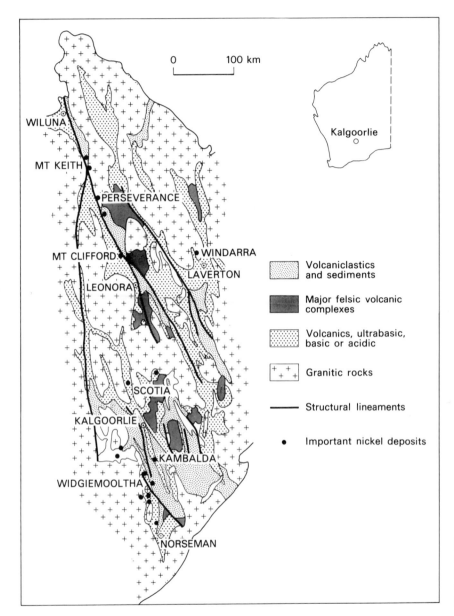

Fig. 13.3 Generalized geological map of the Eastern Goldfields Province of the Yilgarn Block showing some of the important nickel deposits.

Examples of deposits associated with **tholeiites** [1(i)] occur at **Pechenga**, R.F., in, or related to, either large intrusions showing an upward change from peridotite to gabbro or smaller relatively undifferentiated masses. Massive ores with nickel as high as 10–12% and copper varying from very low values to 13% occur together with lower grade disseminated ores. The deposits at **Lynn Lake**, Manitoba occur in sill-like bodies ranging from peridotite through norite to diorite. Massive and disseminated ores are present.

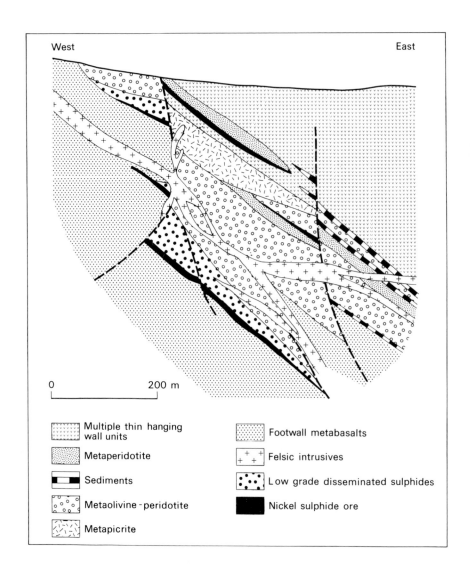

West   East

0   200 m

Multiple thin hanging wall units	Footwall metabasalts
Metaperidotite	Felsic intrusives
Sediments	Low grade disseminated sulphides
Metaolivine-peridotite	Nickel sulphide ore
Metapicrite	

Fig. 13.4 Section through the Lunnon and neighbouring ore shoots, Kambalda, Western Australia.

13.7.2 Plutonic association

(a) *Sudbury, Ontario.* The Sudbury Structure is a unique crustal feature lying just north of Lake Huron near the boundary of the Superior and Grenville Provinces of the Canadian Shield. The structure is about 60×27 km (Fig. 13.5) and its most obvious feature is the **Sudbury Igneous Complex** (1 849 Ma old), which consists of a Lower Zone of augite–norite, a thin Middle Zone of quartz-gabbro and an Upper Zone of granophyre. The Complex is believed to have the shape of a deformed funnel. At the base of the norite there is a discontinuous zone of inclusion and sulphide-rich norite and gabbro known as the Sublayer. In the so-called Offsets (steep to vertical, radial and concentric dykes that appear to penetrate downwards

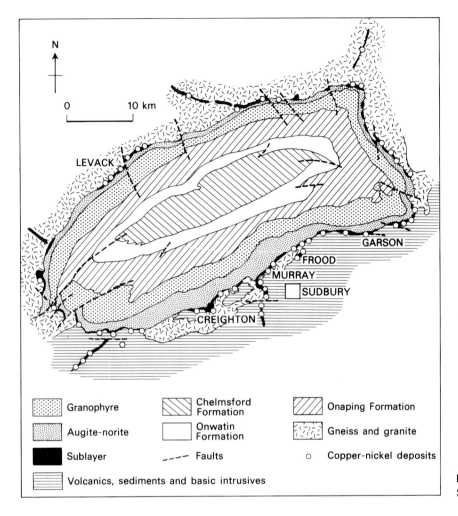

N

0   10 km

LEVACK

GARSON

FROOD

MURRAY

SUDBURY

CREIGHTON

Granophyre

Augite-norite

Sublayer

Volcanics, sediments and basic intrusives

Chelmsford Formation

Onwatin Formation

Faults

Onaping Formation

Gneiss and granite

Copper-nickel deposits

Fig. 13.5 Geological map of the Sudbury District.

into the footwall from the base of the Complex) the inclusion-rich sulphide-bearing rock is a quartz-diorite. The **Sublayer** and **Offsets are at present the world's largest source of nickel** as well as an important source of copper, cobalt, iron, platinum and eleven other elements. Inside the Complex is the Whitewater Group of clastic deposits that has suffered only slate grade metamorphism. Outside the Complex there are higher grade and migmatitic Proterozoic and Archaean rocks. Structural analysis of the rocks inside the Complex suggests that it was shortened by 30% in a north-westerly direction and was formely less elliptically shaped.

The **origin** of the Sudbury Structure and the structure along which the Complex was intruded have been debated since the area was first mapped about the turn of the century. There is now an important school of thought that regards the Structure as having resulted from a **meteoritic impact**. Shock metamorphic features, including shatter cones, are common in the rocks around the Complex for as much as 10 km from its footwall, and in the breccias of the Onaping Formation, which is regarded by this school as being a **fall-back breccia** resulting from the impact. (The Onaping Formation is the principal deposit at the base of the tripartite Whitewater Group.) Geologists arguing against the meteorite impact hypothesis interpret the Onaping Formation as **an ignimbrite deposit** and have mapped a quartzite unit, believed to be a **basal quartzite**, lying beneath the Onaping Formation. This school generally regards the Complex as having been intruded along an unconformity at the base of the Whitewater Group. Another controversial rock type is the 'Sudbury Breccia'. This consists of zones, a few centimetres to several kilometres across, of brecciated country rocks which are **sometimes hosts for the orebodies.**

The discovery of mineralization during the construction of the Canadian Pacific Railway in 1883 has led to the development of over forty mines and the total declared ore reserves of the district from the time of the original discovery to the present are of the order of 930 Mt. Of these, about 500 Mt have been exploited but new reserves are constantly being blocked out, just about keeping pace with production. The grade of ores worked in the past was about 3.5% nickel and 2% copper. Today, with **large scale mining methods**, the **grade** worked is **about 1% for both metals**. Most of the orebodies occur in the Sublayer whose magma was rich in sulphides and inclusions of peridotite, pyroxene and gabbro. In some places the Sublayer appears to be older than the main mass of the Complex and in other places younger and presumably intruded along the footwall of the main mass. It appears to have acquired its high content of inclusions during its passage through an underlying hidden basic igneous complex whose existence is suggested by geophysical data. The sulphides tended to sink into synclinal embayments in the footwall giving a structural control of the mineralization. The Creighton ore zone has the greatest number of ore varieties. It plunges north-westward down a trough at the base of the Complex for at least 3 km (Fig. 13.6) and consists of a series of ore types. The hanging wall quartz–norite above the Sublayer occasionally contains enough **interstitial sulphide** to form low grade ore. In the upper part of the Sublayer, **ragged disseminated sulphide ore** occurs, consisting of closely packed inclusions (several millimetres to 10 cm in size) in a matrix of sulphides and subordinate norite. The sulphide content increases downwards as does the ratio of matrix to inclusions, with a concomitant increase in inclusion size up to 1 m, resulting in an ore called **gabbro–peridotite inclusion sulphide**. This ore changes towards the footwall into **massive sulphide** containing fragments of footwall rocks. It is called **inclusion massive sulphide** and it is discontinuous, and also forms stringers and pods in the footwall. The **Frood–Stobie Orebody** is an example of an orebody in an offset dyke. This parallels the footwall of the Complex and it has been suggested that it

Surface

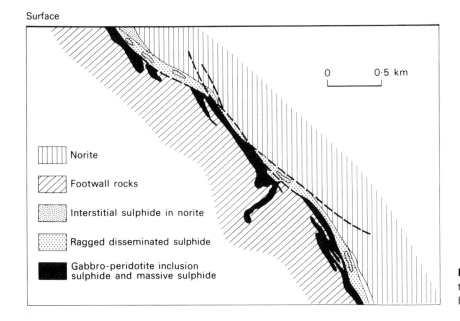

Norite

Footwall rocks

Interstitial sulphide in norite

Ragged disseminated sulphide

Gabbro-peridotite inclusion sulphide and massive sulphide

0   0·5 km

Fig. 13.6 Generalized section through the Creighton ore zone looking west.

might once have been continuous with the Complex at a higher level. It is a huge orebody, 1.3 km long, 1 km deep and nearly 300 m across at its widest point. It consists of a wedge-shaped body of inclusion-bearing quartz–diorite with disseminated sulphide partially sheathed by inclusion massive sulphide.

(b) *Deposits in intrusions related to flood basalts.* **Rifting of cratons** has resulted in the development of flood basalts at various times during the earth's history, and some of the associated basic intrusions carry important nickel sulphide mineralization. These include the Jurassic Insizwa Complex, R.S.A., which is part of the Karoo Magmatic Event; the Lower Triassic ores of the Noril'sk-Talnakh Region in Siberia, and the 1100 Ma mineralization of the Duluth Complex, Minnesota plus the Great Lakes deposit of Ontario, which are both related to Keweenawan magmatism.

The **Noril'sk-Talnakh Region** (Fig. 13.7) has the largest reserves of copper–nickel ore in the R.F. It lies deep in northern Siberia near the mouth of the Yenisei River. The country rocks are carbonates and argillaceous sediments of the early and middle Palaeozoic overlain by Carboniferous rocks with coals, Permian and a Triassic basic volcanic sequence. The associated gabbroic intrusions form sheets, irregular masses and trough-shaped intrusions depending on their location in the gentle folds of the sediments lying beneath the volcanic sequence.

The Noril'sk I deposit occurs in a differentiated layered dominantly gabbroic intrusion which extends northwards for 12 km and is 30–350 m

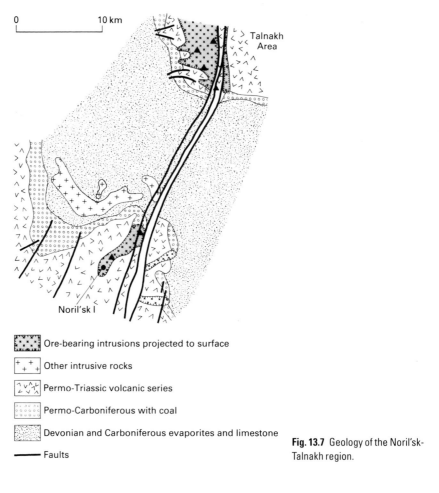

0   10 km

Talnakh
Area

Noril'sk I

⊠	Ore-bearing intrusions projected to surface
⊞	Other intrusive rocks
⊡	Permo-Triassic volcanic series
⊡	Permo-Carboniferous with coal
⊡	Devonian and Carboniferous evaporites and limestone
——	Faults

Fig. 13.7 Geology of the Noril'sk-Talnakh region.

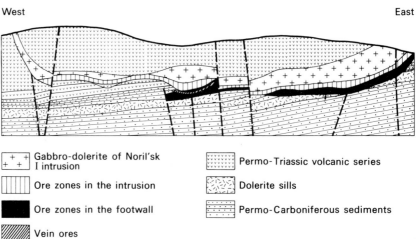

West   East

⊞ Gabbro-dolerite of Noril'sk I intrusion	⊡ Permo-Triassic volcanic series
⊞ Ore zones in the intrusion	⊡ Dolerite sills
■ Ore zones in the footwall	⊡ Permo-Carboniferous sediments
▨ Vein ores	

Fig. 13.8 Cross section through the Noril'sk I deposit.

thick. In cross section it is lensoid with steep sides (Fig. 13.8). The **copper–nickel sulphides** form breccia and disseminated and massive ores at the base of the intrusion, and vein orebodies developed in the footwall rocks and the basal portion of the intrusion. Like Sudbury there is a high Cu/Ni ratio and (Pt + Pd)/Ni = 1:500.

13.8 Environmental concerns

Nickel sulphide ores contain **8 t sulphur for each tonne of nickel** and nickel smelters used to emit hundreds of thousands of tonnes of sulphur. A century of relatively legislation-free smelting at Sudbury produced 100 km² of **barren land** in which vegetation had been killed and much of the topsoil eroded away. This area was surrounded by a further 360 km² of stunted birch and maple woodland. Soil near the smelters had a pH of nearly 3 and that of nearby lakes was as low as 4. Volatile metals such as Hg, As and Cd were dispersed over a much larger area. Sulphur emissions from Sudbury have been cut dramatically since the 1960s particularly as a result of the introduction of **flash smelting**, a facility that has yet to be introduced at Noril'sk whose smelters are responsible for widespread pollution in Fenno-Scandinavia.

13.9 Further reading

Naldrett A.J. (1989) *Magmatic Sulphide Deposits*. Oxford University Press, New York. This book provides an excellent and authoritative coverage of the subject.

14: The Skarn Environment

14.1 Introduction

Skarn is an old Swedish mining term for gangue, and who might want to mine gangue except perhaps for use as an aggregate?! But some skarns are economically mineralized and these, with examples, are the subject of this chapter.
- Skarns that make the headlines are either big and/or high grade or were found dramatically like Marmoraton, Ontario and Sarbai, Kazakhstan by aeromagnetic anomalies soon after W.W.2.
- Skarns mined for industrial minerals rarely achieve mention but the Trimouns Talc Mine 1800 m up in the Pyrenees is a notable example, as are the graphite orebodies within Norway's section of the Arctic Circle.

Skarns were formed at **elevated temperatures** with the addition and subtraction of material (**metasomatism**). Their general morphology and nature have been summarized in Section 4.2.2(b). They are developed most often, but not invariably, at the **contact of intrusive plutons and carbonate country rocks**. The latter are converted to marbles, calc-silicate hornfelses and/or skarns by contact metamorphic effects. The majority of skarns are devoid of economic mineralization.

The calc–silicate minerals, such as diopside, andradite and wollastonite, which are often the principal minerals in these ore-bearing skarns, attest to the high temperatures involved, and various lines of evidence suggest a range of 650–400°C for initial skarn formation, but in some skarns, particularly Zn–Pb, lower temperatures appear to have obtained. Skarns are classified according to their dominant mineralogy: as magnesian if they contain an important component of Mg silicates such as forsterite, or as calcic when Ca silicates, e.g. andradite, diopside, are predominant. **The majority of the world's economic skarn deposits occur in calcic skarns.** Skarn deposits are usually described according to the dominant economic metal or mineral present, e.g. copper, iron, tungsten, zinc–lead, molybdenum, tin, talc etc. These deposits are generally smaller than many other deposit types, such as porphyry coppers, porphyry molybdenums and lead–zinc, sediment-hosted sulphide deposits, but they are very important sources of tungsten. In some countries, e.g. Kazakhstan, they are of considerable importance for their iron production. Some particularly rewarding copper skarn deposits, especially those with by-product Au and/or Ag, have been worked in various parts of the world and notably large deposits may occur associated with porphyry copper deposits, e.g. Twin Buttes, Arizona, 500 Mt of 0.8% Cu. Zinc–lead skarn deposits occur throughout the world but are seldom of large tonnage. Skarn deposits of molybdenum and tin are of little importance compared with other deposits of these metals, apart from the San Antonio Mine, Santa Eulalia District, Mexico.

14.2 Some examples of skarn deposits

(a) *The Memé Mine, Northern Haiti—a copper skarn.* Frequently at the contacts of skarns and intrusions there is a **completely gradational contact** and this is the case at the Memé copper mine where a large block of Cretaceous limestone has been surrounded by monzogranite. Mineralization was preceded by extensive magmatic assimilation that formed zones of syenodiorite and granodiorite around the limestone. Following the crystallization of the magma, the limestone and neighbouring parts of the intrusion were replaced by skarn. **Mineralization followed skarn formation** and consisted of the introduction of hematite, magnetite, pyrite, molybdenite, chalcopyrite, bornite, chalcocite and digenite, in that paragenetic order. These occur **as replacement zones**. The main skarn and ore development is along the lower contact with the limestone block (Fig. 14.1). Skarn formation took place at between 480°C and 640°C and exsolution textures suggest that the minimum temperature of copper–iron sulphide deposition exceeded 350°C and the youngest ore minerals crystallized about 250°C. The grade is about 2.5% Cu.

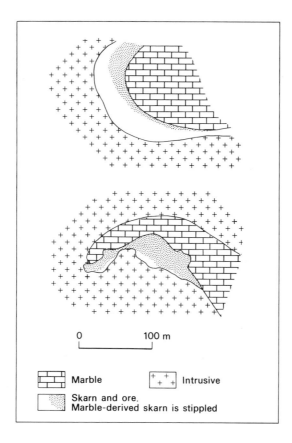

Fig. 14.1 Geological map of the 1 500 ft (457.2 m) level (above) and an east–west section (below) of the Memé Mine, Haiti. Note the concentration of skarn and ore beneath the marble.

0   100 m

Marble   Intrusive

Skarn and ore.
Marble-derived skarn is stippled

(b) *Iron skarn deposits*. Skarns have long been important sources of iron ore and the magnetite mine at Cornwall, Pennsylvania, supplied much of the iron used during the industrial revolution in the U.S.A. It is the oldest continuously operated mine in North America. Mining commenced in 1737 and by 1964, 93 Mt of ore had been produced with an average mill feed grade in 1964 of 39.4% Fe and 0.29% Cu, from which minor by-product cobalt, gold and silver were obtained. A pyrite concentrate was used to produce sulphuric acid, and up to 1953, when open pit operations ceased, limestone overburden was crushed and sold as aggregate — a good example of 'waste not, want not'! Cornwall was **a calcic iron skarn and such skarns are associated with intrusives ranging from gabbro through diorite to syenite, whilst magnesian iron skarns are normally associated with granites or granodiorites.** The largest known deposits in either class occur in the R.F. and Kazakhstan. With magnetite as the major ore mineral, these deposits are often marked by pronounced magnetic anomalies, and the detection of these by aeromagnetic surveys led to the discovery *inter alia* of Sarbai in Kazakhstan and of Marmoraton, Ontario. These deposits typically run 5–200 Mt with a grade of about 40% Fe.

Sarbai is the giant of skarn deposits with 725 Mt grading 46% Fe. It lies in the Turgai Iron Ore Province of Kazakhstan, i.e. the south-western part of the Siberian Platform to the east of the southern end of the Urals and about 500 km south-east of the famous and long-worked iron skarns of Magnitogorsk. The orebodies (Fig. 14.2) occur in a succession of metamorphosed pyroclastics, marbles and skarns developed from a Carboniferous volcaniclastic-sedimentary succession in the western limb of an anticline. At Sarbai, as elsewhere in the Turgai Province, there is a marked development of chlorine-bearing scapolite associated with the iron skarn orebodies, indicating the presence of important amounts of **brine solution** during the metasomatism. The orebody dimensioins are impressive: the eastern ore has a strike length of 1.7 km and a thickness of up to 185 m; it has been traced down dip for more than 1 km and the western orebody for 1.8 km. There is a further possible reserve of 775 Mt!

(c) *Tungsten skarns*. Tungsten skarns, veins and stratiform deposits provide most of the world's annual production of tungsten, with skarns being predominant. In the market economy countries (M.E.C.) most skarn tungsten comes from a few relatively large deposits: King Island, Tasmania; Sangdong, Korea; Canada Tungsten (NW.T) and MacMillan Pass (Yukon), Canada; and Pine Creek, California, U.S.A. To put 'relatively large' in perspective, it must be remarked that annual M.E.C. production of tungsten is only about 10 000 t, i.e. the equivalent of the production of just one modestly sized copper mine, which in turn is small compared with an iron mine! China is the world's main producer and in 1994 produced about

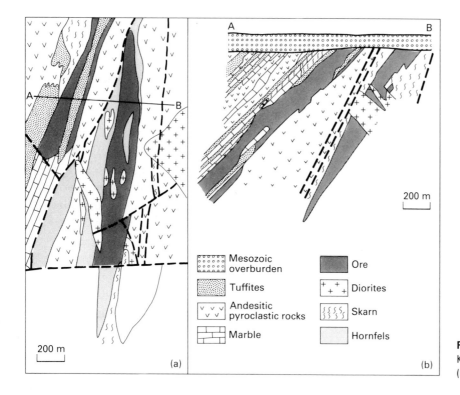

Mesozoic overburden

Tuffites

Andesitic pyroclastic rocks

Marble

Ore

Diorites

Skarn

Hornfels

200 m

200 m

(a)

(b)

Fig. 14.2 Sarbai iron skarns, Kazakhstan: (a) plan of 80 m level; (b) cross section.

32 000 t. The R.F. ranked next with 8 300 t. The other leading M.E.C. countries in 1994 were Bolivia with 540 t, Peru with 330 t, and Portugal with 350 t.

A good example of a tungsten skarn is the **King Island Scheelite Mine**. King Island, which lies between Australia and Tasmania at the western approach to Bass Strait, contains a number of important tungsten deposits. These are **scheelite-bearing skarns** developed in Upper Proterozoic to Lower Cambrian sediments that have been intruded by a granodiorite and a monzogranite of Devonian age. Mined ore and reserves at 1980 were put at 14 Mt averaging 0.8% WO_3. Scheelite-bearing andradite skarns were formed by the **selective replacement of limestone beds** and as a result they form stratiform orebodies 5–40 m thick. There is a great deal of mineralized skarn below ore grade. The orebodies lie either in the exocontact of the granodiorite or the monzogranite (Fig. 14.3). Irregular relics of marble occur in the skarn demonstrating its metasomatic origin, and it has been shown that the **replacement was a volume-for-volume process with massive addition of silica, iron and aluminium and subtraction of calcium and CO_2**. Stable isotopic, paragenetic and petrological investigations indicate the following stages in the development of these orebodies. Intrusion of the granitoids into a sequence of interbedded impure carbon-

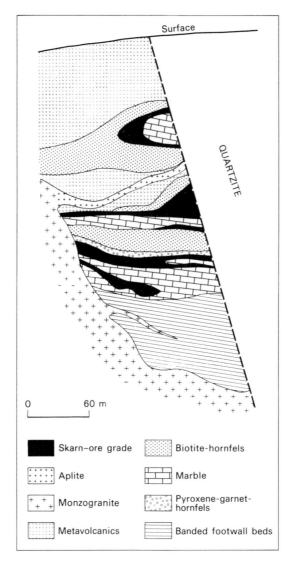

Fig. 14.3 Section looking north through the Bold Head orebodies, Tasmania.

ates, shales and volcanic rocks gave rise to contact metamorphism during which the shales became highly impermeable metapelites, but the interbedded dolomitic carbonates probably became more permeable when they were changed into marbles. Pre-existing faults and faults and fractures created by the intrusions channelled the flow of high temperature fluids (up to 800°C) into the permeable marbles creating massive calc–silicate replacement skarns and a chemical environment favouring the precipitation of scheelite.

(d) *Talc and graphite skarns*. Talc-bearing skarns and similarly altered carbonate and other metasedimentary rocks provide about 70% of world talc production. A good and important example of these deposits occurs in France. The Trimouns Mine, an open pit operation, lies towards the eastern end of the French Pyrenees and at an altitude of 1 800 m. Talc production is over 300 000 t p.a. and reserves of at least 20 Mt are present. The ores occur along the boundary between a basement of high grade metamorphic rocks and migmatites of the St Barthelemy Massif and an overthrust cover of lower grade metamorphic rocks of upper Ordovician to Devonian age (Fig. 14.4). The lower part of these hanging wall rocks contains discontinuous dolomitic lenses 5–80 m thick, overlying but also intercalated with mica schists in which small bodies of cross cutting leucogranites, aplites, pegmatites and quartz veins occur.

During the overthrusting intense shearing of the dolomites, schists and other rocks permitted **extensive hydrothermal circulation which produced talc-rich ore (80–97% talc) in the dolomites and chlorite-rich ore (10–30% talc) in the silicate rocks**. Many residual blocks of the primary rocks are present within the ore. The main orebody is 10–80 m thick and dips eastwards at 40–80°. Rock volumes appear to have remained roughly constant during the metasomatism. From a study of mineral assemblages and compositions it has been shown that the metasomatism took place at about 400°C under a pressure of about 0.1 GPa. **Highly saline, CO_2-poor solutions** with high Ca and Mg concentrations played an important role in this metasomatism but their source is still conjectural.

A minor amount of **graphite production** comes **from skarns**, e.g. the Norwegian Skaland Mine deep within the Arctic Circle just south of Tromsø, where lenses of skarn up to 200 m long and usually 5–6 m thick (maximum 24 m), carry 20–30% graphite and occur within mica schists surrounded by metagabbros and granites. Gangue minerals include diopside, hornblende, labradorite, sphene, garnet, scapolite and wollastonite.

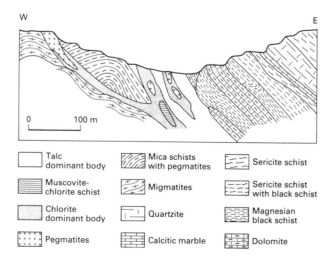

Fig. 14.4 Sketch section through the Trimouns Talc Mine, Pyrenees, France.

Talc dominant body	Mica schists with pegmatites
Muscovite-chlorite schist	Migmatites
Chlorite dominant body	Quartzite
Pegmatites	Calcitic marble
	Sericite schist
	Sericite schist with black schist
	Magnesian black schist
	Dolomite

The deposits are thought to have resulted from the **concentration of pre-existing carbon** in the sediments and these could be calc–silicate hornfelses or reaction skarns.

14.3 Genesis of skarn deposits

A common pattern in the evolution of proximal skarns (skarns near or at an igneous contact) has been recognized which takes the form of:
1 initial isochemical metamorphism;
2 multiple stages of metasomatism; and
3 retrograde alteration.

14.3.1 Stage 1

This involves the recrystallization of the country rocks around the causative intrusion, producing marble from limestone, hornfels from shales, quartzites from sandstones etc. Reaction skarns may form along lithological contacts. If the marbles are impure then various calcium and magnesian silicates may form and we have a calc–silicate hornfels that might contain **minerals of economic interest**, such as **talc and wollastonite**. The principal process involved in this isochemical metamorphism is diffusion of elements in what can be an essentially stationary fluid, apart from the driving out of some metamorphic water. The rocks as a whole may become more brittle and more susceptible to the infiltration of fluids in stage 2.

14.3.2 Stage 2

The infiltration of the contact rocks by hydrothermal-magmatic fluids leads to the conversion of pure and impure marbles, and other rock types, into skarns and the modification of calc–silicate hornfelses of stage 1. This is a prograde metamorphic and metasomatic process operating at temperatures of about 800–400°C during which an ore fluid evolves, **initial ore deposition takes place** and the pluton begins to cool. The new minerals developed are dominantly anhydrous. Deposition of oxides (magnetite, cassiterite) and sulphides commences late in this stage **but generally peaks during stage** 3.

14.3.3 Stage 3

This is a retrograde (destructive) stage accompanying cooling of the associated pluton and involving the **hydrous alteration** of early skarn minerals and parts of the intrusion **by circulating meteoric water**. Calcium tends to be leached and volatiles introduced with the development of minerals such as low-iron epidote, chlorite, actinolite, etc. Declining temperatures lead to the precipitation of sulphides.

The origin of all the introduced material in certain skarns, e.g. vast tonnages of iron, has been much debated. The great majority of workers who have investigated these deposits consider that in most cases the pluton responsible for the contact metamorphism was also the source of the metasomatizing solutions. Whilst it is conceivable that a granitic pluton might supply much silica, it might be thought unlikely that it could have supplied the amount of iron that is present in some deposits. However, it has been shown that it is probable that in natural magmatic systems, the concentration of iron in chloride solutions coexisting with magnetite or biotite is very high. This high solubility may explain the large quantities of iron in some skarns associated with granitic intrusions. On the other hand, where the pluton concerned is basic, the supply of iron does not present such great problems.

14.4 Further reading

Einaudi M.T., Meinert L.D. and Newberry R.J. (1981) Skarn Deposits. *Econ. Geol.*, 75th Anniversary Volume, 317–91. A very comprehensive account of skarn deposits.

Kwak T.A.P. (1987) *W–Sn Skarn Deposits*. Elsevier, Amsterdam. Includes important references to the King Island deposits and much fluid inclusion work.

For useful short summaries of skarn deposits see Sawkins F.J. (1990) *Metal Deposits in Relation to Plate Tectonics*, Springer-Verlag, Berlin, and Edwards R. and Atkinson K. (1986) *Ore Deposit Geology*, Chapman & Hall, London.

15: Disseminated and Stockwork Deposits Associated with Plutonic Intrusives

15.1 Introduction

In this chapter we meet some real elephants among orebodies and discuss their origin.
- Mining engineers learnt the craft and benefits of bulk mining methods by exploiting cupriferous examples of these deposits in the U.S.A. in the 1920s.
- Besides producing over 50% of the world's Cu, this deposit type yields most of our Mo and some Sn and W.

We are concerned in this chapter with **low grade, large tonnage deposits** which are mined principally for **copper, molybdenum and tin**. These deposits are normally intimately associated with intermediate to acid plutonic intrusives and all are characterized by intense and extensive hydrothermal alteration of the host rocks. The ore minerals in these deposits are scattered through the host rock either as what is called **disseminated mineralization**, which can be likened to the distribution of seeds through raspberry jam, or they are largely or wholly restricted to quartz veinlets that form a ramifying complex called a **stockwork** (Fig. 15.1). In many deposits or parts of deposits both forms of mineralization occur (Fig. 15.2).

The first copper deposits of this type to be mined on any scale are in some of the south-western states of the U.S.A. and it was here that the **cost effectiveness of bulk mining methods** was first demonstrated in the 1920s and the mining of much lower grade copper ores than had hitherto been exploited became possible. These American copper deposits are **associated with porphyritic intrusives**, often mapped as porphyries. The deposits soon came to be called copper porphyries, the name by which they are still generally known. On the other hand, more or less identical molybdenum deposits have been known as disseminated molybdenums, although they are also called molybdenum stockworks or porphyry molybdenums. Similar tin deposits are more usually called tin stockworks, though the term porphyry tins has been used. The student will find all these names in present use. Whereas economic **porphyry coppers and molybdenums are usually extremely large orebodies** (50–500 Mt is the common size range), **tin stockworks are much smaller**, 2–20 Mt being the common size range. All three types of metal deposit may yield **important by-products**. Amongst these are molybdenum and gold from porphyry coppers; tin, tungsten and pyrite from the Climax molybdenum deposit (other porphyry molybdenums tend to be without useful by-products); and tungsten, molybdenum, bismuth and fluorite from tin stockworks.

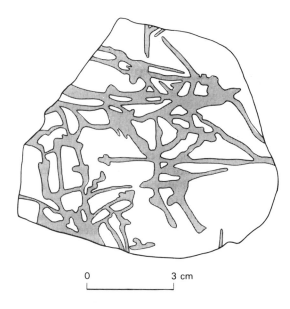

Fig. 15.1 Stockwork of molybdenite-bearing quartz veinlets in granite that has undergone phyllic alteration. Run of the mill ore, Climax, Colorado.

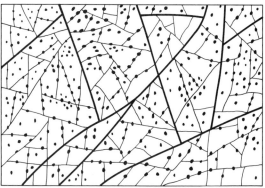

Fig. 15.2 Schematic drawing of a stockwork in a porphyry copper deposit. Sulphides occur in veinlets and disseminated through the highly altered host rock.

Porphyry coppers annually provide over 50% of the world's copper and there are many deposits in production. They are situated in orogenic belts in many parts of the world. **Porphyry molybdenums** in production are far fewer, about ten, and **account for over 70% of world production**. Tin stockworks are much less important, most tin production coming from placer and vein deposits. Porphyry copper and molybdenum deposits are closely related and, although the next section is mainly devoted to porphyry coppers mention will be made of some salient points concerning porphyry molybdenum.

15.2 Porphyry copper deposits

15.2.1 General description

As has been indicated above, these are large low grade stockwork to disseminated deposits of copper that may also carry minor recoverable amounts of molybdenum, gold and silver. Usually they are copper–molybdenum or copper–gold deposits. They must be amenable to bulk mining methods — that is, open pit or, if underground, block caving. **Selective mining** is of course **impossible** and host rock, stockwork and disseminated mineralization have to be extracted *in toto*. In this way, some of the largest man-made holes in the crust have come into being. **Most deposits have grades of 0.4–1% copper and total tonnages range up to 1000 million** with a few giants being even larger than this. The typical porphyry copper deposit occurs in a cylindrical stock-like, composite intrusion having an elongate or irregular outcrop about 1.5×2 km, often with an outer shell of equigranular medium-grained rock. The central part is porphyritic—implying a period of rapid cooling to produce the finer grained groundmass—the porphyry part of the intrusion.

15.2.2 Petrography and nature of the host intrusions

The most common hosts are acid plutonic rocks of the granite clan ranging from granite through granodiorite to tonalite, quartz monzodiorite and diorite. However, diorite through monzonite (especially quartz-monzonite) to syenite (sometimes alkalic) are also important host rock types. Many authors agree that porphyry copper deposits are **normally hosted by I-type granitoids**, within which category it is important to distinguish the I(Cordilleran) from the I(Caledonian) intrusives, as the latter rarely carry economic mineralization. Host intrusions in island arc settings have primitive initial strontium isotope ratios of 0.705–0.702 and are presumably derived from the upper mantle or recycled oceanic crust. The same ratios from mineralized intrusions in continental settings are generally higher indicating derivation from or, more probably, contamination by, crustal material. **Multiple intrusive events** are common in areas with porphyry copper mineralization, **with the host intrusions** normally being the most differentiated and **youngest of those present**. The host intrusions usually appear to be passively rather than forcefully emplaced, stoping and assimilation being the principal mechanisms. Economic hosts may be isolated stocks or a late stage unit of a composite, co-magmatic intrusion, often batholithic in dimensions.

15.2.3 Hydrothermal alteration

In 1970 Lowell and Guilbert described the San Manuel–Kalamazoo orebody (Arizona) and compared their findings with twenty-seven other porphyry copper deposits. From this study they drew up what is now known as the **Lowell–Guilbert model**. In this invaluable and fundamental work they demonstrated that the best reference framework to which we can relate all the other features of these deposits is the nature and distribution of the **zones of hydrothermal wall rock alteration**. They claimed that generally **four** alteration **zones** are **present** as shown in Fig. 15.3. These are normally centred on the porphyry stock in coaxial zones that form concentric but often incomplete shells and they are frequently used as a guide to ore in exploring porphyry copper deposits. In the Lowell–Guilbert model they are as follows.

(a) *Potassic zone.* This zone is not always present. When present it is characterized by the development of **secondary orthoclase and biotite** or by orthoclase–chlorite and sometimes orthoclase–biotite–chlorite. Sericite may also be present. These secondary minerals replace the primary orthoclase, plagioclase and mafic minerals of the intrusion. Anhydrite may be prominent in this zone.

(b) *Phyllic zone.* This is alteration of the type known in other deposits as sericitization and advanced argillic alteration. It is characterized by the assemblage **quartz–sericite–pyrite** and usually carries minor chlorite, illite and rutile. The sericitization affects the feldspars and primary biotite, alteration of the latter mineral producing the minor rutile. These are silica-generating reactions, so much secondary quartz is produced (silicification). The contact with the potassic zone is gradational over tens of metres. When the phyllic zone is present it possesses the greatest development of disseminated and veinlet pyrite.

(c) *Argillic zone.* This zone is not always present. Clay minerals are prominent with kaolin being dominant nearer the orebody, and montmorillonite further away.

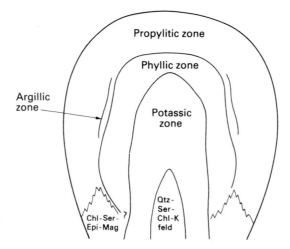

Fig. 15.3 Hydrothermal alteration zoning pattern in the Lowell–Guilbert model of porphyry copper deposits.

(d) *Propylitic zone.* This outermost zone is never absent. **Chlorite** is the most common mineral. **Pyrite, calcite and epidote** are associated with it. Primary mafic minerals (biotite and hornblende) are altered partially or wholly to chlorite and carbonate. Plagioclase may be unaffected. This zone fades into the surrounding rocks over several hundreds of metres.

15.2.4 Hypogene mineralization

The **ore may be found in three different situations**. It may be: (i) totally within the host stock; (ii) partially in the stock and partially within the country rocks; or (iii) in the country rocks only. The most common shape for the orebody in the examples analysed by Lowell and Guilbert is that of a steep-walled cylinder (Fig. 15.4) Stubby cylindrical to flat conical forms and gently dipping tabular shapes are also known. The orebodies are usually surrounded by a **pyrite-rich shell**. Like the alteration, the **mineralization** also tends to occur **in concentric zones** (Fig. 15.4). There is a central barren or low grade zone with minor chalcopyrite and molybdenite, pyrite usually forming only a few per cent of the rock, but occasionally ranging up to 10%. Passing outwards, there is an increase first in molybdenite and then in chalcopyrite as the **main ore shell** is encountered. Pyrite mineralization also increases in intensity outwards to form a peripheral pyrite-rich halo with 10–15% pyrite but only minor chalcopyrite and molybdenite. **The shells show a spatial relationship to the wall rock alteration zones** (see figures) with the highest copper values often being developed at and near the boundary between the potassic and phyllic zones. Weak, non-economic mineralization continues outwards into the propylitic zone.

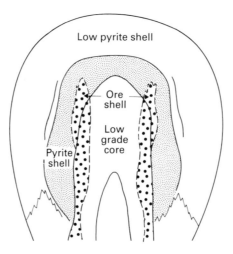

Fig. 15.4 Schematic diagram of the principal areas of sulphide mineralization in the Lowell–Guilbert model of porphyry copper deposits. Solid lines represent the boundaries of the alteration zones shown in Fig. 15.3.

15.2.5 The Diorite model

Subsequent to Lowell's and Guilbert's classic work it has been recognized that some porphyry copper deposits are associated with intrusives having low silica to alkali ratios. Various names have been suggested for this type. The one that has won general recognition is **diorite model**, although the host pluton may be a syenite, monzonite, diorite or alkalic intrusion. Diorite model deposits differ in a number of ways from the Lowell–Guilbert model; one of the main reasons appears to be that **sulphur concentrations were relatively low in the mineralizing fluids**. Consequently, not all the iron oxides in the host rocks are converted to pyrite and much iron remains in the chlorites and biotites while excess iron tends to occur as magnetite, which may be present in all alteration zones.

(a) *Alteration zoning*. The phyllic and argillic alteration zones are usually absent so that the **potassic zone is surrounded by the propylitic zone**. In the potassic zone, biotite may be the most prominent potassium mineral and when orthoclase is not well developed, plagioclase may be the principal feldspar.

(b) *Mineralization*. The main difference from the Lowell–Guilbert model is that **significant amounts of gold** may now occur and **molybdenum/copper is usually low**. The fractures containing gangue silicate minerals and copper sulphides may be devoid of quartz. On the other hand, chlorite, epidote and albite are fairly common.

15.2.6 Regional distribution of porphyry deposits

The distribution of porphyry copper and molybdenum deposits is shown in Fig. 15.5. From this map it can be seen that the majority of porphyry deposits are associated with Mesozoic and Cenozoic orogenic belts in two main settings—island arcs and continental margins. The major exceptions are the majority of the R.F. and Kazakhstan deposits and the Appalachian occurrences of the U.S.A. These exceptions belong to the Palaeozoic. Only a few porphyry deposits have so far been found in the Precambrian. These facts are of great importance from the exploration point of view.

15.2.7 Genesis of porphyry copper deposits

The principal arguments over recent years have been concerned with a **magmatic versus a meteoric derivation for the mineralizing fluids** and the origin of the metals and sulphur. In considering the formation of these deposits we must remember that the most striking characteristic of

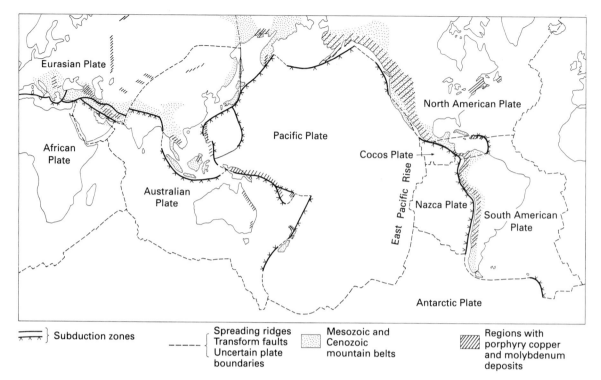

Fig. 15.5 The principal porphyry copper and molybdenum regions of the world. Also shown are present plate boundaries and Mesozoic–Cenozoic mountain belts.

porphyry copper deposits when compared with other hydrothermal ore-bodies is their enormous dimensions. The size and shape of these deposits imply that the hydrothermal solutions permeated very large volumes of rock, including country rocks, as well as the parent pluton. That at least some of these solutions originated in the host pluton is suggested by the existence of crackle brecciation.

(a) *Crackle brecciation and its origin.* Crackle brecciation is the name given to the fractures that are usually healed with veinlets to form the stockwork mineralization (Fig. 15.1). The zone of crackle brecciation is usually circular in outline, always larger than the orebodies and it fades out in the propylitic zone. It is often less well developed near the centre of the deposit, particularly if potassic alteration is present. This brecciation is thought to be due to the expansion resulting from the release of volatiles from the magma. The host magmas of porphyry copper deposits appear to have reached to within 0.5–2 km of the surface before equigranular crystallization commenced in their outer portions. The intrusions would then be stationary and the confining pressure would not fluctuate. With the steady development of crystallization, however, anhydrous minerals form and the liquid magma becomes richer in volatiles, leading to an increase in the vapour pressure. If the vapour pressure rises above the confining pressure, then what is called **retrograde boiling** will occur and a rapidly boiling liquid will separate. If retrograde boiling occurs in a largely consolidated rock, the vapour pressure has to overcome the tensile strength of the rock as well as rising above the confining pressure. This will result in expansion and extensive and **rapid brecciation** (Fig. 15.6). The reason for this is that water released at a depth of about 2 km at 500°C would have a specific volume of 4 and, if 1% by weight formed a separate phase, it would produce an increase in volume of about 10%. At shallower depths the

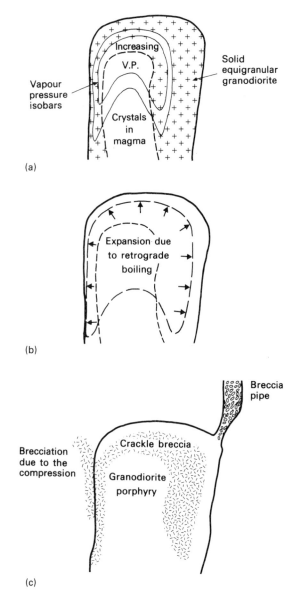

Fig. 15.6 Three stages in the development of crackle brecciation. (a) Vapour pressure building up in and around upper portion of magmatic fraction. (b) Retrograde boiling occurs, causing expansion. (c) Distribution of resulting brecciation.

increase would be even greater and the degree of fracture intensity higher. **Evidence** for the development **of retrograde boiling** in porphyry copper deposits is common in the form of the **widespread** occurrence of liquid-rich and gas-rich fluid inclusions in the same thin section (see Section 5.4).

(b) *Some chemical processes in the formation of porphyry copper deposits.* Retrograde boiling produces an **aqueous phase** (hydrothermal solution) in a porphyry system and chloride ion is partitioned strongly into it as is bisulphide ion, provided a sulphide mineral, such as pyrrhotite, is not stable. The presence of **chloride ion supplies a transporting mechanism** for the base metals that also fractionate strongly into the aqueous phase, and the bisulphide provides the sulphur for the eventual precipitation of sulphides.

(c) *Evidence from isotopic and geochemical investigations.* Further evidence of a magmatic derivation of at least some of the hydrothermal solutions comes from stable isotope investigations. This shows that **waters in equilibrium with potassium silicate alteration** assemblages and formed at 550–700°C are isotopically indistinguishable from **primary magmatic waters**. On the other hand, **waters** associated with sericites **from the phyllic zone of alteration** appear to have been **formation and meteoric waters**. This suggests, as does the field and microscopic evidence, that the phyllic and argillic alterations were later and superimposed upon the potassium silicate and propylitic alterations. These two stages of development are depicted in Fig. 15.7. When the intrusion cools, this meteoric-formation hydrothermal system may encroach upon and mix with the waning magmatic system leading to the development of lower temperature minerals: sericite, pyrophyllite and clay minerals. These would replace in particular the feldspar and biotite of the outer part of the original potassium silicate zone. The relatively **rapid gradients** in pH, temperature, salinity etc. across the interface between these two hydrothermal systems probably account for the **concentration of copper** around the boundary zone between the potassium silicate and phyllic zones. With this second stage of alteration, a Lowell–Guilbert model deposit comes into being.

(d) *Fluid inclusion evidence.* A consideration of the extensive fluid inclusion data on porphyry coppers emphasizes the remarkable uniformity of the three characteristics of **high maximum temperatures** (as high as 725°C), **high salinities** (up to 60 wt % alkali chlorides) and **evidence of boiling** for all deposits. These data clearly favour a magmatic hypothesis of origin for the mineralizing fluids. The highest temperature inclusions, which are rich in copper and other metals, characterize the central portions of porphyry systems and their distribution patterns mimic the zonal alteration–mineralization patterns. Fluid inclusion temperatures and

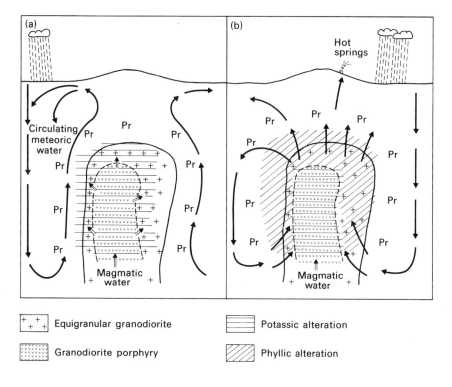

Fig. 15.7 Diagrammatic sections through a porphyry copper deposit showing two stages in the development of the hydrothermal fluids leading to the formation of a Lowell–Guilbert model deposit (b). Pr, propylitic alteration.

salinities decrease both away from the central portions and with time of formation. All this evidence agrees well with the concept depicted in Fig. 15.7.

15.3 Porphyry molybdenum deposits

15.3.1 General description

These have many features in common with porphyry copper deposits and some of these features have been touched on above. Average grades are 0.1–0.45% MoS_2 (molybdenum grades are *more usually* given as MoS_2) and one deposit produces by-product tin and tungsten. Host intrusions vary from quartz monzodiorite through granodiorite to granite. **Stockwork mineralization is more important** than disseminated mineralization and the orebodies are associated with simple, multiple or composite intrusions or with dykes or breccia pipes. There are three general orebody morphologies (Fig. 15.8), and tonnages range from 50 to 1500 Mt. Supergene enrichment, which can be very important in porphyry coppers, is generally absent or minor.

15.3.2 Hydrothermal alteration

The alteration patterns are very similar to those found in porphyry copper

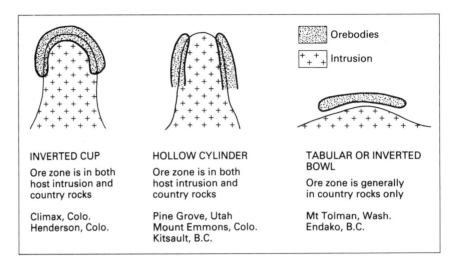

Fig. 15.8 Porphyry molybdenum orebody morphologies.

deposits, with potassic alteration and silicification being predominant. The lower temperature zones of alteration envelop intrusive centres rather than single stocks. They have been summarized as quartz–topaz, phyllic, argillic and propylitic zones. A prominent pyrite zone, carrying 6–10% pyrite, is developed around the Henderson orebody and a similar less distinct zone is present at Climax; both deposits are in Colorado. Peripheral pyrite zones, which appear to coincide mainly with the phyllic zone, have been reported from a number of other porphyry molybdenums.

15.3.3 Genesis of porphyry molybdenum deposits

The close spatial association of the orebodies of most deposits with the potassic zone of alteration in a number of small stocks suggests a magmatic source. However, the volumes of the host intrusions are considered by some workers to be far too small to have supplied the large tonnages of molybdenum present in these deposits, and they suggest that lower unexposed master reservoirs of magma that fed the host intrusions were the source. This view is supported by isotopic evidence which also indicates that meteoric water played a minor or no role in the mineralization.

15.4 Porphyry tin and porphyry tungsten deposits

Much primary tin has been won in the past from stockworks at Altenberg in Germany and Cínovec in the Czech Republic, from many deposits in Cornwall, England, and from deposits in New South Wales, Tasmania and South Africa. Eroded stockworks in Indonesia and Malaysia have provided much of the alluvial cassiterite in those countries. Exploitation of such deposits in recent times has been mainly in Germany and New South Wales. At Ardlethan in New South Wales a quartz–tourmaline stockwork occurs in altered granodiorite carrying secondary biotite, sericite and siderite. With a grade of 0.45% tin, this deposit is economic. The **majority of tin stockworks** in the world, however, only run about 0.1% tin and they are **not at present mineable at a profit**. Tin stockworks in the countries mentioned above belong to the plutonic environment. Porphyry tin deposits also occur in the subvolcanic section of the Bolivian tin province south of Oruro and these deposits have much in common with porphyry copper deposits—large volumes of rock grading 0.2–0.3% tin. There is pervasive sericitic alteration that grades outwards into propylitic alteration and pyrite halos are present in two deposits. Major differences include the absence of a potassic zone of alteration, the association with stocks having the form of inverted cones rather than upright cylinders and the presence of swarms of later vein deposits. **Porphyry tungsten** (usually W–Mo) deposits have now been described from various parts of the world, especially North America and China. Hydrothermal alteration is not always present as pervasive zones and when present may not show a regular concentricity and grades are low.

15.5 Further reading

McMillan W.J. and Panteleyev A. (1989) Porphyry Copper Deposits. In Roberts R.G. and Sheahan P.A. (eds), *Ore Deposit Models*, 45–58. Geological Association of Canada, Memorial University, Newfoundland.

Wallace S.R. (1991) Model Development: Porphyry Molybdenum Deposits. *Econ. Geol. Monogr.* **8**, 207–24.

Both these references are excellent and concise reviews of their subjects.

16: Stratiform Sulphide, Oxide and Sulphate Deposits of Sedimentary and Volcanic Environments

16.1 Introduction

- Most of the world's Cu not produced from porphyry copper deposits comes from stratiform deposits which are also important producers of Pb, Zn, Ag, Au, Fe, W and baryte.
- These deposits have either a sedimentary or a volcanic setting, the former are divisible into dominantly copper or lead–zinc; the latter are normally polymetallic.
- The sediment-hosted copper deposits were probably formed from post-diagenetic, hydrothermal solutions that leached metals from subjacent rocks but all the others are probably of exhalative origin.

In this chapter we are concerned with a class or classes of deposit whose origin is at present hotly debated. As was pointed out in Section 4.3, there are many types of stratiform deposit. This chapter is mainly devoted to sulphide deposits and the related oxide deposits can only be mentioned *en passant*. The latter do not include bedded iron and manganese deposits, placer deposits and other ores of undoubted sedimentary origin, which are dealt with in Chapter 19. The related oxide deposits are of iron, tin and uranium with which may be grouped certain tungsten deposits.

Concordant deposits referred to in Section 4.3 which belong to this class include the Kupferschiefer of Germany and Poland; Sullivan, British Columbia; the Zambian Copperbelt, and the large group of volcanic-associated massive sulphides. There appears to be a possible gradation in type and environment from deposits such as the Kupferschiefer, which are composed dominantly of normal sedimentary material and which occur in a non-volcanic sedimentary environment, through deposits such as Sullivan, which are richer in sulphur and may have some minor volcanic formations in their host succession, to the volcanic-associated massive sulphide deposits which are composed mainly of sulphides and occur in host rocks dominated by volcanics. Some writers consider that these and other deposits form a single class. Other workers have felt it better to divide them into **two classes, the first** class being those **developed in a sedimentary environment** where sedimentary controls are important, and the **second** being the **volcanic-associated massive sulphide deposits** in which exhalative processes were important during genesis. Such a division introduces difficulties when dealing with deposits showing only a weak link with volcanism but where exhalative processes may have been important, e.g. Sullivan. However, this division will be followed in this chapter as the author feels that deposits such as the Kupferschiefer and the Zambian Copperbelt are sufficiently different from the volcanic-associated massive sulphide deposits to warrant some differentiation. Among the sediment-hosted, stratiform, base metal deposits two major groups are recognized by the majority of workers: **sediment-hosted copper and sediment-**

hosted lead–zinc deposits. The latter are often termed **sedex** (i.e. sedimentary exhalative) **deposits**.

16.2 Stratiform sulphide deposits of sedimentary affiliation

16.2.1 General characteristics

The majority of these deposits occur in non-volcanic marine or deltaic environments. They are widely distributed in space and time, i.e. from the Proterozoic to the Tertiary, and can vary in tonnage from several hundred millions down to subeconomic sizes. In shape, they are broadly **lensoid to stratiform** with the length at least ten times the breadth. There is often more than one ore layer present. **Feeder zones** (cf. Fig. 4.6) have been identified below some deposits and may be present below many more, but as mining operations rarely penetrate into footwalls on a large scale, they will probably never be seen. The degree of deformation and metamorphism varies with that of the host rocks, suggesting a pre-metamorphic formation. They are frequently organic rich, particularly those in shales, and usually contain a less complex and variable suite of minerals and recoverable metals than volcanic-associated massive sulphide deposits. The **sulphides have a small grain size** so that fine and often **costly grinding is necessary** to liberate them from the gangue. They may show a shore to basinward zoning of $Cu + Ag \rightarrow Pb \rightarrow Zn$. There is a tendency for copper and lead–zinc deposits to be separate from each other and to have markedly different metal ratios.

The **geological settings** of these deposits are **mostly intracratonic** and the majority do not appear to be related directly to orogenic events or plate margin activity. Regional settings include: (i) first marine transgressions over continental deposits (Kupferschiefer, Zambia, White Pine, U.S.A.); (ii) carbonate shelf sequences (Ireland); and (iii) fault controlled, sedimentary basins (Selwyn Basin, Yukon; Belt-Purcell Basin, British Columbia). Some of these environments appear to be aulacogens.

Economically, both the copper and lead–zinc deposits are of great importance on a worldwide scale; indeed, sediment-hosted, stratiform copper deposits are second only to porphyry coppers as producers of the metal.

16.2.2 Copper deposits

In addition to the general points listed above, it should be noted that this deposit type is the world's most important source of **cobalt** (from the Central African Copperbelt) and is becoming an important producer of **byproduct silver** (Poland and U.S.A.). Most grades of recently worked and working deposits vary from 1.18% to 5% Cu, but the lower grade deposits have sweeteners. **Tonnages can be enormous**, e.g. Lubin, Poland 2600 Mt running about 2% Cu, 30–80 g t^{-1} Ag and 0.1 g t^{-1} Au.

Most major deposits occur in reduced, pyritic, organic-rich, calcareous shales or their metamorphic equivalents, but the remaining approximately one third occur in sandstones. These immediate host rocks occur in anoxic, paralic marine (or large scale, saline lacustrine) sediments immediately above typically red, oxidized, continental clastic sediments, and deposits of this type are found in rock sequences post-dating the first appearance of red beds (*c.* 2400 Ma ago) and range in age to Recent. The most important and abundant deposits are in Upper Proterozoic and Upper Palaeozoic rocks that were deposited in arid and semi-arid areas within continental rift environments not further than 20–30° from the palaeoequator. In many areas these rocks are interbedded with evaporites. At the oxidation–reduction boundary the ascending sequence of minerals in the mineralized ground contains all or some of the following: hematite, native copper, chalcocite, bornite, chalcopyrite, galena, sphalerite and pyrite. These occur in mineral zones that overlap upward and outward.

16.2.3 Some examples of copper deposits

(a) *The European Kupferschiefer.* This is probably **the world's best known copper-rich shale**. It is of late Permian age and has been mined at Mansfeld, Germany for almost 1000 years. The Kupferschiefer underlies about 600 000 km^2 in Germany, Poland, Holland and England (Fig. 16.1). **Copper** concentrations greater than 0.3% occur in about 1% and **zinc** concentrations greater than 0.3% in about 5% of this area. Thus, although all the Kupferschiefer is anomalously high in base metals, ore grades are only

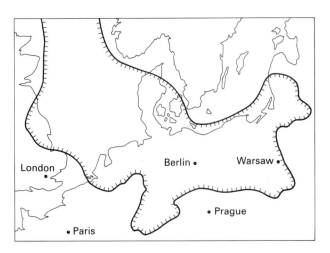

Fig. 16.1 Extent of the Zechstein Sea in central Europe. The Kupferschiefer occurs at the base of the Upper Permian Zechstein.

encountered in a few areas. The most notable recent discoveries have been in southern Poland where deposits lying at a depth of 600–1500 m have been found during the last two decades. Here, the Kupferschiefer varies from 0.4 to 5.5 m in thickness. Average copper content is around 1.5% and reserves at 1% Cu amount to nearly 3000 Mt making Poland the leading copper producer in Europe. The area underlain by these deposits is approximately 30 × 60 km.

The Kupferschiefer consists of thin alternating layers of carbonate, clay and organic matter with fish remains which give it a characteristic dark grey to black colour. The **Kupferschiefer is the first marine transgressive unit** overlying the non-marine Lower Permian Rotliegendes, a red sandstone sequence, and it is succeeded by the Zechstein Limestone which in turn is overlain by a thick sequence of evaporites. The Kupferschiefer and the Zechstein evaporites may represent a tidal marsh (sabkha) environment which developed as the sea transgressed desert sands. Sulphide mineralization occurs across the contact between the Upper Permian Zechstein marine sequence and the Lower Permian Rotliegendes red beds. Ore is contained in the Kupferschiefer, the overlying limestone and the underlying sandstone. The copper and other metals are disseminated throughout the matrix of the rock as **fine-grained sulphides** (principally bornite, chalcocite, chalcopyrite, galena, sphalerite), **commonly replacing earlier calcite cement, lithic fragments and quartz grains** as well as other sulphides. Typical features of the mineralization are shown in Fig. 16.2. Horizontal and vertical veinlets of gypsum, calcite and base metal sulphides are common in the Lubin district and increase the grade of the ore significantly. They have been interpreted as resulting from **hydraulic fracturing** (see Section 6.4). A zone of superposed diagenetic reddening known as the **Rote Fäule facies** transgresses the stratigraphical horizons. Copper miner-

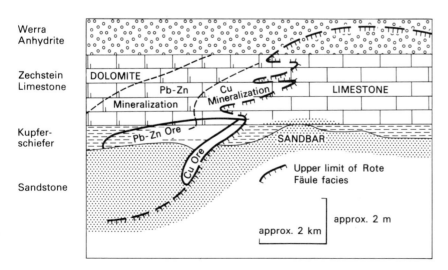

Werra Anhydrite
Zechstein Limestone
Kupfer-schiefer
Sandstone

DOLOMITE
Pb-Zn Mineralization
Cu Mineralization
LIMESTONE
Pb-Zn Ore
Cu Ore
SANDBAR
Upper limit of Rote Fäule facies
approx. 2 km
approx. 2 m

Fig. 16.2 Diagrammatic section through orebodies in the basal Zechstein with the Rote Fäule facies alteration gently transgressing the bedding above an area of sandbars formed by marine working of the Rotliegendes sandstone. Sulphide mineralization occurs in the unoxidized zone adjacent to the Rote Fäule with copper nearest to it and lead–zinc further away.

alization lies directly above the Rote Fäule and the copper zone is overlain, in turn, by lead–zinc mineralization. This relationship to the Rote Fäule has meant that the delineation of this facies is the most important feature of the search for new orebodies. The Rote Fäule copper zones are coincident with underlying highs in the buried basement, and the metal zoning dips away from the highs toward the basin centres. A recently discovered but as yet unexploited aspect of the Polish Kupferschiefer is the occurrence of **Pt-rich** (>10 ppm) **shales** along strike lengths in excess of 1.5 km and values of >200 ppm Pt have been found over 50 m strike lengths. (P.G.M.-rich shales have recently been recognized in Canada, China, central Europe and the U.S.A. and may represent a possible source of these metals.)

(b) *The Zambian Copperbelt*. This is part of the larger Central African Copperbelt of Zambia and Shaba (Zaïre) which produced about 17% of the western world's copper in the early 1980s. In 1984 Zambia produced 531 000 t of copper and the mill grade at the principal mines varied from 1.49% to 2.81%, with appreciable **by-product cobalt** from some. The industry in Zambia is contracting rapidly. The harsh realities are that the mines are old, with declining grades and an average of ten years' life ahead of them, and the country simply does not have the cash for investment to increase their efficiency and operate them profitably. Shaba produced about 500 000 t copper in 1984 from ore averaging around 4% Cu + Co; but here too the industry is suffering and production is down.

Almost all the copper mined in 1984 came from restricted horizons within the late **Proterozoic Katangan sediments** of the Lufilian Arc (Fig. 16.3). The Katangan rests unconformably on a granite–schist–quartzite basement and the lowermost Katangan sediments fill in the valleys of the pre-Katangan land surface. Most mineralization in Zambia and south-eastern Shaba occurs in the **Ore Formation** which lies a few metres above the level at which the pre-Katangan topography became filled in. **Shale or dolomitic shale forms the host rock for about 60% of the mineralized ground**, and the shale orebodies form a linear group to the south-west of the Kafue Anticline (Fig. 16.4). **Arkose-arenite hosted ores occur** mainly to the north-east of the anticline, e.g. Mufulira. The footwall succession consists of quartzites, feldspathic sandstones and conglomerates of both aquatic and aeolian origin. The Ore Formation, generally 15–20 m thick, is succeeded by an alternating series of arenites and argillites which, with the rocks below them, make up the Lower Roan Group. All the rocks and their contained copper minerals have suffered low to high grade greenschist facies metamorphism and many of the so-called shales are biotite-schists. In places they are tightly folded (Fig. 16.5). **Copper**, together with minor amounts of iron and cobalt, occurs mainly in the lower part of the Ore Formation as disseminated bornite, chalcopyrite and chalcocite. In places, mineralization passes for short distances into the

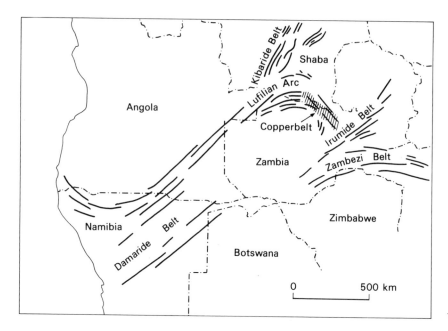

Fig. 16.3 Location of the Copperbelt in relation to the main tectonic trends of central Africa.

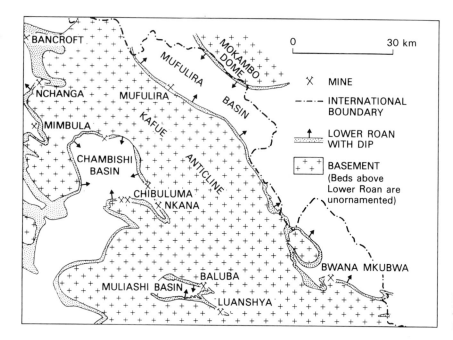

Fig. 16.4 Location map for the Zambian Copperbelt showing the regional geology.

underlying beds. Both the upper and lower limits of mineralization are usually sharply defined.

16.2.4 Genesis

Work on deposits of this type all over the world has led to the general

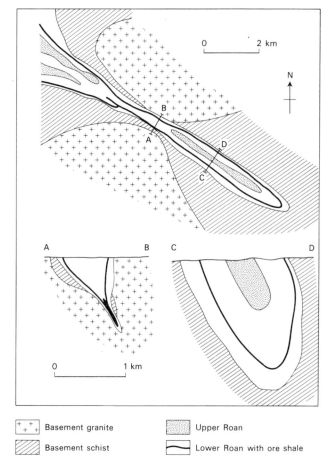

Fig. 16.5 Sketch map and sections of the Luanshya Deposit, Zambia.

Basement granite

Basement schist

Upper Roan

Lower Roan with ore shale

conclusion that copper and associated **metals** have been **added to their host rocks after sedimentation** and after at least very early syndiagenetic accumulations of sulphate and sulphide were formed, some of which, particularly pyrite, have been replaced by later copper and cobalt minerals. Features such as this and the slightly transgressive nature of the mineralization have now been reported from many deposits. In the Kupferschiefer the clear spatial relationship between sulphide and hematite mineralization suggests that the ore genesis was closely related to processes responsible for the Rote Fäule facies formation. Metal zoning transgressing the strata, localization of copper deposits around the Rote Fäule facies and the coexistence of hematite grains, hematite pseudomorphs after pyrite, metal sulphides and copper sulphide replacements of pre-existing pyrite within the outer part of the Rote Fäule facies suggest **a post-sedimentary origin** of the Rote Fäule ore system. It has been postulated that the ore solutions were metalliferous formation waters expelled from the Rotliegendes after leaching much of its copper content. Much evidence indicates that the

Central African Copperbelt sedimentation and mineralization took place in a **rift zone** and that the mineralization was due to hydrothermal leakage from the bounding fractures of saline formation waters that had leached Fe, Co and Cu, plus a wide range of minor elements, from basement rocks particularly basalts. Isotopic research indicates that the **sulphur** in the sulphides was **derived from sea water sulphate** already present in the arenites in the form of anhydrite and released from this mineral by high-temperature (>250°C) inorganic reduction. In the shales the **metals** were probably **precipitated under the influence of bacteriogenically reduced sulphur** at temperatures of 140–215°C.

16.2.5 Sediment-hosted lead–zinc deposits (sedex deposits)

These have a worldwide distribution and are very important metal producers. The grades and tonnages of a number of deposits are given in Fig. 16.6. For multi-lensoid deposits these represent the average grade and the total tonnage for all component lenses. Figure 16.6 demonstrates the impressive size (average of 70 Mt) and grade (average Pb + Zn = 12%) of these deposits, some of which have **important by-products** such as Ag, Au, Cd, Cu, Sn or baryte. They appear to form a distinctive group of ores **formed in local basins on the sea floor** as a result of protracted hydrothermal activity accompanying continental rifting and they constitute the important ore deposit type known as **sedex deposits**. Their host rocks are generally shales, siltstones and carbonates. In Ireland at the Silvermines and Tynagh deposits, **fossil hydrothermal chimneys** have been found which are similar to those known from present day hydrothermal vents on the East Pacific Rise and other constructive plate margins. These discoveries, together with the evidence of **feeder zones** and other observations described above, suggest that some (probably all) of these deposits have been formed by hydrothermal solutions venting into restricted basins on

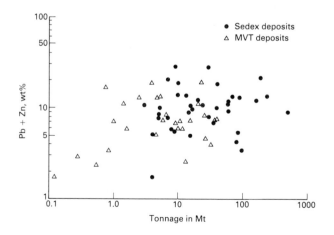

Fig. 16.6 Grade–tonnage plot for sedex and Mississippi Valley-type deposits.

the sea floor. The environment was not, however, that of the deep ocean, but more like that of the Gulf of California, where sulphide and baryte deposits are forming today in the axial rifts of a region of active sedimentation. However, the sea depths are still much greater than those postulated for the formation of these giant lead–zinc deposits—about 50–800 m. **How were these metal-bearing, hydrothermal brines generated?** Attention has been drawn to the similarity of temperature, salinity and pH of the solutions and the sulphur isotopic compositions of the sulphide of both sedex and Mississippi Valley-type deposits with the inference that the **solutions** responsible for the formation of both deposit types originated in, and were **expelled from clastic sedimentary basins**—a mode of genesis illustrated in Fig. 6.3. Other workers contend that the **exhalations** were **formed by sea water convection cells** similar to those shown in Fig. 16.7, which dissolved base metals from the rocks they traversed [see Section 6.2.3(a)]. Such cells would only have to penetrate a few kilometres into the oceanic crust in order to be heated to and sustained at the necessary temperatures, *if* supplied with the latent heat of crystallization from an underlying magma chamber.

For an intracratonic situation, like that of Ireland, it is of course necessary to postulate much deeper penetration of sea water—a penetration of about 10 km if the deposit was mainly Pb–Zn, with a depth of about 15 km being necessary to permit the leaching of copper. In a terrane having a high geothermal gradient under rifting conditions, continued cooling of the rocks would allow convection to reach deeper and deeper into the crust as shown in Fig. 16.7. In the early stages of convection the shallow penetration would lead in the main to leaching of iron, manganese and silica, which would account for the manganese enrichment in the footwall rocks of many deposits. With deeper circulation the temperature and the time for

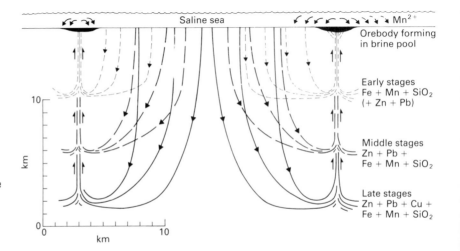

Fig. 16.7 Diagram showing how sea water circulation through the crust might give rise to the formation of an exhalative, sediment-hosted, stratiform ore deposit.

Saline sea

Mn^{2+}

Orebody forming in brine pool

Early stages
$Fe + Mn + SiO_2$
$(+ Zn + Pb)$

Middle stages
$Zn + Pb +$
$Fe + Mn + SiO_2$

Late stages
$Zn + Pb + Cu +$
$Fe + Mn + SiO_2$

10

km

0

0   10
km

water–rock interaction would increase and **lead and zinc** also would be leached. For many deposits this might be the limit of the convective system, leading to the formation of essentially copper free deposits. With still deeper penetration **copper** too might be leached to make that element an important by-product in the deposit.

16.3 Volcanic-associated massive sulphide (V.M.S.) deposits

Some attention has already been paid to these deposits in Sections 4.3.2(a) and 6.3.1. This section will therefore be used to amplify and add to what has already been written.

16.3.1 Size, grade, mineralogy and textures

The majority of world deposits are small and about 80% of all known deposits fall in the size range 0.1–10 Mt. Of these about a half contain less than 1 Mt. Average figures of this type tend to hide the fact that **this is a deposit type that can be very big or rich or both** and then very profitable to exploit. Examples of such deposits are given in Table 16.1.

The **mineralogy** of these deposits is fairly **simple** and often consists of over 90% iron sulphide, usually as pyrite, although pyrrhotite is well developed in some. Chalcopyrite, sphalerite and galena may be major constituents, depending on the deposit class, bornite and chalcocite are occasionally important and arsenopyrite, magnetite and tetrahedrite–tennantite may be present in minor amounts. **With increasing magnetite content these ores grade to massive oxide ores.** The **gangue** is principally quartz, but occasionally carbonate is developed and chlorite and sericite

Mined ore + reserves	(Mt)	Cu%	Zn%	Pb%	Sn%	Cd%	Ag g t⁻¹	Au g t⁻¹
Kidd Creek, Ontario	155.4	2.46	6.0	0.2	r‡	r	63	—
Horne, Quebec	61.3	2.18	—	—	—	—	—	4.6
Rosebery, Tasmania	19.0	0.8	15.7	4.9	—	—	132	3.0
Hercules, Tasmania	2.3	0.4	17.8	5.7	—	—	179	2.9
Rio Tinto, Spain*	500	1.6	2.0	1.0	—	—	r	r
Azalcollar, Spain	45	0.44	3.33	1.77	—	—	67	1.0
Neves-Corvo,	30.3	7.81	1.33	—	—	—	—	—
Portugal†	2.8	13.42	1.35	—	2.57	—	—	—
	32.6	0.46	5.72	1.13	—	—	—	—

Table 16.1 Tonnages and grades for some large or high grade V.M.S. deposits.

* Rio Tinto, originally a single stratiform sheet, was folded into an anticline whose crest cropped out and vast volumes were gossanized. The base metal values are for the 12 Mt San Antonio section.

† The three lines of data represent three ore types and not particular orebodies of which there are four. Ag is present in some sections.

‡ r indicates that this metal is recovered, but the average grade is not available.

may be important locally. Their mineralogy results in these deposits having a high density and some, e.g. Aljustrel and Neves-Corvo in Portugal, give marked **gravity anomalies**, a point of great exploration significance.

The **vast majority** of massive sulphide deposits are **zoned**. Galena and sphalerite are more abundant in the upper half of the orebodies whereas chalcopyrite increases towards the footwall and grades downward into **chalcopyrite stockwork** ore (see Figs 4.10 & 6.9). This zoning pattern is only well developed in the polymetallic deposits.

Textures vary with the degree of recrystallization. The dominant original textures appear to be colloform banding of the sulphides with much development of framboidal pyrite, perhaps reflecting colloidal deposition. Commonly, however, **recrystallization**, often due to some degree of metamorphism, has destroyed the colloform banding and produced a **granular ore**. This may show banding in the zinc-rich section, whereas the chalcopyrite ores are rarely banded. Angular inclusions of volcanic host rocks are occasionally present and soft sediment structures (slumps, load casts) are sometimes seen. Graded bedding has also been reported from some deposits.

16.3.2 Wall rock alteration

Wall rock alteration is usually confined to the footwall rocks. **Chloritization** and **sericitization** are the two most common forms. The alteration zone is pipe-shaped and contains within it and towards the centre the chalcopyrite-bearing stockwork. The diameter of the alteration pipe increases upward until it is often coincident with that of the massive ore.

16.3.3 Classifications

The geochemical division into iron, iron–copper, iron–copper–zinc and iron–copper–zinc–lead deposits has been touched on already [see Sections 4.3.2(a), 6.3.1], but it must be emphasized that while we may find pyrite deposits without any appreciable copper, copper is never found on its own. Similarly, if we find lead, we will have zinc and at least accessory copper too. With zinc will come copper and perhaps lead. Using a different approach from the simple chemical one, it can be shown that there are just **two major groups**, **Cu–Zn** and **Zn–Pb–Cu**. Indeed, there are few so-called copper deposits without some zinc. Some of the names commonly given to these different types have been mentioned in Section 6.3.1. Although for a number of years the dominantly iron–copper–zinc deposits of the Canadian Archaean were considered by many workers to be a variant of the **Kuroko type**, it is now generally agreed that they are best considered as a separate type, termed **Primitive**. A summary of the nature of the different types of V.M.S. deposits is given in Table 16.2. It is important to note that **precious metals** are also produced from some of these deposits; indeed in some Canadian examples of the Primitive type they are the prime products. Both **Besshi** and Kuroko types may also produce silver and gold whilst the **Cyprus type** may have by-product gold.

16.3.4 Some important field occurrence features

(a) *Association with volcanic domes.* This frequent association is stressed in the literature and the Kuroko deposits of the Kosaka district, Japan, are a good example (Fig. 16.8). Many examples are cited from elsewhere, e.g. the Noranda area of Quebec and some authors infer a genetic connexion. Certain Japanese workers, however, consider that the Kuroko deposits all formed in depressions and that the domes are late and have uplifted many of the massive sulphide deposits. In other areas, e.g. the Ambler District, North Alaska no close association with rhyolite domes has been found.

(b) *Cluster development.* Although the Japanese Kuroko deposits occur over a strike length of 800 km with more than a hundred known occurrences, these are clustered into eight or nine districts. Between these districts lithologically similar rocks contain only a few isolated deposits and this tends to be the case, with a few notable exceptions, for massive sulphide occurrences of all ages.

Table 16.2 Volcanic-associated massive sulphide deposit types.

Type	Associated volcanic rocks	Clastic sedimentary rocks	Depositional environment	Plate tectonic setting	Known age range
Besshi Cu–Zn ± Au ± Ag	Within plate (intraplate) basalts	Continent-derived greywackes and other turbidites	Deep marine sedimentation with basaltic volcanism	Epicontinental or back-arc rifting	Early Proterozoic, Palaeozoic
Cyprus Cu (± Zn) ± Au	Ophiolitic suites, tholeiitic basalts	Minor or absent	Deep marine with tholeiitic volcanism	Oceanic rifting at accreting margin	Phanerozoic
Kuroko Cu–Zn–Pb ± Au ± Ag	Bimodal suites, tholeiitic basalts, calc-alkaline lavas and pyroclastics	Shallow to medium-depth clastics, few carbonates	Explosive volcanism, shallow marine to continental sedimentation	Back-arc rifting, caldera formation	Early Proterozoic, Phanerozoic
Primitive Cu–Zn ± Au ± Ag	Fully differentiated suites, basaltic to rhyolitic lavas and pyroclastics	Immature greywackes, shales, mudstones	Marine, <1 km depth. Mainly developed in greenstone belts	Much debated, major subsidence, fault-bounded troughs, back-arc basins?	Archaean–early Proterozoic

Sulphide deposit

Gypsum deposit

Volcanic breccia

Stockwork or
dissemination deposit

Structure contours on the lava
domes, mapped and assumed

Fig. 16.8 Distribution of dacite lava domes and Kuroko deposits, Kosaka District, Japan.

(c) *Favourable horizons.* The deposits of each cluster often occur within a limited stratigraphical interval. For Primitive and Kuroko types, this is usually at the top of the felsic stage of cyclical, bimodal, calc-alkaline volcanism related to high level magma chambers rather than to a particular felsic rock type. Sometimes this favourable horizon is hosted by relatively quite thin developments of volcanic rocks, as in the Iberian Pyrite Belt, where the volcanic–sedimentary complex underlying the many enormous orebodies of this region is only 50–800 m thick. The Japanese Kuroko mineralization and associated volcanism occurred during a limited period of the Middle Miocene in the Green Tuff volcanic region.

(d) *Deposit stratigraphy.* As has already been indicated massive sulphide deposits tend to have a well developed **zoning** or layering. Kuroko deposits have the best and most consistent stratigraphical succession of ore and rock types, and an idealized deposit (Fig. 16.9) contains the following units:

1   hanging wall—upper volcanics and/or sedimentary formation;
2   ferruginous quartz zone—chiefly hematite and quartz (chert);
3   baryte ore zone;
4   Kuroko or black ore zone—sphalerite–galena–baryte;
5   Oko or yellow ore zone—cupriferous pyrite ores; about this level, but often towards the periphery of the deposit, there may be the Sekkoko zone of anhydrite–gypsum–pyrite;
6   Keiko or siliceous ore zone — copper-bearing, siliceous, disseminated and/or stockwork ore;
7   footwall—silicified rhyolite and pyroclastic rocks.

16.3.5 Genesis

The genesis of these deposits is discussed in Section 6.3.1.

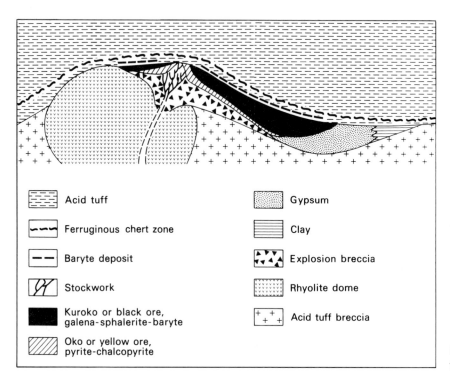

Acid tuff		Gypsum
Ferruginous chert zone		Clay
Baryte deposit		Explosion breccia
Stockwork		Rhyolite dome
Kuroko or black ore, galena-sphalerite-baryte		Acid tuff breccia
Oko or yellow ore, pyrite-chalcopyrite		

Fig. 16.9 Schematic section through a Kuroko deposit.

16.4 Volcanic-associated oxide deposits

16.4.1 Iron deposits

It has been suggested that stratiform oxide ores such as the magnetite–hematite–apatite ores of **Kiruna** and Gällivare, northern Sweden, are oxide end-members of the massive sulphide group. These ores sometimes contain **as much as 15% P** in the form of apatite, and magnetite (± hematite)–apatite ores are often referred to as **Kiruna-type** ores, but some in the type area are virtually free from apatite. There does indeed appear to be a range of ore types from massive sulphide deposits containing minor magnetite, through magnetite–pyrite ores with minor chalcopyrite and trace sphalerite such as Savage River, Tasmania, to stratiform iron oxide deposits, which may, or may not, carry appreciable phosphorus. Some possibly related deposits, e.g. in Iran, South Australia and the R.F., appear to have the form of **mineralized pipes**. These oxide deposits can be immense. The Kiruna orebody crops out over a strike length of 4 km, extends down dip for at least 1 km and is 80–90 m thick. Phosphorus-poor ore runs 67% Fe and phosphorus-rich 2–5% P and about 60% Fe. Production is now entirely from underground workings and at 18 Mt p.a. makes **Kiruna the world's largest underground mine**. At Savage River, ore reserves in 1975 were 93 Mt in one of many deposits in a zone up to 23 km long. Among the pipe deposits there are fourteen in the Bafq district of Iran together totalling more than 1000 Mt, the single deposit of Korshunovsk in Irkutsk, R.F. carries 428 Mt of ore and the enormous Cu–U–Au–Ag Olympic Dam deposit in South Australia contains over 2000 Mt of mineralized hematite breccias having an average grade of 1.6% Cu, 0.06% U_3O_8, 3.5 g t^{-1} Ag and 0.6 g t^{-1} Au. The amount of iron in the hematite and copper–iron sulphides is huge. Here we have space only for a few remarks on the **geology and genesis** of these deposits. They all occur in volcanic or volcanic-sedimentary terrane, but there is no suggestion of some correlation of deposit type with volcanic rock type, as may be the case with the massive sulphides. A sketch map of the Kiruna area is given in Fig. 16.10. The stratiform and concordant nature of the orebodies and their volcanic setting are apparent. The ores are both massive in part and well banded elsewhere—a banding that in places looks very much like bedding, sometimes with cross stratification. At Luossavaara, what appears to be a stockwork underlies the orebody and explosive volcanic activity has formed a hanging wall breccia containing fragments of the ore. All this evidence suggests that these are **exhalative deposits**.

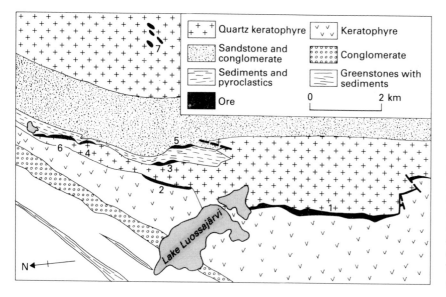

Fig. 16.10 Sketch map of the geology of the Kiruna area showing the location of the iron deposits. 1, Kirunavaara; 2, Luossavaara; 3, Rektorn; 4, Henry; 5, Haukivaara; 6, Nukutusvaara; 7, Tuollovaara. The formations dip and young eastwards.

16.4.2 Other metalliferous deposits

Other possible exhalative oxide deposits include certain uranium deposits in Labrador and some tin ores. **Exhalative tin ores** were first described from the Erzgebirge of Germany. However, the most important deposits so far found are probably those of Changpo in Dachang, China. **Exhalative tungsten ores** in Austria, have also come into prominence during recent years and Sb–W–Hg mineralization has been traced along the Eastern Alps through the whole length of Austria. The mineralization is stratiform, occurs in Lower Palaeozoic inliers and is associated with metatholeiites. The scheelite mineralization is mostly present in fine-grained metachert bands that probably represent an original exhalite and which is very finely banded. One of the world's largest tungsten mines (Felbertal, to the south of Salzburg) exploits this mineralization and mineralization of this type has now been reported from a number of areas of the world.

16.4.3 Exhalative baryte deposits

Stratiform baryte deposits occur in both dominantly sedimentary settings, often as part of, or associated with, sedex deposits, and volcanic environments where they may form part of a volcanic-associated massive sulphide deposit.

(a) *Sediment-hosted stratiform baryte deposits.* Some of these deposits, e.g. Ballynoe at Silvermines, Ireland, Meggen and Rammelsberg, Germany, are associated so intimately with sedex sulphide deposits that they must have been generated by the same period of hydrothermal activity. Deposits of this type have a worldwide distribution and they provide much of the baryte used in oil and gas drilling rigs. This is because generally they are insufficiently pure to be used for applications requiring highly refined products, e.g. the paint industry. However, the deposits themselves provide large tonnages that can often be mined in open pits to yield a cheap product.

(b) *Volcanic-associated stratiform baryte deposits.* Baryte is found with some volcanic-associated massive sulphide deposits, particularly those of Kuroko type, and in a number of these it is mined. Being a low priced product (about $U.S.50 t^{-1}$ for oil fluid grade in August 1996) the deposits must be amenable to cheap mining methods to be economic. This is the case in the Hokuroku and Hokkaido districts of Japan, which have supplied much of that country's baryte, and Peru, where the deposits are conveniently near the port of Callao and the Talara Oilfields.

16.5 Further reading

Boyle R.W., Brown A.C., Jefferson C.W., Jowett E.C. and Kirkham R.V. (eds) (1989) *Sediment-hosted Stratiform Copper Deposits.* Geological Association of Canada Special Paper 36, Memorial University, Newfoundland. A mine of information!

Lydon J.W. (1989) Volcanogenic Massive Sulphide Deposits Parts 1 & 2. In Roberts R.G. and Sheahan P.A. (eds), *Ore Deposit Models*, 145–181. Geological Association of Canada, Memorial University, Newfoundland. An invaluable discussion of the nature and origin of V.M.S. deposits.

Morganti J.M. (1989) Sedimentary-type Stratiform Ore Deposits: Some Models and a New Classification. In Roberts R.G. and Sheahan P.A. (eds), *Ore Deposit Models*, 67–78. Geological Association of Canada, Memorial University, Newfoundland. This useful summary covers both copper, sedex and other related mineralization giving the reader an overall view of stratiform, sediment-hosted base metal deposits.

17: The Vein Association and Some Other Hydrothermal Deposit Types

17.1 Introduction

Vein, manto and pipe deposits have already received considerable attention in some earlier chapters. Their morphology and nature were described in Section 4.2.1 when such subjects as pinch-and-swell structure, ribbon ore shoots, mineralization of dilatant zones along faults, vein systems and orebody boundaries were outlined. Similarly, in Chapter 5 discussion of precipitation from aqueous solutions, crustiform banding (that characteristic texture of veins) was described, as were fluid inclusions and wall rock alteration, whose relationship to veins is most marked. It is generally agreed that vein filling minerals were deposited from hydrothermal solutions, and in Sections 6.2.3 and 6.2.4 some consideration was given to the origin and nature of hydrothermal solutions, including metamorphic processes such as lateral secretion. In Section 6.4 the role of hydraulic fracturing in the formation of veins and other mineral deposits is reviewed. In Section 7.2 paragenetic sequence and zoning were discussed with special reference to vein deposits.

Veins used to be one of the most important types of mineral deposit. However, improved mining methods involving large scale mining operations now permit the economic exploitation of large low grade deposits such as are described in Chapter 15. Veins are nevertheless still important for their production of gold, tin, tungsten, uranium and a number of other metals and industrial minerals, such as fluorspar and baryte. Of course thick, high grade, base metal veins are still well worth finding and a good example is the El Indio Mine high in the Chilean Andes, where massive sulphide veins can be over 10 m thick with average grades of 2.4% Cu.

Vein deposits and the allied, but less frequent, tubular orebodies show a great variation in all their properties. For example, thickness can vary from a few millimetres to more than 100 m and so far as environments are concerned, veins can be found in practically all rock types and situations, though they are often grouped around plutonic intrusions as in Cornwall, England. Mineralogically, they can vary from monominerallic to a mineral collector's paradise, such as the native silver–cobalt–nickel–arsenic–

- Although the importance of veins as orebodies has declined markedly over the last eighty years, rich base metal veins are still worth seeking and veins remain important for their production of Au, Ag, Sn, W, U and some industrial minerals.
- Hypothermal Archaean Au veins can be very profitable as can the epithermal precious metal veins of volcanic regions.
- Disseminated hydrothermal Au deposits occur in many environments and include the noseeum gold of the Carlin-type deposits only recognized just over thirty years ago.
- Unconformity-related uranium deposits, also newly recognized deposits, are now prodigious producers of low cost uranium from very rich orebodies.

uranium association of the Erzgebirge and Great Bear Lake, Northwest Territories.

17.2 Kinds of veins

The majority of veins have probably formed from uprising hydrothermal solutions which precipitated metals under environments extending from near magmatic high temperature, high pressure conditions, to near surface low temperature, low pressure conditions. Mineralogically, **gangue minerals** can be the dominant constituents as in auriferous quartz veins. **Quartz and calcite are the commonest**, with quartz being dominant when the host rocks are silicates and calcite with carbonate host rocks. This suggests **derivation of gangue material from the wall rocks**. Sulphides are most commonly the important metallic minerals but in the case of tin and uranium oxides are predominant. Here and there, native metals are abundant, especially in gold- and silver-bearing veins.

17.3 Some important vein deposit types

At the present time gold veins are much sought after by explorationists. There are many types and among the more important are the **volcanic-associated and intrusion-associated gold ores of the Archaean greenstone belts and the epithermal gold deposits in volcanic terranes and Carlin-type deposits**. Vein uranium has a worldwide significance. The classic vein deposits of Joachimsthal in the Erzgebirge (*Jáchymov* in Czech), from which the Curies separated radium, produced large tonnages of uranium concentrates for U.S.S.R. consumption during the two decades following the explosion of the first atomic bombs. These are the famous Ag–Co–Ni–As–U veins described by Georgius Bauer, who wrote under the Latin name of Agricola, in his book *De Re Metallica* (1556). Agricola's works laid the foundation of modern mining science and of the geology of ore deposits. Vein uranium is important throughout the Hercynian Massifs of central and western Europe and indigenous deposits make a significant contribution to the French nuclear energy industry.

17.4 Some examples

These examples have been chosen to illustrate not only different vein types but also the classification into hypothermal, mesothermal and epithermal classes—see margin note in Section 7.2.2.

17.4.1 Hypothermal Archaean vein gold deposits

These big (and small) gold producers are found and sought for in Archaean greenstone belt terrane all over the world. Many famous names belong here, such as the Golden Mile at Kalgoorlie, Western Australia, the Kolar Goldfield of India and the Kirkland Lake and Timmins area of Ontario.

(a) *Size and grade.* Those greenstone belts that are well mineralized and lie within major Archaean provinces, e.g. the Superior Province of Canada and the Yilgarn Block of Western Australia, contain hundreds to several thousands of individual gold deposits. The majority of these have, or had, less than 1 t Au. However, such belts often contain several large deposits, sometimes with one or more giant gold districts (or camps), e.g. the Superior Province with twenty-five orebodies that each contain (or contained) 45 t Au, including the Timmins (Porcupine) Camp which has produced more than 1530 t Au. Grades of mined ore vary widely even within individual camps and show a dramatic decrease with time, e.g. in the Yilgarn Block from 40 ppm in the 1890s to 5 ppm in 1988. This is largely owing to improved mining and mineral processing technology and changes in the gold price. At the present day most underground deposits run 4–8 ppm but some grade 10–15 ppm. Open pit mines are exploiting grades as low as 1–2 ppm.

(b) *Host rocks and structural control of mineralization.* These deposits are usually present in greenschist facies terranes (Fig. 17.1) in structures of brittle–ductile transition regime. The original rocks are very variable but the **dominant hosts** in all the world's Archaean cratons are **tholeiitic pillowed basalts** and komatiites and their pyroclastic equivalents or felsic to mafic intrusives. The physical properties of these rocks have favoured **hydraulic fracturing** and **fluid access**, and their mineralogy and geochemistry have controlled gold deposition within the veins. At least some of the gold-bearing **hydrothermal solutions** appear to be **of metamorphic origin** as described in Section 6.2.5(b). The gold mineralization is associated with laterally zoned, wall rock alteration haloes whose formation involved the metasomatic addition of SiO_2, K_2O, CO_2, H_2O and Au. As the structural control of mineralization changes from more widely spaced faults or fractures to closely spaced minor fractures, these deposits shade into one of the types of disseminated gold deposits described below.

17.4.2 Butte, Montana—a mesothermal orefield

This is one of the world's most famous vein mining districts. From 1880 to

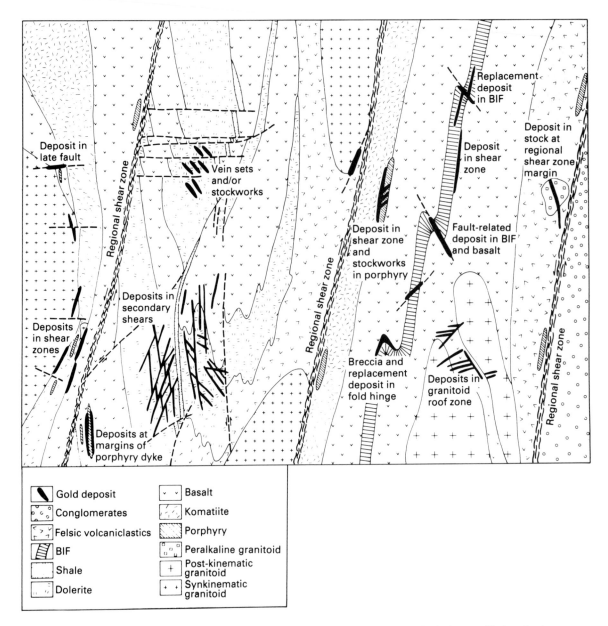

Fig. 17.1 A schematic representation of the nature of epigenetic, Archaean gold mineralization showing many of the variable structural styles and host rocks. (This figure is based largely on Western Australian examples and is taken with permission from Groves *et al.*, 1988, *Geology Department and University Extension, University of Western Australia, Publication 12.*)

1964 Butte produced 300 Mt of ore yielding 7.3 Mt Cu, 2.2 Mt Zn, 1.7 Mt Mn, 0.3 Mt Pb, 20 000 000 kg Ag, 78 000 kg Au, together with significant amounts of bismuth, cadmium, selenium, tellurium and sulphuric acid. Because of this wealth of mineral production from a very small area (little more than

$6 \times 3 \text{ km}$), with more than a score of mines, Butte has been aptly called the **richest hill on earth**, for the monetary value of its production has been exceeded only by the much larger Witwatersrand Goldfield of South Africa. Cut-and-fill stoping of the veins was the main mining method up to 1950, then it was joined by the block caving of veined ground, and in 1955 large scale, open pit working of low grade porphyry-type mineralization commenced. Reserves are still extensive, of the order of 10 Mt of high grade vein copper and silver ore and 500 Mt of low grade copper mineralization.

The Butte field is in the south-western corner of the Cretaceous Boulder Batholith. Soon after the emplacement of the intrusion, large parts of the area were covered by rhyolitic and dacitic eruptions (Fig. 17.2). The veins occur in a granodiorite which has given a radiometric age of 78 Ma. The main mineralization stage occurs in several vein systems of which the most important are the easterly trending Anaconda and the later north-westerly trending Blue veins. The Anaconda veins were the major producers in the western third of the mineralized zone and also in the eastern third, where they divide into myriads of closely spaced south-easterly trending minor veins. This is called **horse-tailing** and gives rise to porphyry-type mineralization suitable for mass mining methods. The Anaconda veins were the

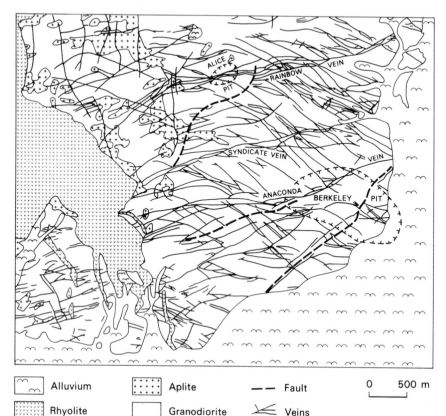

Alluvium	Aplite	— — Fault	0 500 m
Rhyolite	Granodiorite	Veins	

Fig. 17.2 Surface geology and veins of the Butte District, Montana.

largest and most productive, averaging 6–10 m in thickness with local ore pods up to 30 m thick. The Blue veins usually offset the Anaconda veins with a sinistral tear movement and sometimes ore has been dragged from the Anaconda into the Blue veins by this fault movement. Individual oreshoots persist along strike and in depth for hundreds of metres.

All the veins contain similar mineralization and this is **strongly zoned**. There is a Central Zone of copper mineralization which is particularly rich in chalcocite-enargite ore. This gradually gives way outwards to ore dominated by chalcopyrite and containing minor amounts of sphalerite — the Intermediate Zone. The Peripheral Zone is principally sphalerite–rhodochrosite mineralization with small quantities of silver. All the veins are bordered by **zones of alteration**, which usually consist of sericitization next to the vein followed outwards by intermediate argillic alteration and then by propylitization. The district zoning suggests that the main mineralization was effected by hydrothermal solutions which passed upwards and outwards during a long period of time through a steadily evolving fracture system.

17.4.3 The Llallagua Tin Deposits, Bolivia

Bolivia possesses the greatest known reserves of tin outside the countries of south-east Asia, most of her reserves being in **vein and disseminated deposits**. There is a long history of mining dating back to the Spanish colonial days of the sixteenth century when most of the important tin deposits were discovered and were mined initially for the fabulously rich silver ores that had formed, partly by supergene enrichment processes, in the upper parts of the vein systems. Some of these deposits are still being mined, mainly for tin ores beneath the silver-rich zones. In common with other tin-producing countries whose output came largely from vein deposits, Bolivia has been hit by the low tin prices of recent years. Tin production fell from 16 472 t in 1985 to 4 039 t in 1989 but by 1994 had recovered to 16 100 t, much of this production, however, being **a by-product of zinc and silver mining**.

The Bolivian Tin Belt extends along the Andean ranges east of the high plateau of Bolivia from the Argentine border northwards past Lake Titicaca and into Peru (Fig. 17.3). North-west of Oruro the deposits are mainly **tin and tungsten veins** associated with granodioritic batholiths. The batholiths range in age from Triassic to Miocene, with the Miocene ones being best mineralized.

South of Oruro there is **a tin–silver association** spatially related to high level subvolcanic intrusions. At some of these volcanic centres both the intrusives and the coeval volcanics are preserved, at others erosion has removed the volcanics completely leaving only the intrusives. This is the

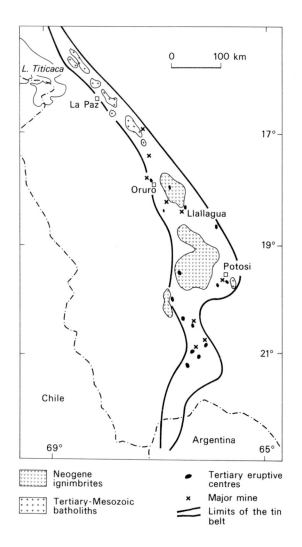

Fig. 17.3 The Bolivian Tin Belt.

0   100 km

L. Titicaca

La Paz

17°

Oruro

Llallagua

19°

Potosi

21°

Chile

Argentina

69°   65°

Neogene ignimbrites

Tertiary-Mesozoic batholiths

Tertiary eruptive centres

Major mine

Limits of the tin belt

case at Llallagua, the world's largest tin mine which worked primary tin deposits estimated to have produced over 500 000 t of tin since the beginning of this century. The mine occurs in the Salvadora Stock, which occupies a volcanic neck cutting the core of an anticline in Palaeozoic rocks. The stock, which is made up of xenolithic and highly brecciated porphyry, narrows with depth and has suffered **pervasive hydrothermal alteration**. The alteration has produced a host rock consisting of primary quartz, tourmaline, sericite and secondary quartz. There is a network of veins in and around the stock, some of which are shown in Fig. 17.4. The **major veins** trend about 030° and **appear to form part of a conjugate system of normal faults**. They are typified by the San José type, which generally has a dip of 45–80°, a good width and strike persistence, and an

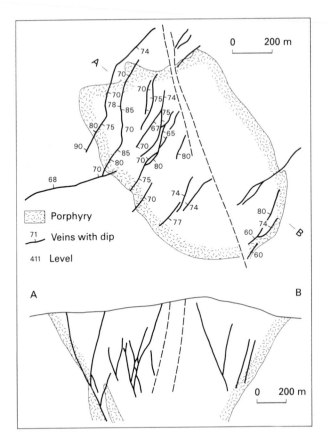

Fig. 17.4 Plan and section of the major veins at Llallagua, Bolivia.

average width of about 0.6 m, although widths of up to 1.8 m are known. These are the richest veins and can contain up to 1 m of solid cassiterite. Clay gouge is very common. The Serrano vein type is much thinner (average, 0.3 m), nearly vertical and impersistent. These veins can be as richly mineralized as the San José type, but **with such narrow veins, dilution with country rock** material occurs during mining. There is little or no clay gouge. These veins may have formed in vertical tension gashes associated with the normal faulting.

17.4.4 Epithermal precious metal deposits in volcanic rocks

Some examples of these deposits take the form of discrete veins but others are stockwork or breccia hosted (bulk-tonnage deposits) that could be included among the disseminated gold deposits described later in this chapter. This precious metal deposit type may carry important quantities of silver as well as gold and it has been important during various historical periods. The quartz–gold veins of Krissites and other areas in northern

Greece were the backbone of the Ancient Greek economy and those of Dacia (Romania) helped to sustain the Roman economy. At the present day deposits of this type are the goal of major exploration programmes around the Pacific rim, especially in Japan, the south-western Pacific and western North America. Two recent finds illustrate the range of their grade-tonnage spectrum. Hishikari in Kyushu, Japan, discovered in 1980, consists of a vein system, with veins up to a few metres wide having an average grade of $70\,g\,t^{-1}$ and high silver values, that forms an ore reserve of 1.4 Mt—a real bonanza find which during 1988 was the world's lowest cost gold mine. By contrast, the Ladolam deposit, Lihar Island, P.N.G., is a breccia hosted, disseminated deposit with at least 167 Mt of ore grading $3.43\,g\,t^{-1}$ and a mineable reserve of 14.6 million troy ounces. Bulk tonnage open pit mining with modern cyanide extraction and heap leach techniques have made even the large, low grade deposits attractive exploration targets.

Using their dominant wall rock alteration assemblages, these deposits have been divided into two classes: **acid sulphate** and **adularia–sericite**. **Acid sulphate deposits** formed from acid hydrothermal solutions, containing much CO_2, SO_2 and HCl, and are characterized by **advanced argillic alteration** [see Section 5.5.1(a)]. They are mainly associated with acid volcanic activity and have a limited vertical extent. **Adularia–sericite deposits** are much more common and their orefields and deposit dimensions larger. They were formed from near neutral solutions and the wall rock **alteration is sericitic to intermediate argillic** [see Sections 5.5.1(b) & (c)] **with abundant adularia**. These deposits are widespread in western North America, the western Pacific island arcs, Japan and New Zealand. They are associated with both acid and intermediate volcanic activity. Both classes possess gold- and silver-rich deposits and the acid sulphate class may have significant **copper** production. For both classes, **calderas are important as structural settings** and they are interpreted as forming from **near surface geothermal systems** (Fig. 17.5). It is possible that these two different classes of deposit reflect in particular their distance from their heat source, the acid sulphate type being spatially and temporally related to a shallow hydrothermal system in the core of a volcanic dome (Fig. 17.5). These **domes** often occur **along or near the ring fracture faults of calderas** and as a result of this location the mineralized zone is usually small, the heat source being the volcanic conduit. Mineralization may be capricious but the El Indio acid sulphate deposit, Chile is a spectacularly rich example with the ore in well-developed veins several metres in width. Grades averaging $225\,g\,t^{-1}$ Au, $104\,g\,t^{-1}$ Ag and 2.4% Cu have been reported. This deposit and the ores of Goldfield, Nevada have been mined using selective methods, but those of the Peublo Viejo district, Dominican Republic are being mined successfully using bulk mining methods.

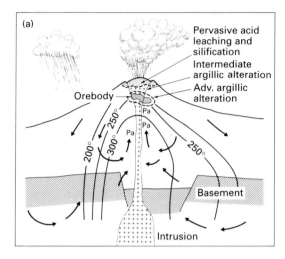

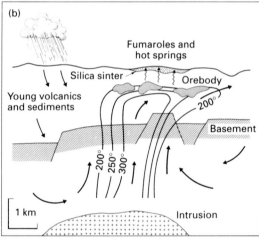

Fig. 17.5 Schemata for the formation of two types of precious metal deposits in volcanic terranes. (a) Acid sulphate type. Pa, propylitic alteration. Note that the mineralization occurs within the heat source. (b) Adularia–sericite type. The upwelling plume of hydrothermal fluid is outlined by the 200°C isotherm. The mushroom-shaped top reflects fluid flow in the plane of major fracture systems, a much narrower thermal anomaly would be present perpendicular to such structures. The heat source responsible for the buoyancy of the plume is shown as an intrusion several kilometres below the mineralized zone. In both schemata the arrows indicate the circulation of meteoric water. (a) Is at the same scale as (b).

17.5 Disseminated gold deposits

The range of deposit subtypes that can be described under this title is considerable; for reasons of space only three can be touched on here.

17.5.1 Disseminated and stockwork gold–silver deposits in igneous intrusive bodies

Some of these are referred to as porphyry gold deposits and some are clearly copper-deficient, gold-rich porphyry copper deposits. Orebodies of this general type are commonly in the range 5–15 Mt with grades around 8–16 ppm, but there are many larger deposits of lower grade. The mineralization occurs in highly fractured zones of irregular outline which have been healed by veinlets, veins and stringers of auriferous quartz. These zones are marked by considerable hydrothermal alteration of the host rock. Deposits of this subtype occur in orogenic belts in all the continents and range from Archaean to Phanerozoic in age.

17.5.2 Disseminated deposits in tuffaceous rocks and iron formations

Gold deposits in tuffs and other pyroclastic rocks are common in the green-stone belts of the Precambrian. A well described example is the Madsen Mine, Ontario. The orebodies took the form of echelon ore zones hosted by heterogeneous, sheared and highly altered tuffs occurring in the lower parts of a tholeiitic–komatiitic sequence (Fig. 17.6). The orebodies, which were delineated by assay boundaries and had an average grade of about 8 ppm, were localized by rolls (open folds) in the hanging wall and footwall contacts.

A common feature of the greenstone belts of Archaean cratons is the presence of **several large gold deposits** (>50 t Au) in areas containing numerous smaller deposits. These large and many smaller deposits occur

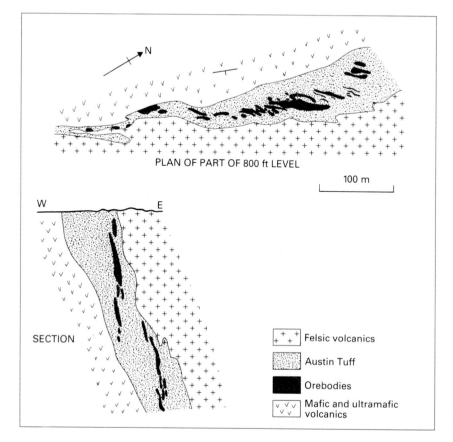

PLAN OF PART OF 800 ft LEVEL

100 m

W   E

SECTION

Felsic volcanics

Austin Tuff

Orebodies

Mafic and ultramafic volcanics

Fig. 17.6 Generalized geology of part of the Madsen Mine, Ontario.

in various facies of iron formation of which **lean sulphide-bearing carbonate and silicate facies** are perhaps the most important. These have been termed **'Homestake-type' deposits** after the famous large and long lived mine in South Dakota, and included in this group are Morro Velho, Brazil and the Kolar Goldfield of India, where similar facies and mineral associations to those at Homestake occur. Many of the Zimbabwean deposits are of this type, as well as many in the Murchison Province, Western Australia. There is a continuing debate as to whether these deposits are syngenetic and formed from fluids of deep metamorphic origin debouching into marine basins in which banded iron formation was being precipitated, or whether the older epigenetic hypothesis is still correct for at least some of these deposits.

17.5.3 Carlin-type deposits

These deposits are responsible for the majority of newly mined gold production in the U.S.A., principally in Nevada and Utah. Typically the gold occurs as micron-sized grains which are invisible to the naked eye (**micron or 'noseeum' gold**), within impure limestones or calcareous siltstones. The host rocks appear to be little different from the surrounding rocks and the **margins of the deposit are only distinguishable by assay**. Grades vary from 1 to 13.4 ppm Au and mercury and silver are often won as by-products, although in some Au/Ag can be very high. Orebody tonnages are commonly in the range 5–80 Mt.

Individual deposits occur mainly within major linear trends, up to 34 km in length, the most important of which is the **Carlin Trend** in north-east Nevada. This trend and type of deposit are named after the Carlin Mine, which was the first significant deposit of this type to be recognized, although others had previously been mined. The deposits vary from broadly tabular to highly irregular within favourable beds and adjacent to structures such as faults that have acted as fluid pathways. The faulting and fluid flow has often resulted in brecciation and several deposits are hosted in breccias.

Silicification is the commonest form of alteration and the occurrence of **jasperoid**, which **replaces carbonate**, is common in many deposits. **Gold** is thought to occur predominantly as native metal although its grain size is very small, usually less than 1 μm, and the metal's exact mineralogical form remains unknown in some mines. Gold forms films on sulphide and amorphous carbon and is particularly associated with arsenian pyrite. Pyrite is the most common sulphide and may be accompanied by marcasite and arsenic, antimony and mercury sulphides. The arsenic sulphides realgar and orpiment, which are readily distinguishable by their bright colour, occur in many deposits, as does stibnite and cinnabar. The general

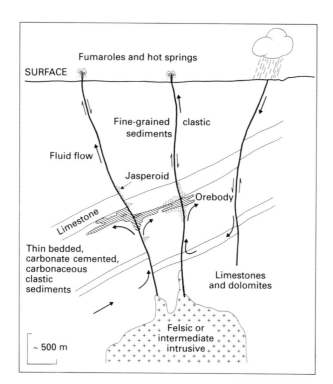

Fig. 17.7 Schema for the formation of Carlin-type deposits.

geochemical association is of As, Sb, Tl, Ba, W and Hg in addition to Au and Ag. The deposits are **spatially associated with granites** and many authors favour **intrusion driven hydrothermal systems** (Fig. 17.7) with a mixing model in which magmatic fluids have been mixed with meteoric water causing precipitation in favourable host rocks.

17.6 Unconformity-associated uranium deposits

Deposits of this type have come into prominence only during the last twenty years or so and are currently the world's main source of high grade uranium. The first to be discovered was the Rabbit Lake deposit in Saskatchewan in 1968, and the initial discoveries in the East Alligator River Field of Australia were made in 1970. As a result, there is no agreed name for this deposit type, and in addition to the above title they have been called *inter alia* 'unconformity vein-type deposits' and 'Proterozoic vein-like deposits'. Some are in no way vein-like but are pod shaped and a name omitting the word 'vein' is to be preferred. Since 1970, more deposits have been found and are now known to include (after Olympic Dam, South Australia) the largest (Jabiluka, Northern Territory, 200000 t contained U_3O_8) and highest grade (Cigar Lake, Saskatchewan, 9.04% U) uranium deposits in the world. The Athabasca Basin (see Fig. 17.9) in Saskatchewan and the Northern Territory of Australia host the main deposits so far found and these **contain about half the world's low cost uranium reserves**. Large deposits have also been found in the Northwest Territories of Canada and in west Africa.

Orebodies of this deposit type range from very small up to more than 50 Mt (Jabiluka II) and **grade from 0.3 to over 9%**. They may also have many **metal by-products**, e.g. Jabiluka II has 15 ppm Au, Cluff Lake produces by-product gold and Key Lake contains large quantities of nickel, but when smelted the nickel contains too much uranium to be marketable. Other elements of note are Co, As, Se, Ag and Mo. The geological environment is that of middle Proterozoic, **sandstone-dominated sequences unconformably overlying** older metamorphosed Proterozoic **basement rocks**. Most of the mineralization occurs at or just below the unconformity, where it is intersected by faults passing through carbonaceous schists in the basement (Fig. 17.8). The orebodies tend to be tubular to flattened cigar-shaped. The high grade orebodies grade outwards into stratiform disseminations and fracture fillings. Important **zones of wall rock alteration** are developed around the orebodies and these zones greatly **broaden the exploration target**. The dominant types of alteration are chloritization, argillization, carbonitization (commonly dolomitization), silicification, pyritization and tourmalinization.

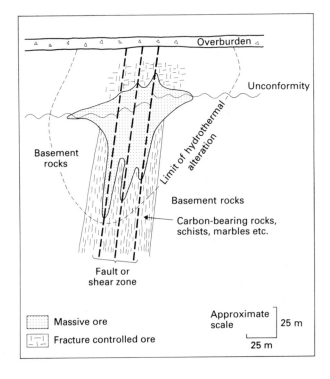

Fig. 17.8 Generalized diagram of an unconformity-associated uranium deposit.

Although most of the known Saskatchewan deposits are found along the basin rim (Fig. 17.9), the presence of the Cluff Lake Mine and other deposits in the Carswell Circular Structure, which brings the unconformity to the surface, and Cigar Lake at 400 m depth, indicates that other orebodies probably await discovery in deeper parts of the basin. The Carswell Circular Structure may have resulted from a meteoritic impact or a diapiric cryptoexplosion in Cambro-Ordovician time. **Fluid inclusion studies** indicate that both the uranium deposits and the Athabasca sandstones have been permeated by brines and CO_2-rich fluids at about 160–200°C. **Stable isotopic H and O data** suggest that these were one and the same fluid. (Similar fluids have been found in inclusions in the Australian deposits.) Isotopic studies of the McClean Lake deposit permit a positive correlation of the $d^{13}C$ from graphite in the host metasediments and from siderite intergrown with uranium mineralization, suggesting that the carbon of the siderite has been derived from the graphite. Similarly, work on **sulphur isotopes** showed that the sulphur in the ore zone sulphides could have been derived from the basement rocks. All the deposits have a largely epigenetic aspect, and radiometric dating at Key Lake shows that the mineralization is about 300 Ma younger than the Athabascan sediments and their major period of diagenesis.

All this evidence has led most geologists to postulate uranium deposition from low to medium temperature (100–300°C) **geothermal systems driven by regional heating events**. In both the Canadian and Australian fields there are uranium-rich rocks and uranium deposits present in the basement. Source rocks for the uranium are not therefore a problem in elucidating the genesis of these deposits.

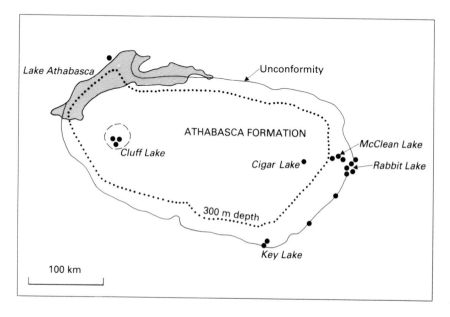

Fig. 17.9 Outline of the Athabasca Basin and distribution of the associated uranium deposits. The basement rocks outside the basin are older Proterozoic and Archaean in age.

17.7 Further reading

Berger B.R. and Bagby W.C. (1991) The Geology and Origin of Carlin-type Gold Deposits. In Foster R.P. (ed.), *Gold Metallogeny and Exploration*, 210–48. Blackie, Glasgow. An excellent summary of these noseeum gold deposits.

Evans A.M. (1993) *Ore Geology and Industrial Minerals: An Introduction*. Blackwell Scientific Publications, Oxford. Much of the above chapter is based on Chapter 16 of this book which will provide the reader with references and an amplified discussion of many points.

Henley R.W. (1991) Epithermal Gold Deposits in Volcanic Terranes. In Foster R.P. (ed.), *Gold Metallogeny and Exploration*, 133–64. Blackie, Glasgow. A good summary of precious metal deposits in volcanic rocks but note that Henley prefers the term alunite–kaolinite rather than acid sulphate.

18: Strata-bound Deposits

18.1 Introduction

The term strata-bound is applied, irrespective of their morphology, to those deposits which are restricted to a fairly limited stratigraphical range within the strata of a particular region. For example, the vein, flat and pipe lead–zinc–fluorite–baryte deposits of the Pennine orefields of Britain are restricted to the Lower Carboniferous and therefore spoken of as being strata-bound. To take a very different example, the stratiform deposits of the Zambian Copperbelt are all developed at about the same stratigraphical horizon in the Roan Series [see Section 16.2.3(b)] and these too may be described as strata-bound. Clearly, stratiform deposits can be strata-bound but strata-bound ores are not necessarily stratiform. Two important associations will be dealt with here: carbonate-hosted base metal deposits and sandstone–uranium–vanadium base metal deposits.

18.2 Carbonate-hosted base metal deposits

These are important producers of lead and zinc and also, sometimes principally, of fluorite and baryte. Copper is important in some fields, notably that of central Ireland.

18.2.1 Distribution in space and time

Most of the lead and zinc produced in Europe and the U.S.A. comes from this type of deposit. In Europe there are important fields in central Ireland, the Alps, southern Poland and the Pennines of England. In the U.S.A. there are the famous Appalachian, Tri-State (south-west Missouri, north-east Oklahoma and south-east Kansas), South-east Missouri and Upper Mississippi Districts (the locations of some of these fields are shown on Fig. 29.8). There are also important fields in north Africa (Tunisia and Algeria) and Canada. Few deposits of this type have as yet been located in the Precambrian. Minor Mississippi Valley-type mineralization, at least 2.3 Ga old, occurs in the Transvaal but the Gayna River deposit, Northwest Territory of Canada, has more than 50 Mt running up to 10% combined metals. It is,

In this chapter we learn:
- that strata-bound does **not** mean stratiform;
- that the long exploited, carbonate-hosted Mississippi Valley deposit type is still an important base metal producer for Pb, Zn and the industrial minerals fluorite and baryte, although the origin is still much debated;
- that known occurrences are almost entirely restricted to the Phanerozoic;
- that the host rocks are shallow water carbonates laid down in low latitudes, which host orebodies in various settings from faults through supratenuous folds to karstic cavities;
- that Zn tends to dominate Pb;
- of another strata-bound deposit type—the sandstone U-V-base metal deposit which, by contrast, occurs in terrestial sediments.

however, only 0.9 Ga old. Substantial deposits first appear in the **Cambrian** of south-east Missouri and important deposits occur in all systems, except the Silurian, **up to the Cretaceous**.

18.2.2 Environment

We are not concerned in this group of deposits with skarns and the skarn environment, but with deposits which occur mainly in dolomites and, to a lesser extent, in limestones. The most important feature is the presence of **a thick carbonate sequence** because thin carbonate layers in shales seldom contain important deposits of this type. The fauna and lithologies of the limestone and dolomite hosts show that they were mostly formed in **shallow water**, near shore environments of warm seas peripheral to intra-cratonic (or craton centre) basins, and a plot of major carbonate-hosted deposits on palaeolatitude maps shows **a grouping of these deposits in low latitudes**. Some ore districts, e.g. Nanisivik, Canada and the Eastern Alpine District straddling Austria, Italy, Slovenia and Croatia, occur in former rift zones and others occur close to carbonate-shale facies changes, e.g. Pine Point, Canada, suggesting a near shelf-edge setting. The warmer climate of low latitudes encourages the development of reefs and so the frequent, but by no means universal, association of these deposits with reefs (e.g. South-east Missouri) and carbonate mudbanks (e.g. Ireland) is not surprising. The occurrence of **reefs and carbonate mudbanks** is related to ancient shorelines and sea bed topographies. Nowadays, along the shorelines where carbonate deposition is occurring we often find arid zones, as in the Persian Gulf, with desert-prograding supratidal flats or coastal sabkhas. There, gypsum and anhydrite are precipitated from marine-derived ground waters to form evaporites. These may be of considerable significance. The isotopic composition of the **sulphur** or sulphides from a number of carbonate-hosted deposits suggests origination **from sea water sulphate**, particularly sea water of the same age as the limestone country rocks. **Sulphate evaporites** are known to be interbedded with the limestones in relatively close regional proximity to many carbonate-hosted deposits. Thus, the primary regional control of such deposits is palaeogeographical.

Environments such as those described above developed in the past along the margins of marine basins that formed in stable cratonic areas, such as the Devonian Elk Point Basin of western Canada in which the Pine Point deposits occur. A negative but important point is that these lead–zinc deposits are remote from post-host rock, igneous intrusives which might be the source for mineralizing solutions. Other regional controls, which are well exemplified by the Mississippi Valley region and the British deposits, are the presence of the orefields **in positive areas of shallow** water sedimentation and separated from each other by **shale-rich basins**. Such positive areas in the British Isles are often underlain by older granitic masses, suggesting that these very competent rocks fractured easily to produce channelways for uprising solutions which, on reaching the overlying limestones, gave rise to the mineralization. In addition, a large number of deposits are clearly related spatially to faults, sometimes of a regional character (see Fig. 7.2) up which the ore solutions may have passed.

18.2.3 Terminology

These deposits are frequently, some would say loosely, described as **Mississippi Valley-type** (M.V.T.). Critics of this usage have divided them into two types: M.V.T. and Alpine-type, or into three types according to the enthalpy involved in their formation, or restricted the term in the main to North American deposits, or to those deposits containing J-type lead (see below)! The present writer prefers to use the term Mississippi Valley-type to include all low temperature deposits containing one or more of the minerals baryte, fluorite, galena and sphalerite, that are dominantly **carbonate-hosted** *and epigenetic* in origin. Thus, as many authors do, I keep the epigenetic, fracture-controlled deposits, such as those of the English Pennines, in the Mississippi Valley class, but assign the syngenetic, carbonate-hosted deposits, such as Navan, Silvermines and Tynagh in Ireland, to the sedex class (see Section 16.2.5).

18.2.4 Orebody types and situations

As has been made clear by the above comments, the orebodies are very variable in type. In the British Pennines, vein orebodies with ribbon ore shoots occupying normal faults are the main deposit type in the northern field (see Fig. 4.2). In the southern Pennines, veins are again the most important orebodies but there they occupy tear (wrench) faults. The orebodies in the Tri-State Field are **in solution and collapse structures**, caves and underground channelways connected with karst topography and these are common features in many ore districts. They arise from the fact that the host rocks formed in very shallow waters, such that even minor uplifts led to elevation above sea level, resulting in the common development of disconformities, non-conformities and the associated subaerial weathering that produced the solution-collapse breccias that host so many orebodies. At Pine Point, Northwest Territory, Canada the ores are **in interconnected small-scale solution cavities**, with may have resulted from the dissolution of carbonate rocks by corrosive fluids generated by a reaction between petroleum and sulphate ion. Some of the geological situations in which these deposits occur are shown in Fig. 18.1. They may be listed as follows.

1 Above unconformities in environments, such as permeable reefs and facies changes (A-1); supratenuous folds (A-2); above the pinch-outs of permeable channelway horizons (A-3); above or in mudbank complexes (A-4).

2 Below unconformities in solution-formed open spaces (caves etc.) related to a karst topography predating the unconformity (B-1); or in collapse structures formed by the dissolution of underlying beds by subsurface drainage (B-2 and B-3).

3 At a facies change in a formation, or between basins of deposition (C-1).

4 In regional fracture systems (D-1).

18.2.5 Grade, tonnage and mineralogy

Average ore grades range mainly from 3 to 15% combined Pb + Zn with individual orebodies running up to 50%. **Tonnages** generally range from a few tens of thousands up to 20 Mt—see Fig. 16.6 for a grade–tonnage plot

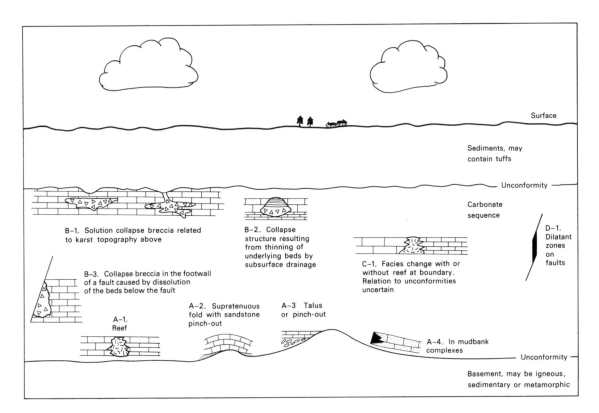

Fig. 18.1 Idealized vertical section illustrating the range of geological situations in which carbonate-hosted base metal deposits are known to occur.

for these deposits. Lead and/or zinc are the elements that commonly determine economic viability. In a few mines silver and copper are important **by-products**, as are cadmium and germanium. **Fluorite and/or baryte** may be important by-products or the prime products that are mined. From Fig. 16.6 it can be seen that the majority of M.V.T. deposits are greater than 10 Mt in size, but orefield totals, made up of a number of deposits, often lie in the range 50–500 Mt, e.g. Pine Point with over eight separate orebodies and the Upper Mississippi Valley with nearly three hundred. Most M.V.T. orefields have a high Zn/Pb ratio and a few produce very little or no lead. **Lead-dominated fields are rare** and limited largely to the very productive Old and New Lead Belts of the South-east Missouri District (the New Lead Belt is now referred to more commonly as the Viburnum Trend), unless one includes the Laisvall–Largentière type deposits of Sweden and France as suggested by some workers.

The characteristic minerals of this ore association are galena, sphalerite, fluorite and baryte in different ratios to one another varying from field to field. Pyrite and especially marcasite may be common and chalcopyrite is important in a few deposits. Calcite, dolomite, other carbonates and various forms of silica usually constitute the main gangue material.

18.2.6 Isotopic characteristics

(a) *Sulphur*. Sulphur isotopic abundances have been studied in both the sulphide and sulphate minerals. These tend to vary from field to field (Fig. 18.2) but generally show a range of positive $\delta^{34}S$ values. This range may be explained in terms of fractionation as a function of mineral species, temperature of chemical environment or by the mixing of sulphur from different sources. A comparison with values for crustal rocks (Fig. 18.2) indicates **a crustal source for the sulphur** of these deposits. Thus, the Pine Point values suggest that the sulphur was derived from marine evaporites but the low values of the Cave-in-Rock District are thought to indicate a significant contribution during mineralization of H_2S from petroleum and possibly igneous or crustal sulphur from the basement.

(b) *Lead*. Like sulphur, lead isotopic abundances are variable in nature. This variation results from the addition to existing lead of varying amounts of ^{206}Pb, ^{207}Pb and ^{208}Pb produced by the radioactive decay of uranium and thorium over geological time. ^{204}Pb is a stable isotope, possibly the decay product of a now unknown primordial radioelement. As ^{204}Pb amounts have remained unchanged, isotopic ratios are usually stated by reference to it, e.g. $^{206/204}Pb$, $^{207/204}Pb$ and $^{208/204}Pb$. Two distinct categories of lead, ordinary lead and anomalous lead, have been recognized. **Ordinary lead** has isotopic ratios that increase steadily with time so long as

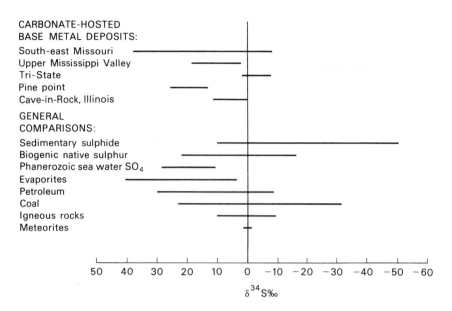

Fig. 18.2 The range of δ^{34}S for some carbonate-hosted, base metal deposits and the range for major sources of sulphur that could have contributed to the ore deposits.

it remains in uniform source rocks (probably the mantle) in contact with constant amounts of uranium and thorium. Once it has been removed from its source and separated from the elements producing radiogenic lead ($^{206-208}$Pb), its isotopic composition is fixed and if a mathematical model is assumed for the rate of addition of radiogenic leads in the source region then **a model lead age** can be calculated. Generally, such ages are in reasonable agreement with other radiometric age determinations, but sometimes they are grossly incorrect and such leads are defined as **anomalous**. Some give negative ages, e.g. the model lead age says that they have not yet been formed! These leads must have had a more complicated history, presumably within the crust, during which they acquired extra amounts of radiogenic lead. They are sometimes called **J-type leads** after Joplin, Missouri from where the first examples to be found were collected. Other leads, on a simple mathematical model, are older than the rocks in which they occur. These could be leads which were first removed from the mantle, then 'stored' in some older rocks before being remobilized and redeposited in younger rocks, e.g. the Silesian and Alpine deposits.

Leads from the various fields of the Mississippi Valley have been found to be notably enriched in radiogenic lead (i.e. having ^{206}Pb/^{204}Pb ratios of 20 or greater) compared to ordinary lead. They are all J-type leads. The strange thing is that although **J-type lead is ubiquitous in the Mississippi Valley fields**, most similar carbonate-hosted deposits, such as those of Pine Point, the British Pennines, central Ireland and southern Poland, contain ordinary lead. These facts suggest that the Mississippi Valley lead was derived from a crustal source relatively high in uranium and thorium, which could have provided it with anomalous amounts of radiogenic lead. A highly probable source of this nature would be the Precambrian basement.

18.2.7 Origin

There is little doubt that the majority of deposits of this class have been formed from **epigenetic hydrothermal solutions**. Some of the mechanisms of transport and deposition of lead and zinc in hydrothermal solutions have been discussed in Chapter 6. The **source(s)** of these solutions and their metallic constituents is **very problematical**. As there is generally no obvious spatial association with igneous intrusions that could be regarded as ore fluid sources, and geothermometry indicates temperatures of below 270°C for the source regions of these fluids, then the mineralizing fluids must be either formation water or heated circulating surface water. For many fields outside the U.S.A., lead isotopic studies suggest a deep source for the metals, but in some cases basinal brines may have played an important role and many workers have argued strongly for such a mechanism and this may have been the case for the Mississippi Valley districts. Again, the source of sulphur may have been partially deep-seated (Cave-in Rock?) or from marine evaporites (Pine Point?) or from sea water.

As the reader can discern from Chapter 6, we have today four general models for the genesis of carbonate-hosted lead–zinc deposits:
1 transport of the metals as bisulphide complexes, one fluid carrying metals and sulphur with precipitation by boiling, cooling by contact with ground water etc.;
2 transport of metals as chloride complexes (more favoured hypothesis) and precipitation when this solution meets one carrying H_2S — mixing model;
3 transport of metals as chloride complexes and sulphur as sulphate in the same solution, precipitation when the sulphate is reduced by encountering organic material;
4 organometallic complexes as carriers of the metals, H_2S in the same solution with precipitation by cooling.

Much more research will have to be carried out before we can decide whether one general model is applicable or whether we have a number of processes at work with different combinations being operative during the genesis of different deposits.

18.3 Sandstone uranium–vanadium–base metal deposits

These deposits are found in terrestrial sediments, frequently fluviatile, which were generally laid down under arid conditions. As a result, the host rocks are often red in colour and for this reason copper deposits of this type are commonly referred to as **'red bed coppers'** whilst uranium-rich examples are called sandstone–uranium type. In deposits of this group, one or two metals are present in economic amounts, whilst the others may be present in minor or trace quantities. Thus, copper mineralization (with chalcocite, bornite and covellite) is widespread in red bed successions, though it is not often up to ore grade and the same applies to silver and lead–zinc mineralization. Uranium mineralization (± vanadium) may be accompanied by trace amounts of the above metals but usually occurs as separate deposits.

18.3.1 Sandstone–uranium-type deposits

Uranium deposits of this general type are widespread in the U.S.A. and they have provided over 95% of its domestic production of uranium and vanadium. They can be divided into four classes but there is only space for a generalized view in this chapter. In the U.S.A., these deposits are well developed in the **Colorado Plateau** region and in **Wyoming** (Fig. 18.3). Metals occurring in these deposits in significant quantities are: uranium,

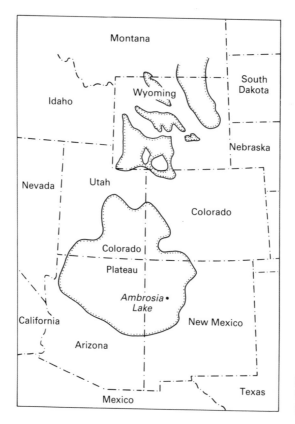

Fig. 18.3 Map showing the Colorado Plateau and Wyoming Basin.

vanadium, copper, silver, selenium and molybdenum. A deposit may contain any one or more of these metals in almost any combination except vanadium and copper, which are usually mutually exclusive. Amounts of uranium, vanadium and copper vary enormously within and between deposits and many orebodies fluctuate so much in grade that a single overall average figure is not informative. Generally, **grades vary from 0.1 to 1% U_3O_8,** but locally can be much higher, with such phenomena as whole tree trunks entirely replaced by uranium minerals. The usual range in mineral deposits is 1 000–10 000 t of contained U in ores grading 0.1–0.2% U. Many deposits contain less than 1 000 t U, but some contain more than 30 000 t U. Some American deposits carry up to 1.5% V_2O_5 and others up to 0.2% Mo, but Mo in many cases is deleterious.

Most of the **orebodies** are **similar. Small irregular pods** are common and are sporadically distributed within a favourable rock unit. The larger deposits form **mantos** hundreds of metres long, about 100 m wide and a few metres thick; these may also be referred to as tabular bodies. They can be mined by open pit or underground methods whilst the smaller bodies can be exploited by *in situ* leaching. The elongate orebodies follow buried stream courses or lenses of conglomerate material. The most common forms of deposit are termed: (a) **roll-front** (Fig. 18.4); (b) **blanket** or peneconcordant (Fig. 18.5a); and (c) **stack** or tectolithological (Fig. 18.5b), which are often related to permeable fault zones. These different morphologies can be related to the flow of mineralizing fluids through the host rock. The deposits are epigenetic, in the sense that they were formed in their present position after the host sediment was deposited — how much later is very debatable. The typical orebody represents an addition of less than 1% of ore minerals, which are accommodated in pore spaces where they form thin coatings on the detrital grains, whilst, in the case of high grade deposits, they may entirely fill the pore spaces. The disseminated

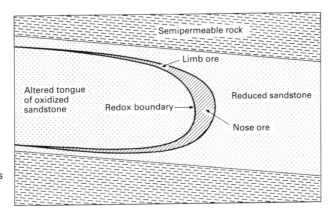

Fig. 18.4 Idealized cross section of a roll-front uranium deposit.

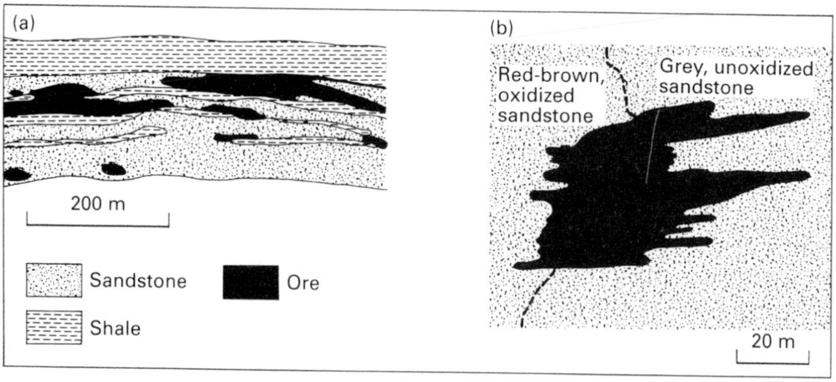

Fig. 18.5 Cross sections through two forms of sandstone uranium-type deposits in the Ambrosia Lake Field, New Mexico (Fig. 18.3). (a) Blanket or peneconcordant. (b) Stack or tectolithological.

form and microscopic size of the ore minerals increase the susceptibility to subsequent oxidation and remobilization by both alteration and weathering. The principal primary uranium minerals are pitchblende and coffinite [$(USiO_4)_{1-x} (CH)_{4x}$]. Vanadium, if present, is generally in the form of roscoelite (vanadium mica) and montroseite [$VO(OH)$].

Sedimentological studies have shown that the usual immediate host rocks of these ores are fossil stream channel deposits. These consist of linear formations of permeable sandstone and conglomeratic sandstone containing much decayed vegetation and authigenic iron disulphides and enclosed by relatively impermeable rocks—shales, mudstones etc. Petrographical studies show that the sediment was often derived from a granitic or acidic tuff source area and during weathering the trace content of uranium would be oxidized to the hexavalent state and taken into solution. This uranium would migrate through the basin of deposition to be lost in the sea unless it came into contact with reducing conditions in the organic-rich sediment a short distance beneath the sediment–water interface, in which case it would enrich the sediment. As the oxygen-rich waters encroached upon the reducing organic-rich environment, an irregular tongue-shaped zone of oxidized rock was formed. The interface, or redox boundary, between the oxidized and reduced rocks has, in cross section, the shape of a crudely crescentic envelope or roll, the leading edge of which cuts across the host strata (Fig. 18.4) and points down dip towards reduced organic-rich ground that still contains authigenic iron disulphides. The reduced ground is generally grey in colour, the oxidized ground is drab yellow to orange or red owing to the development of limonite and hematite by the alteration of the sulphides. Upon encountering reducing conditions, the uranium became reduced to the insoluble tetravalent state and was precipitated.

18.4 Further reading

Adams S.S. (1991) Evolution of Genetic Concepts for Principal Types of Sandstone Uranium Deposits in the United States. *Econ. Geol. Monogr.* **8**, 225–48. A very good coverage of modern ideas.

Anderson G.M. and Macqueen R.W. (1989) Mississippi Valley-type Lead–zinc Deposits. In Roberts R.G. and Sheahan P.A. (eds), *Ore Deposit Models*, 79–90. Geological Association of Canada Memorial University, Newfoundland. An attractively concise discussion of an age old mineralization problem.

19: Sedimentary Deposits

■ Economically valuable sediments are very variable in type and embrace many rock types or concentrations of one or a few minerals.

■ Mechanical accumulations of heavy minerals produce placer deposits of which alluvial and beach are the most important. Alluvial deposits are today substantial Au and Sn producers, beach placers significant sources of diamond, magnetite, rutile and zircon.

■ Au–U conglomerates are outstanding examples of fossil placer deposits forming extensive orefields.

■ Our greatest reserve of iron ore exists as Precambrian banded iron formation, about 90% of it formed 2 500–1 900 Ma ago. It, and enriched sections (see Section 20.3.2), are now the world's main source of iron ore. Phanerozoic ironstones are now of no economic importance.

■ Mn, the most important ferroalloy element, also has a multitude of industrial mineral end uses. It is won from sedimentary deposits of shallow water, shelf origin related to anoxic shale basins and from residual deposits (see Section 20.3.3).

Broadly speaking sediments can be divided into two large groups, allochthonous deposits and autochthonous deposits. The **allochthonous deposits** are those which were transported into the environment in which they are deposited and they include the terrigenous (clastic) and pyroclastic classes. The **autochthonous sediments** are those which form within the environment in which they are deposited. They include the chemical, organic and residual classes. Table 19.1 shows the relationship between these various terms. Some sediments are sufficiently rich in elements of economic interest to form orebodies and examples from both sedimentary groups will be described in this chapter. The student must realize, however, that the total range of sedimentary material of economic importance is much greater than can be included in this small volume, although some more will be covered in Chapter 21. Organic deposits will be dealt with in Chapters 24 and 25 and residual deposits in the next chapter together with other ores in whose formation weathering has played an important role.

19.1 Allochthonous deposits

Allochthonous sediments of economic interest are usually referred to by ore geologists as mechanical accumulations or **placer deposits**. They belong to the terrigenous class formed by the ordinary sedimentary processes that concentrate heavy minerals. Usually, this natural gravity separation is accomplished by moving water, though concentration in solid and gaseous mediums may also occur. The dense or **heavy minerals** so concentrated must first be freed from their source rock and **must possess a high density**, **chemical resistance to weathering and mechanical durability**. Placer minerals having these properties in varying degrees include: cassiterite, chromite, columbite, copper, diamond, garnet, gold, ilmenite, magnetite, monazite, platinum, ruby, rutile, sapphire, xenotime and zircon. Since **sulphides break up readily**, oxidize and decompose, they are rarely concentrated into placers. There are, however, some notable Precambrian exceptions (perhaps due to a non-oxidizing atmosphere) and a few small recent examples.

Table 19.1 A classification of sedimentary rocks.

Group	Class
Allochthonous sediments	*Terrestial deposits*—clays, siliclastic sands and conglomerates *Pyroclastic deposits*—tuffs, lapillituffs, agglomerates, volcanic breccias
Autochthonous sediments	*Chemical precipitates*—carbonates, evaporites, cherts, banded iron formation, ironstones, phosphorites *Organic deposits*—coal, lignite, oil shales, tar sands *Residual deposits*—laterites, bauxites

Placer deposits have formed throughout geological time, but **most are of Tertiary and Recent age**. The majority of placer deposits are small and often ephemeral as they form on the earth's surface usually at or above the local base level, so that many are removed by erosion before they can be buried. Most placer deposits are low grade but can be exploited because they are loose, easily worked materials which require no crushing and for which relatively cheap semi-mobile separating or hydraulic mining plants can be used. Mining usually takes the form of **dredging**, about the cheapest of all mining methods. Older placers are likely to be lithified, tilted and partially or wholly buried beneath other lithified rocks. This means that exploitation costs are much higher and then the deposits, to be economic, must be of high grade or contain unusually valuable minerals such as gold, as in the Precambrian Witwatersrand of R.S.A. described below, and diamonds, examples of which are the Cretaceous diamondiferous conglomerate near Estrela Do Sul, Brazil and the Max Resources' prospect at Nullagine, 200 km south-west of Port Hedland, Western Australia. The latter consists of a Tertiary conglomerate 2–3 m thick and 2–3 m beneath the surface, that has a recoverable grade of $0.23\,ct\,m^{-3}$ with approximately 60% of the stone being of gem quality. Placers can be classified in various ways but in this book the simple, traditional, genetic classification shown in Table 19.2 will be used.

Table 19.2 A classification of placer deposits.

Mode of origin	Class
Accumulation *in situ* during weathering	Residual placers
Concentration in a moving solid medium	Eluvial placers
Concentration in a moving liquid medium (water)	Stream or alluvial placers Beach placers Offshore placers
Concentration in a moving gaseous medium (air)	Aeolian placers

19.1.1 Residual placers

These accumulate immediately above a bedrock source (e.g. a gold or cassiterite vein, Fig. 19.1) by the chemical decay and removal of the lighter rock materials and they may grade downwards into weathered veins as in some tin areas of Shaba. In residual placers chemically resistant light minerals (e.g. beryl) may also occur.

19.1.2 Eluvial placers

These are formed upon hill slopes from minerals released from a nearby source rock. The heavy minerals collect above and just downslope of the source and the lighter non-resistant minerals are dissolved or swept downhill by rain wash or are blown away by the wind. This produces **a partial concentration by reduction in volume**, a process which continues with further downslope creep. Obviously, to yield a workable deposit, this incomplete process of concentration requires **a rich source**. In some areas with eluvial placers, the economic material has accumulated in pockets in the bedrock surface, e.g. cassiterite in potholes and sinkholes in marble in Malaysia.

19.1.3 Stream or alluvial placers

These were once the most important type of placer deposit and primitive mining made great use of such deposits. The ease of extraction made them eagerly sought after, in early as well as in recent times, and they have been **the cause of some of the world's greatest gold and diamond rushes.** Our understanding of the exact mechanisms by which concentrations of heavy minerals are formed in stream channels is still incomplete.

It is well known that the **heavy mineral fraction** of a sediment is much **finer grained than the light fraction**. There are several reasons for this. Firstly, many heavy minerals occur in much smaller grains than do quartz and feldspar in the igneous and metamorphic rocks from which they are derived. Secondly, the sorting and composition of a sediment is controlled by both the density and size of the particles, known as their **hydraulic**

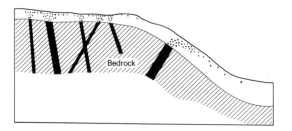

Fig. 19.1 The formation of residual (left) and eluvial (right) placer deposits by the weathering of cassiterite veins.

ratio. Thus a large quartz grain requires the same current velocity to move it as a small heavy mineral. Clearly, if we have a very rapid flow all grains of sand grade will be in motion, but with a slackening of velocity, the first material to be deposited will be large heavy minerals, then smaller heavy minerals, plus large grains of lighter minerals. If the velocity of the transporting current does not drop any further then a heavy mineral concentration will be built up. For this reason, such concentrations are developed when we have **irregular flow** and this may occur in a number of situations —always provided a source rock is present in the catchment area. The first example is that of **emergence from a canyon**. In the canyon itself net deposition is zero. As the stream widens and the gradient decreases at the canyon exit, any heavy minerals will tend to be deposited and lighter minerals will be winnowed away. Again, where we have fast-moving water passing over **projections in the stream bed**, the progress of heavy minerals may be arrested (Fig. 19.2). **Waterfalls** and **potholes** form other sites of accumulation (Fig. 19.3) and the **confluence** of a swift tributary with a slower master stream is often another site of concentration (Fig. 19.4). Most important of all, however, is deposition in rapidly flowing **meandering streams**. The faster water is on the outside curve of meanders and slack water is opposite. The junction of the two, **where point bars form, is a favourable site** for deposition of heavy minerals. With lateral migration of the meander (Fig. 19.5), **a pay streak** is built up which becomes covered with alluvium and eventually lies at some distance from the present stream channel.

Obviously, placer deposits do not form in the meanders of old age rivers because current flow is too sluggish to transport heavy minerals. In

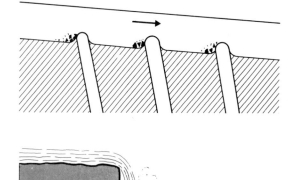

Fig. 19.2 Quartz ribs interbedded with slate serving as natural riffles for the collection of placer gold.

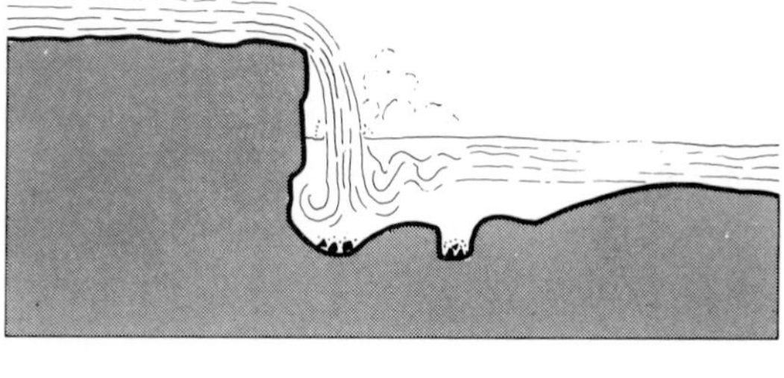

Fig. 19.3 Plunge pools at the foot of a waterfall and potholes can be sites of heavy mineral accumulations.

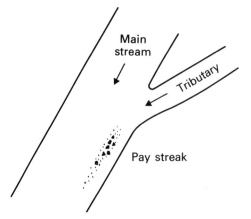

Fig. 19.4 A pay streak may be formed where a fast-flowing tributary enters a master stream.

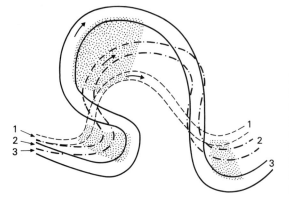

Fig. 19.5 Formation of pay streaks (dotted) in a rapidly flowing meandering stream with migrating meanders. 1, Original position of stream; 2, intermediate position; 3, present position. Note that pay streaks are extended laterally and downstream. Arrows indicate direction of water flow.

the upper reaches, current flow may be too rapid and there may also be a lack of source material. The middle reaches are most likely to contain placer deposits where we have graded streams in which a balance has been achieved between erosion, transportation and deposition. **Gradients** measured on a number of placer gold and tin deposits average out at **a little under 1 in 175**.

Much of the above discussion hinges on the concept of hydraulic equivalence in situations where the mineral grains are being transported either in a series of saltation leaps or in suspension. However, in most fluvial and littoral marine situations the transport of sand and larger particles is largely as part of a **traction carpet**, in which case settling equivalent is unimportant. The important processes are: (i) **entrainment equivalence**—larger light mineral grains stand proud on the carpet and are subject to greater lift and drag and are entrained—and (ii) **interstice entrapment**—the movement of smaller heavy mineral grains into the interstices of coarser sediment, as a result of which gravels will be better traps than sand. This results in fine grains commonly lagging behind coarse grains when both are being transported over a coarse bed and dynamic lag enrichments of heavy minerals are built up. Much of the world's **tin** is won from alluvial placers in Brazil and Malaysia and alluvial placers are also important **gold** producers.

19.1.4 Beach placers

The most important minerals of beach placers are: cassiterite, diamond, gold, ilmenite, magnetite, monazite (an economically important carrier of R.E.E.), rutile, xenotime and zircon. Examples include the gold placer of Nome, Alaska, diamonds of Namibia, ilmenite–monazite–rutile sands of Travencore and Quilon, India, rutile–zircon–ilmenite sands of eastern and western Australia and magnetite sands of North Island, New Zealand. Of course, **a source** or sources **of the heavy minerals** must be present. These may be coastal rocks, or veins cropping out along the coast or in the sea bed, or rivers or older deposits being reworked by the sea. Recent marine placers occur at different topographical levels owing to Pleistocene sea level changes (Fig. 19.6; see Fig. 19.8).

Important beach placers stretch for about 900 km along the eastern coast of Australia and these are particularly important for their rutile and zircon production. They occur in Quaternary sediments that form a coastal strip up to 13 km wide and are usually 30–40 m thick. **Placer deposits occur along the present day beaches and in the Pleistocene sands behind them** (see Fig. 4.8). These stabilized sands are characterized by low arcuate ridges which probably outline the shape of former bays. As the **thickest heavy mineral accumulations** are usually **adjacent to the southern side of headlands**, a reconstruction of the palaeogeography is important during exploration. Thus at Crowdy Head (see Fig. 4.8) the deposit between A and B appears to be related to an old headland, now a bedrock outcrop at B. The point A also appears to mark the site of a former headland and further mineralization may be present to the south-west of this point.

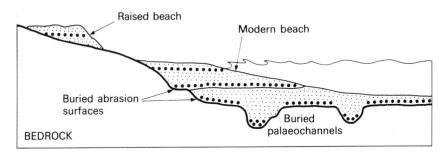

Fig. 19.6 Sketch section to illustrate some sites of beach placer deposits. Placers shown by heavy stipple.

Beach placers are formed along shorelines by the concentrating agency of waves and shore currents. Waves throw up materials on to the beach, the **backwash** carrying out the larger, lighter mineral grains, which are moved away by **longshore drift**, the larger and heavier particles thus being concentrated on the beaches. **Heavy mineral accumulations** can be seen on present day beaches to have **sharp bases** and to form **discrete laminae**. They are especially well developed during **storm wave action**. Inverse grading is present in these laminae. At the base there is a fine-grained and/or heavy mineral-rich layer which grades upwards into coarser and/or heavy mineral-poor sands. These laminations develop during the backwash phase of wave action. The breaking wave moves sand into suspension and carries it beachward and as its velocity drops, its load is deposited. Then the water flow reverses and a surface layer of sand is disturbed, becoming a high particle density bed flow. During such a bed flow the smaller and denser particles sink to the bottom of the flow, producing the reverse grading and also helping to concentrate the heavy minerals. The heavy mineral-poor sand is thus closest to the surface, waiting to be removed by the next wave. **A considerable tidal variation** is also **important** in that it exposes a wider strip of beach to wave action, which may lead to the abandonment of heavy mineral accumulations at the high water mark, where they may be covered and preserved from erosion by seaward advancing aeolian deposits.

Thus, beaches on which heavy mineral accumulations are forming today include many upon which **trade winds impinge obliquely**, and **ocean currents parallel the coast**, these two factors favouring longshore drift. In addition, these beaches face large areas of ocean and so are subjected to fierce storms and large waves. Such situations are found along the eastern and western coasts of Africa and Australia where various important heavy mineral concentrations occur.

But just how do such ephemeral deposits become preserved? The answer is still the subject of considerable debate. Along the present day coast of New South Wales, heavy mineral accumulations formed by storm action are rarely preserved. They are reworked and redeposited in diluted form, possibly because these beaches are now in a stable or slightly erosive stage. In the mining operations inland from the foredune of this area the Holocene and Pleistocene deposits are seen to form overlapping layers separated by heavy mineral-poor quartz sand. These layers dip southeastwards towards the prevailing winds (Fig. 19.7). This suggests that for the **preservation** of heavy mineral deposits **either the shoreline must prograde** because of a previously more abundant sediment supply than at present, **or the sea level must fall** to remove the heavy minerals from the sphere of wave activity.

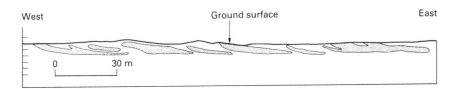

Fig. 19.7 Cross section of heavy mineral concentrations in Quaternary sands at Cudgen, New South Wales. Note seaward (easterly) dip.

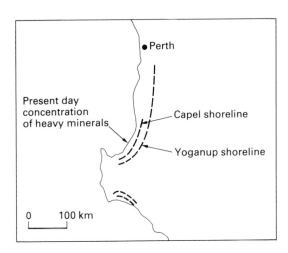

Fig. 19.8 Palaeocoastlines and present day coastline in south-western Australia. The heavy mineral concentrates occur along three former shorelines of Pliocene to early Pleistocene age. The oldest, Yoganup, has six shorelines at 66–26 m elevation and the youngest, Minninup, is at or slightly above present sea level.

Beach placers played an important role in the early days of the titanium industry during the late 1940s and their importance increased with the development of the chloride process, which is preferred to the older sulphate process for the production of titania and titanium as it causes less pollution. Initially, the chloride process required a feedstock with $TiO_2 >$ 70%, but now some plants can use feedstock down to 60% TiO_2. This means that any **ilmenite** in the feedstock **must be largely weathered to leucoxene**. (Pure unaltered ilmenite is about 52% TiO_2.) Only beach placer deposits can supply such highly altered ilmenite and in addition can supply rutile concentrates, which are even more valuable. Economic beach placers with titanium oxide minerals are typically 10 m thick, 1 km wide and over 5 km long. Closely spaced smaller bodies may be workable. Sets of orebodies that are parallel but not contiguous often mark strandlines on a series of marine terraces (Fig. 19.8).

19.1.5 Offshore placers

These occur on the continental shelf, usually within a few kilometres of the coast. They have been formed principally by the submergence of alluvial and/or beach placers (**drowned placers**). Offshore placers are becoming increasingly important as heavy mineral producers and, with the development of more efficient dredges that are capable of working along storm affected coastlines, they will help in prolonging the life of this deposit type.

19.1.6 Aeolian placers

The most important of these have been formed by the **reworking of beach placers** by winds, e.g. the large **titanomagnetite iron sand deposits** of North Island, New Zealand that are estimated to contain more than 1000 Mt of titanomagnetite. High grade aeolian deposits occur at Richards Bay, R.S.A. on the Natal coast facing the Indian Ocean. They are found in a belt along the modern coast up to about 1 km inland and 5–25 km north of Richards Bay. The dunes are locally over 100 m high and rest on a platform at 20–40 m elevation. The deposits average 20 m in thickness and carry 10–14% heavy minerals of which ilmenite, zircon, leucoxene and rutile constitute about a half. By-product monazite is produced.

19.1.7 Fossil placer deposits

The most outstanding examples are the gold–uranium-bearing conglomerates of **late Archaean to middle Proterozoic age**. The principal deposits occur in the Witwatersrand Goldfield of South Africa, the Blind River area along the north shore of Lake Huron in Canada (only trace gold) and at Serra de Jacobina, Bahia, Brazil. Other occurrences are known in many of the other shield areas. The host rocks are **oligomictic conglomerates (vein quartz pebbles)** having a **matrix rich in pyrite** (or more rarely hematitie), **sericite and quartz**. The gold and uranium minerals (principally uraninite) occur in the matrix together with a host of other detrital minerals. In the Witwatersrand Goldfield (Fig. 19.9) the **orebodies** appear to have been formed **around the periphery of** an intermontane, intracrationic lake or **shallow water inland sea** at and near entry points where sediment was introduced into the basin. The individual mineralized areas formed as **fluvial fans** (see Fig. 4.9), which were built up at the entry points. Each fan was the result of sediment deposition at a river mouth that discharged through a canyon and flowed across a relatively narrow piedmont plain before entering the basin. The formation of a **deltaic fan** is shown in Fig. 19.10. Continual uplift of the land along basin–margin faults produced frequent resorting of the sediments, which were further winnowed by longshore currents. **Carbon bands** associated with the conglomerates were once **algal mats** that fringed the deltas. They **contain gold and uranium** that presumably was taken into sea water solution and precipitated by the mats. These processes led to the formation of the world's greatest gold-

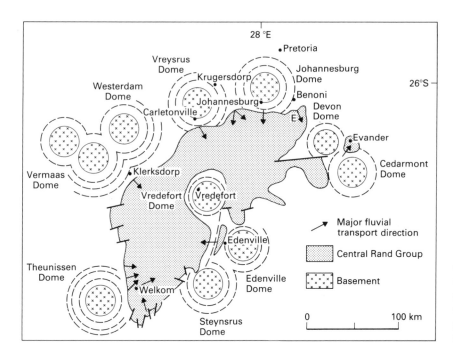

Fig. 19.9 Map showing the distribution of the Central Rand Group, which contains the principal Au–U mineralization within the Witwatersrand Goldfields, together with adjacent granite domes and sites of major fluvial influx. E indicates the position of the East Rand Goldfield (Fig. 4.9).

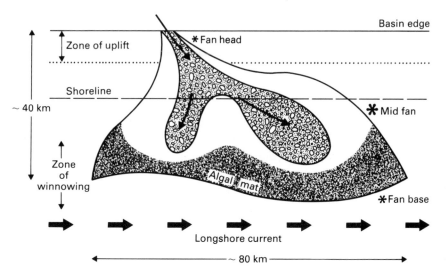

Fig. 19.10 Formation of a deltaic fan in the Witwatersrand Basin. The coarse material is shown by the pebble pattern, the algal mat by the finer pattern. Asterisk sizes indicate the relative gold values. (From Barnes, 1988, *Ores and Minerals*, Open University Press, with permission.)

field, which between its discovery in 1886 and 1983, produced over 35 000 t of gold from ores with an average grade of 10 g t⁻¹. Average mined grades are now below this figure.

The Blind River area has also been closely studied and here the uranium deposits appear to have been laid down in a fluviatile or deltaic environment, perhaps during a wet period preceding an ice age. Unlike the Witwatersrand, the host conglomerates are now at the base of the enclosing arenaceous succession and they appear to occupy valleys eroded in the softer greenstones of the metamorphic basement.

19.2 Autochthonous deposits

In this section we will be concerned with bedded iron and manganese deposits. The iron deposits can be divided conveniently into the Precambrian banded iron formations (B.I.F.) and the Phanerozoic ironstones.

19.2.1 Banded iron formations

These form one of the earth's mineral treasures. They occur in stratigraphical units hundreds of metres thick and hundreds or even thousands of kilometres in lateral extent. Substantial parts of these iron formations are usable directly as a **low grade iron ore** and other parts have been the protores for higher grade deposits (20.3.2). Compared with the present enormous demand for iron ore, now approaching 10^9 t p.a., the **reserves of** mineable ore in the banded iron formations are **very large indeed**. An extraordinary fact emerging from recent studies is that the great **bulk of iron formations** of the world was **laid down** in the very short time interval of **2 500–1 900 Ma ago**. The amount of iron laid down during this period, and still preserved, is enormous—at least 10^{14} t and possibly 10^{15} t, i.e. 90% or more of the total B.I.F. in the Precambrian. Banded iron formations are not restricted to this period, **order and younger examples** being **known**, but the total amount of iron in these is far outweighed by that deposited during the former short time interval and now represented by the B.I.F. of Labrador, the Lake Superior region of North America, Krivoi Rog and Kursk, R.F. and the Hamersley Group of Western Australia.

Banded iron formation is characterized by its **fine layering**. The layers are generally 0.5–3 cm thick and in turn they are commonly laminated on a scale of millimetres or fractions of a millimetre. The layering consists of **silica layers** (in the form of chert or better crystallized silica) alternating with **layers of iron minerals**. The simplest and commonest B.I.F. consists of **alternating hematite and chert layers**. Note that the content of alumina

---

Source and precipitation of the iron

There is little agreement on the source of the iron: one school considers that it was derived by erosion from nearby land masses, another that it is of volcanic exhalative origin. However there is widespread agreement that the iron was chemically or biogenically precipitated. Blue–green algae and fungi have been identified in the Gunflint B.I.F. of Ontario and some of these resemble present day iron-precipitating bacteria which can grow and precipitate ferric hydroxide under reducing conditions, thus fulfilling the requirements of a possibly CO_2 rich, Precambrian atmosphere.

is less than 1% contrasting with Phanerozoic ironstones, which normally carry several per cent of this oxide. There are four important facies of B.I.F.

1 *Oxide facies.* This is the most important facies and it can be divided into the **hematite and magnetite subfacies** according to which iron oxide is dominant. There is a complete gradation between the two subfacies. Hematite in least altered B.I.F. takes the form of fine-grained grey or bluish specularite. An **oolitic texture** is common in some examples, suggesting a shallow water origin, but in others the hematite may have the form of **structureless granules**. The 'chert' varies from fine-grained cryptocrystalline material to mosaics of intergrown quartz grains. In the much less common magnetite subfacies, layers of magnetite alternate with iron silicate or carbonate and cherty layers. Oxide facies B.I.F. typically **averages 30–35% Fe** and these rocks are mineable provided they are amenable to **beneficiation** by magnetic or gravity separation of the iron minerals.

2 *Carbonate facies.* This commonly consists of interbanded chert and siderite in about equal proportions. It may grade through magnetite–siderite–quartz rock into the oxide facies, or, by the addition of pyrite, into the sulphide facies. The siderite lacks oolitic or granular texture and appears to have accumulated as a fine mud below the level of wave action.

3 *Silicate facies.* Iron silicate minerals are generally associated with magnetite, siderite and chert which form layers alternating with each other. Carbonate and silicate facies B.I.F. typically run 25–30% Fe, which is too low to be of economic interest. They also present **beneficiation problems**.

4 *Sulphide facies.* This consists of pyritic carbonaceous argillites — thinly banded rocks with organic matter plus carbon making up 7–8%. The main sulphide is pyrite which can be so fine-grained that its presence may be overlooked in hand specimens unless the rock is polished.

Precambrian B.I.F. can be divided into two principal types.

1 *Algoma type.* This type is characteristic of the **Archaean greenstone belts** where it finds its most widespread development but it also occurs in younger rocks including the Phanerozoic. It shows a greywacke–volcanic association suggesting a **trough environment** and the oxide, carbonate and sulphide facies are present, with iron silicates often appearing in the carbonate facies. Algoma-type B.I.F. generally ranges from a few centimetres to a hundred or so metres in thickness and is rarely more than a few kilometres in strike length. Exceptions to this observation occur in Western Australia where late Archaean deposits of economic importance are found.

2 *Superior type.* These are thinly banded rocks mostly belonging to the oxide, carbonate and silicate facies. They are usually free of clastic material. The **rhythmical banding** of iron-rich and iron-poor cherty layers, which normally range in thickness from a centimetre or so up to a metre, is a prominent feature and this distinctive feature allows **correlation of B.I.F. over considerable distances**. Individual parts of the main Dales Gorge

Member of the Hamersley Brockman B.I.F. of Western Australia can be correlated at the 2.5 cm scale over about 50 000 km², and correlations of varves within chert bands can be made on a microscopic scale over 300 km.

Superior B.I.F. is stratigraphically closely associated with quartzite and black carbonaceous shale and usually also with conglomerate, dolomite, massive chert, chert breccia and argillite. **Volcanic rocks are not always directly associated with this B.I.F.**, but they are nearly always present somewhere in the stratigraphical column. Superior type B.I.F. may extend for hundreds of kilometres along strike and thicken from a few tens of metres to several hundred metres. The successions in which these B.I.F. occur usually lie unconformably on highly metamorphosed basement rocks with the B.I.F., as a rule, in the lower part of the succession. In some places they are separated from the basement rocks by only a metre or so of quartzite, grit and shale and in some parts of the Gunflint Range, Minnesota, they rest directly on the basement rocks. The associated rock sequences and sedimentary structures indicate that Superior B.I.F. formed in **fairly shallow water on continental shelves**, in evaporitic **barred basins**, on flat **prograding coastlines** or in **intracratonic basins**.

THE LAKE SUPERIOR REGION

For an example of B.I.F., we can look briefly at the deposits in the U.S.A. to the west and south of Lake Superior. This was one of the greatest iron ore districts of the world. The western part, which is shown in Fig. 19.11, can be divided into three major units: a basement complex (>2 600 Ma old); a thick sequence of weakly to strongly metamorphosed sedimentary and volcanic rocks (the Marquette Range Supergroup); and later Precambrian (Keweenawan) volcanics and sediments.

Banded iron formation is mainly developed in the Marquette Range Supergroup, but in the Vermilion district it is present in the basement. In this district there is a great thickness of mafic to intermediate volcanic rocks and sediments. Banded iron formation, mainly of oxide facies, occurs at many horizons, generally as thin units rarely more than 10 m thick, but one iron formation (the Soudan) is much thicker and has been mined extensively.

The remaining iron ore of this region comes from the Menominee Group of the Marquette Range Supergroup. All the iron formations of this group in the different districts are of approximately the same age. The Marquette Range Supergroup shows a complete transition from a stable craton to deep water conditions. Clastic rocks were first laid down on the bevelled basement but most of these were removed by later erosion and in many places the Menominee Group rests directly on the basement. Iron formation is the principal rock type of this group. Despite the approximate strati-

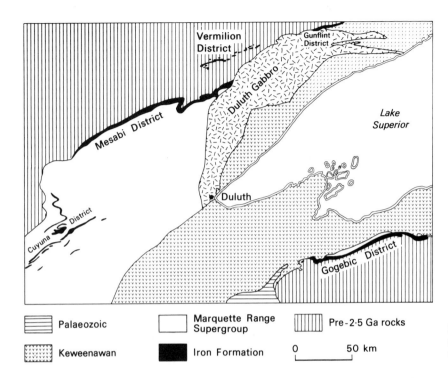

Palaeozoic

Keweenawan

Marquette Range
Supergroup

Iron Formation

Pre-2·5 Ga rocks

0   50 km

Fig. 19.11 Distribution of iron formation in Minnesota and Northern Wisconsin.

graphical equivalence of the **major iron formations, they differ greatly from one district to another in thickness, stratigraphical detail and facies type**. They appear to have been **deposited either in separate basins or in isolated parts of the same basin or shelf area**.

The same iron formation appears in the Mesabi and Gunflint districts. It is 100–270 m thick and consists of alternating units of dark, non-granular, laminated rock and cherty, granule-bearing irregularly to thickly bedded rock. The granules are mineralogically complex containing widely different proportions of iron silicates, chert and magnetite; some are rimmed with hematite. The iron formation of the Cuyuna district consists principally of two facies, thin bedded and thick bedded, which differ in mineralogy and texture. The first is evenly layered and laminated, the layers carrying varying proportions of chert, siderite, magnetite, stilpnomelane, minnesotaite and chlorite, while the second contains evenly bedded and wavy bedded rock in which chert and iron minerals alternately dominate in layers 2–30 cm thick. Granules and oolites are present. In the Gogebic district the iron formation is 150–310 m thick and consists of an alternation of wavy to irregularly bedded rocks characterized by granule and oolitic textures. The iron in the irregularly bedded rocks is principally in the form of magnetite and iron silicates, and granule textures are common. The evenly bedded iron formation is mineralogically complex, consisting of chert, siderite, iron silicates and magnetite. Each mineral may dominate a given layer and may be accompanied by one or more of the other minerals.

19.2.2 Phanerozoic ironstones

These are usually classified into two types, Clinton and Minette, but both are now of very diminished economic importance as they are of low grade and impossible to beneficiate economically on account of their silicate mineralogy. Mining of ironstones in the U.K., once very important, has now ceased.

1 *Clinton type.* This forms massive beds of oolitic hematite–chamosite–siderite rock. The iron content is about 40–50% and they are **higher in Al and P than B.I.F.** They also differ from B.I.F. in the **absence of chert bands**, the silica being mainly present in iron silicate minerals with small amounts as clastic quartz grains. They are common in rocks of Cambrian to Devonian age in eastern North America. One of the best examples is the Ordovician Wabana ore of Newfoundland.

2 *Minette type.* These are the most common and widespread ironstones. The principal minerals are **siderite and chamosite** or another iron chlorite, the chamosite often being oolitic; the iron content is around 30% while lime runs 5–20% and silica is usually above 20%. The high lime content forms one contrast with B.I.F. and often results in these ironstones being **self-fluxing ores**.

Minette ironstones are particularly widespread and important in the Mesozoic of Europe, examples being the ironstones of the English Midlands, the Minette ores of Lorraine and Luxembourg and the Salzgitter ores of Saxony. Unlike the B.I.F., **neither** the Minette nor the Clinton ironstones **show a separation into oxide, carbonate and silicate facies**. Instead the minerals are intimately mixed, often in the same oolite.

19.2.3 Sedimentary manganese deposits

Sedimentary manganese deposits and their metamorphosed equivalents produce the bulk of the world's output of manganese. Residual deposits are the other main source. World production in 1994 was 22.1 Mt and the principal producers were China (~7.0 Mt), Ukraine (2.98 Mt), R.S.A. (2.87 Mt), Brazil (2.5), Australia (1.99 Mt), India (1.6 Mt) and Gabon (1.44 Mt). Weiss wrote that manganese is 'used in feed, food, fertilizer, fungicides, facebricks, frits, flux, fragrances, flavors, foundries, ferrites, fluorescent tubes, fine chemicals, ferric leaching — and ferroalloys. It is an oxidant, deoxidant, colorant, bleach, insecticide, bactericide, algicide, lubricant, nutrient, catalyst, drier, scavenger—and much more'. Manganese for these various purposes must be of a specific type in a particular form and certain producers have become expert in supplying the **special ores** required and the end products derived from them.

An important use is in **dry cell batteries** in which ground manganese oxide acts as an oxidizing agent or depolarizer. Battery grade manganese oxide requires at least 80% MnO_2, less than 0.05% of metals which are electronegative to zinc and no nitrates. Battery grade ore from Gabon is 83–84% Mn; from Ghana, 78% and Greece, 75%.

The geochemistry of iron and manganese is very similar and the two elements might be expected to move and be precipitated together. This is indeed the case in *some* Precambrian deposits; for example, in the Cuyuna District (Fig. 19.11), manganese is abundant in some of the iron formations and forms over 20% of some of the ores. In other areas there seems to have been a complete separation of manganese and iron during weathering, transport and deposition so that many iron ores are virtually free of manganese and many manganese ores contain no more than a trace of iron. The mechanism of this separation is still unknown. But recently a number of workers have proposed that **anoxia** plays a role in the formation of the giant deposits of shallow marine environments **in intracratonic basins**. Anoxia leads to an increase in the dissolved manganese in the deeper sea water of adjacent shale basins and, following a marine transgression on to a nearby platform, the dissolved manganese can be precipitated in the shallow oxic zone of the platform. Shallow marine manganese deposits are thus considered to be **lateral facies of black shales** formed in anoxic basins. Large manganese deposits do not appear until the Proterozoic within which the world's largest known resource of manganese occurs in the Transvaal Supergroup (2500–2100 Ma) of the Kalahari Field. It has been estimated that this field contains 5026.3 Mt of Mn compared with the world's total resources of 6376.1 Mt!

19.2.4 Examples of manganese deposits

Bedded and nodular manganese deposits have been worked over a long period in the Ukraine and the southern European region of the R.F. These deposits formed in a shelf environment under estuarine and shallow marine conditions. On one side they may pass into a non-ore-bearing, coarse, clastic succession, sometimes with coal seams, that lies between the manganese orefields and the source area for the sediments. On the other side they pass into an argillaceous sequence that marks deeper water deposition.

The largest manganese ore basin of this type is the South Ukrainian Oligocene Basin and its deposits include about 70% of the world's reserves of manganese ores. It forms a part of the vast South European Oligocene Basin, which also contains the deposits of Chiatura and Mangyshlak in

Georgia and Varnentsi in Bulgaria. The distribution of ore deposits in the southern Ukraine is shown in Fig. 19.12.

The manganese ore forms a layer interstratified with sands, silts and clays. It is 0–4.5 m thick, averages 2–3.5 m and extends for over 250 km. There are intermittent breaks owing to post-Oligocene erosion. **A shore-**

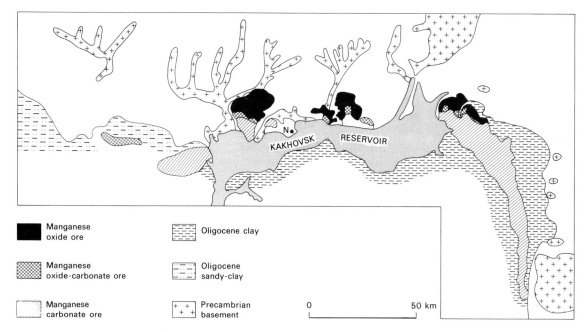

Fig. 19.12 Distribution of manganese ore in the South Ukrainian Basin. The northern and eastern parts of the map area with outcrops of Precambrian basement are largely covered by Quaternary sediments. N, Nikopol.

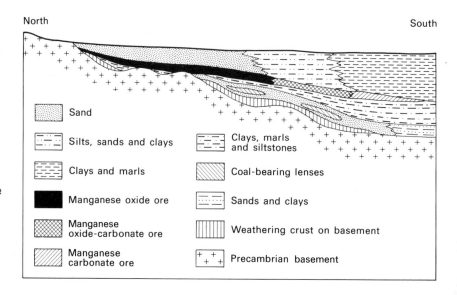

Fig. 19.13 Diagrammatic cross section through the Nikopol manganese deposits showing the zonation of the manganese ores and the transgressive nature of the sedimentary sequence with its overlap on to the Precambrian basement of the Ukrainian Platform.

ward to deeper water zoning is present (Figs 19.12, 19.13). The dominant minerals of the oxide zone are pyrolusite and psilomelane. The principal carbonate minerals are manganocalcite and rhodochrosite. Progressing into a deeper water environment, the ore layer in the carbonate zone grades into green–blue clays with occasional manganese nodules. The Chiatura deposit of Georgia shows a similar zonal pattern. The average ore grade in the Nikopol deposits is 15–25% Mn and at Chiatura it varies from a few to 35%.

An older but similar deposit is the huge Groote Eylandt Cretaceous deposit, Gulf of Carpentaria, Australia, which was also formed in shallow marine conditions just above a basal unconformity during a marine transgression. Recently it has been suggested that these deposits are related to periods of high sea level stands linked to simultaneous anoxic events in adjacent black shale basins. At Imini, Morocco it appears that the ores were formed in a zone of mixing near the Upper Cretaceous coastline where oxidized ground water of meteoric origin bearing Mn (and other metals) entered reducing sea water, itself enriched in Mn derived from the mobilization of Mn from the upper sediment layers in adjoining shale basins. Weathering and oxidation of such primary ores can produce high grade, high quality, supergene ores such as those of Groote Eylandt.

19.3 Further reading

Macdonald E.H. (1983) *Alluvial Mining*. Chapman & Hall, London. A comprehensive coverage of the geological, valuation and mining aspects of the subject. Unfortunately, his deposit classification is not the traditional one and does not appear to have gained general acceptance. For a brief discussion of this point, see Evans A.M. (1993) *Ore Geology and Industrial Minerals: An Introduction*, 245. Blackwell Scientific Publications, Oxford.

Machamer J.F. (1987) A Working Classification of Manganese Deposits. *Min. Mag.*, **157**, 348–51. An invaluable and concise summary of the subject of particular value to exploration geologists.

Pretorius D.A. (1991) Gold and Uranium in Quartz-pebble Conglomerates. *Econ. Geol.*, 75th Anniversary Volume, 117–38. This paper gives us a recent comprehensive discussion of the subject and the history of ideas of the genesis of this type of gold–uranium mineralization.

Roy S. (1988) Manganese Metallogenesis: a Review. *Ore Geol. Rev.*, **4**, 155–70. A strictly academic overview of manganese deposits but nevertheless an important and incisive one.

Weiss S.A. (1977) *Manganese: the Other Uses*. Metal Bulletin Books, London. A wide-ranging and entertaining review of the multitudinous uses of manganese in our civilization.

20: Residual Deposits and Supergene Enrichment

In this chapter you will:
- learn about the role of weathering and groundwater action in upgrading protore to ore and in enriching orebodies,
- look at examples of how these processes have formed bauxite and residual Ni orebodies,
- examine the radical redistribution of elements in mineralized ground that occurs during supergene enrichment,
- briefly survey the supergene enrichment of sulphide, B.I.F. and Mn deposits.

In the previous chapter we considered the concentration into orebodies of sedimentary material removed by mechanical or chemical processes and redeposited elsewhere. Sometimes the material left behind has been sufficiently concentrated by weathering processes and groundwater action to form residual ore deposits. For the formation of extensive deposits, **intense chemical weathering**, such as in tropical climates having a high rainfall, is necessary. In such situations, most rocks yield a soil from which all soluble material has been dissolved and these soils are called laterites. As iron and aluminium hydroxides are amongst the most insoluble of natural substances, laterites are composed mainly of these materials and are, therefore, of no value as a source of either metal. Sometimes, however, residual deposits can be high grade deposits of one metal only, but most iron-bearing laterites are too low in iron to be of economic interest.

20.1 Residual deposits of aluminium

When laterite consists of almost pure aluminium hydroxide, it is called bauxite and is the chief ore of aluminium — the metal used in everything from beer cans to jumbo jets. Total world production of bauxite in 1994 was about 97.2 Mt and most of this was used for the production of aluminium metal. About 4–5 Mt, however, was absorbed by non-metallurgical operations for the manufacture of refractories, abrasives, alumina chemicals and in the cement industry. There are **certain chemical requirements** that bauxites must meet **to be economic**. For refractories the iron content must be low since its presence tends to lower their fusion temperatures. The content of titania, alkalis and alkali earths must also be low and many bauxites do not meet these requirements. High alumina ores from China and Guyana are among the best for this end use. In the manufacture of high alumina cement, bauxite, which replaces the clay or shale in Portland cement, is fused with limestone. This cement resists corrosion by sea water or sulphate-bearing water. As with placer deposits, bauxites are vulnerable to erosion and **most deposits are** therefore **post-Mesozoic**. Older deposits, however are known — for example, those in the Palaeozoic of the C.I.S. Bauxite deposits are usually large deposits worked in open pits. The largest deposit is that of Sangaredi in Guinea, where there are at least 180 Mt forming a plateau up to 30 m thick and averaging 60% alumina, but the largest producer in 1994 was Australia with 42.2 Mt. Next in line in that year were Jamaica (11.6 Mt), Brazil (8.7 Mt), China (~6.5 Mt), India (4.8 Mt), Venezuela (4.8 Mt) and Surinam (3.8 Mt).

High level or upland bauxites generally occur on volcanic or igneous source rocks forming thick blankets of up to 30 m which cap plateaux in tropical to subtropical climates. Examples occur in the Deccan Traps of India, southern Queensland, Ghana and Guinea.

Low level peneplain-type bauxites. These occur at low level along tropical coastlines, such as those of South America, Australia and Malaysia. They are distinguished by the development of pisolitic textures and are often boehmitic in composition. Peneplain deposits are generally less than 9 m thick and are usually separated from their parent rock by a kaolinitic underclay. They are frequently associated with detrital bauxite horizons produced by fluvial or marine activity.

Karst bauxites include the oldest known bauxites—those in the lands just north of the Mediterranean, which range from **Devonian to mid-Miocene**. Other major deposits are the Tertiary ones of Jamaica and Hispaniola. These bauxites overlie a highly irregular karstified limestone or dolomite surface.

Transported or sedimentary bauxites. This is a small class of non-residual bauxites formed by the erosion and redeposition of bauxitic materials.

20.2 Residual deposits of nickel

The first major nickel production in the world came from nickeliferous laterites in New Caledonia where mining commenced in about 1876. It has been calculated that there are about 64 Mt of economically recoverable nickel in land-based deposits. Of this, about 70% occurs in lateritic deposits, although less than half current nickel production comes from these ores.

Residual nickel deposits are formed by the intense tropical **weathering of** rocks rich in trace amounts of nickel, such as **peridotites and serpentinites**, running about 0.25% Ni. During the lateritization of such rocks, nickel passes (temporarily) into solution but is generally quickly reprecipitated either on to iron oxide minerals in the laterite or as garnierite and other nickeliferous phyllosilicates in the weathered rock below the laterite. Cobalt too may be concentrated, but usually it is fixed in wad. Grades of potentially economic deposits range from **1 to 3% Ni + Co** and tonnages from about 10 to 100 Mt.

Nickel deposits of New Caledonia. Much of New Caledonia is underlain by ultrabasic rocks, many of which are strongly serpentinized. A typical environment of nickel mineralization is shown in Fig. 20.1 and a more detailed profile in Fig. 20.2. The nickel occurs in both the laterite and the weathered rock zone. In the latter it forms distinct masses, veins, veinlets or pockets rich in garnierite which occur around residual blocks of unweathered ultrabasic rock and in fissures running down into the underlying rock. The material mined is generally a mixture of the lower parts of the laterite and the weathered rock zone. Above the nickel-rich zone there are pockets of wad containing significant quantities of cobalt. Grades of up to 10% Ni were worked in the past, but today the grade is around 3% Ni. It has been estimated that there are 1.5 Gt of material on the island assaying a little over 1% Ni. All laterites take time to develop and it is thought that those on New Caledonia began to form in the Miocene.

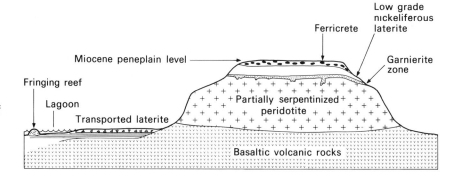

Fig. 20.1 Diagrammatic profile of a peridotite occurrence in New Caledonia showing the development of a residual nickel deposit.

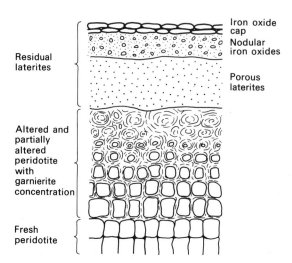

Fig. 20.2 Section through nickeliferous deposits, New Caledonia.

20.3 Supergene enrichment

Although it is applied more commonly to the enrichment of sulphide deposits, the term supergene enrichment has been extended by many workers to include similar processes affecting oxide or carbonate ores and rocks such as those carrying appreciable iron and manganese. In supergene sulphide enrichment, the metals of economic interest are carried down into hypogene (primary) ore where they are precipitated with a resultant **increase in metal content**, whereas in the case of iron and manganese ores it is chiefly the gangue material that is mobilized and carried away to leave behind a purer metal deposit.

20.3.1 Supergene sulphide enrichment

Surface waters percolating down the outcrops of sulphide orebodies **oxidize** many ore minerals and yield solvents that **dissolve** other minerals. **Pyrite** is almost ubiquitous in sulphide deposits and this breaks down to produce insoluble iron hydroxides (limonite) and **sulphuric acid**:

$$2FeS_2 + 15O + 8H_2O + CO_2 \rightarrow 2Fe(OH)_3 + 4H_2SO_4 + H_2CO_3$$
$$\text{and}$$
$$2CuFeS_2 + 17O + 6H_2O + CO_2 \rightarrow 2Fe(OH)_3 + 2CuSO_4 + 2H_2SO_4 + H_2CO_3.$$

Copper, zinc and silver sulphides are soluble and thus the upper part of a sulphide orebody may be oxidized and generally leached of many of its valuable elements right down to the water table. This is called the **zone of oxidation**. The ferric hydroxide is left behind to form a residual deposit at the surface and this is known as a **gossan or iron hat**—such features are eagerly sought by prospectors. As the water percolates downwards through the zone of oxidation, it may, because it is still carbonated and still has oxidizing properties, precipitate secondary minerals, such as malachite and azurite (Fig. 20.3).

Often, however the bulk of the dissolved metals stays in solution until it reaches the water table below which conditions are usually reducing. This leads to various reactions that precipitate the dissolved metals and result in the replacement of primary by secondary sulphides. At the same time, the grade is increased and in this way spectacularly rich bonanzas can be formed. Typical reactions may be as follows:

$$PbS + CuSO_4 \rightarrow CuS + PbSO_4 \text{ (Covellite + anglesite)},$$
$$5FeS_2 + 14CuSO_4 + 12H_2O \rightarrow 7Cu_2S + 5FeSO_4 + 12H_2SO_4 \text{ (Chalcocite)},$$
$$CuFeS_2 + CuSO_4 \rightarrow 2CuS + FeSO_4 \text{ (Covellite)}.$$

This zone of **supergene enrichment usually overlies primary mineralization**, which may or may not be of ore grade. It is thus imperative to

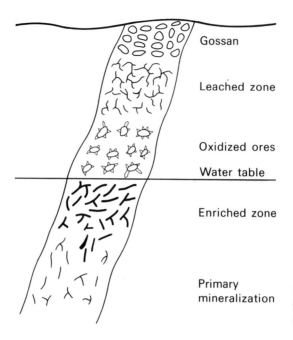

Fig. 20.3 Generalized section through a sulphide-bearing vein showing supergene enrichment.

Gossan

Leached zone

Oxidized ores

Water table

Enriched zone

Primary mineralization

ascertain whether newly discovered near surface mineralization has undergone supergene enrichment, for, if this is the case, a drastic reduction in grade may be encountered when the supergene enrichment zone is bottomed. For this purpose a careful **polished section study** is often necessary.

Clearly, such processes require a considerable time for the evolution of significant secondary mineralization. They also require that the water table be fairly deep and that ground level is slowly lowered by erosion — this usually means that such phenomena are restricted to non-glaciated land areas.

Supergene enrichment has been important in the development of many porphyry copper deposits and a good example occurs in the Inspiration orebody of the Miami district, Arizona (Fig. 20.4). Primary ores of this district are developed along a granite–schist contact with most of the ore being developed in the schist. The unenriched ore averages about 1% Cu and consists of pyrite, chalcopyrite and molybdenite. Supergene enrichment increased the grade up to as much as 5% in some places. The schist is more permeable than the granite and more supergene enrichment occurred within it. The enrichment shows **a marked correlation with the water table** (Fig. 20.4), where it starts abruptly. Downwards, it tapers off in intensity and dies out in primary mineralization. Chalcocite is the main secondary sulphide, having replaced both pyrite and chalcopyrite.

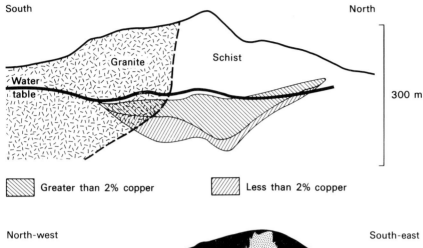

Fig. 20.4 Cross section through the Inspiration orebody at Miami, Arizona, showing the relationship between the low and high grade ores and the position of the supergene enriched zone (Cu over 2%) relative to the water table.

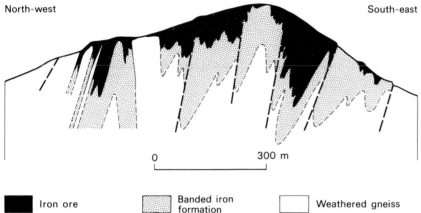

Fig. 20.5 Cross section through the iron orebodies of Cerro Bolivar, Venezuela.

20.3.2 Supergene enrichment of banded iron formation

Most of the world's iron ore is won from orebodies formed by the natural enrichment of banded iron formation (B.I.F.). Through the removal of silica from the B.I.F., the grade of iron may be increased by a factor of two to three times. Thus, for example, the Brookman Iron Formation of the Hamersley Basin in western Australia averages 20–35% Fe but in the orebodies of Mount Tom Price it has been upgraded to dark blue–grey hematite ore running 64–66% Fe.

The agent of this leaching is generally considered to be **descending ground water**, though a minority school in the past has argued the case for leaching by ascending hydrothermal water. In general, these orebodies show such a marked relationship to the present (or a past) land surface that there is little doubt that we are dealing with a process akin to lateritization. This relationship is exemplified clearly by the orebodies of Cerro Bolivar, Venezuela, which are shown in Fig. 20.5. These orebodies are developed in a tropical area having considerable relief and thus the ground water passing through them can be sampled in springs emerging from their flanks. This water carries 10.5 ppm silica and 0.05 ppm iron. Its pH averages 6.1. Clearly, the rate of removal of silica is about 200 times that of iron so that the iron tends to be left behind whilst the silica is removed. The iron is not entirely immobile and this probably accounts for the fact that many orebodies of this type consist of **compact high grade material of low porosity**. Obviously, the voids created by the removal of silica have been filled by iron minerals, normally hematite. This process seems to be taking place at Cerro Bolivar at the present time and it has been calculated that if the leaching process occurred in the past at the same rate as it is proceeding today then the present orebodies would have required 24 Ma for their development. **Enormous orebodies of this type continue to be found**, such as the N4E Mine in the Carajas region of Brazil. This monster contains at least 1 251 Mt of ore averaging 66.13% Fe and only 1% SiO_2 and 0.038% P.

20.3.3 Supergene enrichment of manganese deposits

Deep weathering processes akin to lateritization can also give rise to the formation of high grade manganese deposits. Although not comparable in geographical area with the previously described sedimentary deposits (Chapter 19), they nevertheless form important accumulations of manganese. Large deposits of this type occur at Postmasburg in South Africa, Groote Eylandt, Australia and in Gabon, India, Brazil and Ghana.

20.4 Further reading

Brimhall G.H. and Crerar D.A. (1987) Ore fluids: magmatic to supergene. *Rev. Mineral.*, **17**, 235–321. Contains a good discussion of the development of residual and supergene deposits.

Golightly J.P. (1981) Nickeliferous Laterite Deposits. *Econ. Geol.*, 75th Anniversary Volume, 710–35. A good review of this deposit type.

21: Industrial Minerals

21.1 Industrial minerals

As noted in Section 2.2.1, industrial minerals have been defined as including any rock, mineral or other naturally occurring substance of economic value but excluding metallic ores, mineral fuels and gemstones. Nevertheless, many metallic ores have non-metallurgical end uses and these may be remarkably diverse and important. **Industrial minerals dominate metals in tonnage produced and total product value** (Table 2.1) and have more **rapid growth rates**. Recycling and substitution are of little general importance in the industrial minerals field compared with that of metals and **prices** are **more stable**.

For some industrial minerals, the **unit value** is so low that transport costs over any appreciable distance can make them non-competitive and an efficient bulk transport infrastructure is essential; consequently, deposits of these minerals must be exploited close to a market. For many industrial minerals, particularly those of low unit value, the first essential to be discovered is **a market** rather than a deposit. This is the converse of the situation for metals and industrial minerals of high unit value, as these can be traded internationally. Examples of the value of market creation will be given in the section on olivine.

Some recent trends should be noted. **Product purity**, **grain size** and other specifications have become, and will continue to become, more rigid, making stringent process and quality control practices necessary. For example, in the manufacture of high quality paper, only the finest grades of filler and coating material can be used. The mineral processing techniques for industrial minerals continue to assume more elaborate forms, e.g. the processing of kaolin may now involve grinding and airfloating, washing, delamination, calcining, magnetic separation etc.; mica and talc may be micronized (ultra-fine grinding) and so on. New applications are frequently developed, often as a result of research by mineral producers themselves.

Although industrial minerals are generally thought of as having a wide occurrence, **a small number are rare**. Thus, for boron, garnet, iodine, lithium minerals, natural sodium carbonates, nitrates and vermiculite,

In this chapter we can achieve no more than a scanty survey of the vast spectrum of industrial minerals; therefore, seven main topics have been chosen to illustrate the diversity of physical and chemical properties and geological occurrence that bear upon the exploration for, and the exploitation of, industrial minerals.

■ Readers may find it rewarding to summarize and/or tabulate the significant properties and geology of the products discussed in this chapter to obtain a greater awareness of the range of attributes for which industrial minerals are utilized in our civilization and their variety of occurrence.

about 90% of world production is concentrated in only three countries. In addition, for a total of thirteen industrial minerals, more than 95% of world production comes from only five countries. Besides the commodities listed above, these include bromine, industrial diamonds, R.E.E., rutile, wollastonite and zirconium and hafnium. Much of the supply of another twelve commodities is mined in only five countries. There is thus room for **much exploration and competition**; particularly if new markets can be developed!

The variety of industrial minerals and their uses is so vast that only a book like that by Harben and Bates can begin to do justice to their diversity. I have therefore chosen seven main topics or minerals in an attempt to illustrate this great variation and the importance of **physical** or **chemical properties**, or a combination of both, in the uses to which these resources can be put. **These topics are of course supplemented by the industrial minerals and the descriptions of non-metallurgical uses of metallic ores covered in previous chapters.** Subjects touched upon already are: bauxite, baryte, beryl, chromite, diamonds, fluorite and baryte, graphite, lithium, manganese, phosphate, talc, titania and vermiculite.

21.2 Abrasives

The most important points in choosing an abrasive (grit) are: (i) that it must be harder than the work piece; and (ii) that it is imperative to match the abrasive to the job to be done. As an example of the second point, we can look at the household cleaning of brass and silver. Brass cleaners often carry very fine-grained quartz and mica. The quartz will leave very fine but largely imperceptible scratches on brass ornaments; however, such scratches cannot be tolerated on silver and only mica is added to the cleaning fluid to act as an abrasive.

21.2.1 Some abrasives

Quartz will be present in a sand or crushed sandstone in various sizes and can be sieved to obtain the grade required; it is, however, only about as hard as steel. **Garnet** is a little harder followed by **emery** which is a mixture of corundum + magnetite ± hematite ± spinel ± plagioclase, but the use of these minerals is largely superseded by artificial abrasives — **carborundum** etc. However, the bulk of **diamond** production, both natural and artificial, is used in the abrasives industry. Softer materials include **diatomite, pumice, various metal oxides** and **whiting** (ground and washed chalk).

21.3 Aggregates and constructional materials

21.3.1 Coarse aggregates

Industrial usage separates coarse and fine aggregates since they are used for different purposes. Coarse aggregates have **rock particles greater than 5 mm in diameter**. In the U.S.A. most of the particles should be retained on an A.S.T.M. No. 4 (3/8 in, 9.5 mm) sieve; in the U.K. aggregate grading must comply with B.S. regulations using B.S. sieves. Of course the volume and size of particles retained by a particular sieve are governed by the particles' shapes. This means that, although coarse aggregates are sold in nominal size grades (e.g. 14, 20 mm etc.) produced by screening processes, **undersize and oversize particles** are present in each size grade. **Limits on the permitted amounts** of undersized and oversized particles are set in most countries.

Particles above 5 mm in size may occur naturally in glacial, alluvial, beach or marine gravels or may be produced by crushing igneous, sedimentary or metamorphic rocks. In desert regions it may be necessary to use carbonate sands of coastal origin or material in active wadis in mountainous areas. In tropical regions, near surface rocks may be weakened by weathering but occasional hills of fresh rock may be present.

Aggregates must be tested carefully to assess their suitability for various functions. If they are to be embedded in bitumen or cement they must react favourably with them. Resistance to heavy loads, high impacts and severe abrasion, together with durability, are all important. **Many properties have to be tested**. Normally, **coarse aggregate producers are sited very close to their markets**, but in special circumstances they can export over considerable distances, e.g. the Glensanda Quarry, Scotland where shallow draft, bulk carrier ships can load alongside the granite quarry. In this case, the benefits of large scale production and cheap transport permit profitable export of aggregate to markets as far removed as Texas.

21.3.2 Fine aggregates

With an upper size limit of 5 mm, these can be much coarser than the geolo-

---

Some properties to be assessed

Compressive strength, water absorption, aggregate crushing value, aggregate impact value, aggregate abrasion value, polished stone value, flakiness, resistance to weathering and so on!

gist's sand grade. They may be derived from natural loose materials, beach sands, dune sands etc. by crushing rock or by crushing artificial materials, such as slag. Fine aggregates may be used in concrete making; road construction; precast concrete products, such as beams and other large sections for buildings and bridges; drainage pipes; tiles; mortars; plasters; trench fillings on which to lay cables, water pipes etc.; filter materials; glass manufacture; foundry sand and many other purposes. With such a variety of uses **many different specifications and tests** must be met and passed by producers.

In many of the above uses the fine aggregate acts as a **filler**, cutting down the amount of **binding material** (cement, bitumen) that is required or imparting stiffness to a mixture of materials. Thus, because the packing behaviour of the grains is important, grain size analysis and shape assessment (roundness and sphericity) are essential in determining the suitability of fine aggregates for particular uses.

21.3.3 Sources of coarse and fine aggregates

Naturally occurring aggregates—sand and gravel—are common in many parts of the world. Difficulties in locating such deposits may obtain in lower latitudes where chemical weathering is dominant and lateritization removes silica from many source rocks, so that quartz sands may be uncommon. In higher latitudes, especially those which have been glaciated, sand and gravel deposits are common. The major mineral in these sands is usually quartz. Some of the locations in which sand and gravel deposits may be found are shown in Fig. 21.1.

21.3.4 Structural clay products

Besides **bricks**, important clay products used by the building industry are **vitrified clay pipes and floor and wall tiles**. Most brickclays are won from unconsolidated or soft sediments. Quartz may form up to 90% of the clay, but satisfactory bricks can be made from quartz-poor clay. The **clay minerals** are usually mixtures of kaolinite, illite, smectite and chlorite. A high proportion of smectite will impart a higher drying shrinkage than the other clay minerals and may give rise to other difficulties. After being ground and moistened with water, a brickclay must be capable of taking a good shape by moulding or pressure and of **retaining** that **shape** without detrimental shrinkage, warping or cracking when the bricks are dried and then fired. A high clay content will make the material too plastic and 20% is about right.

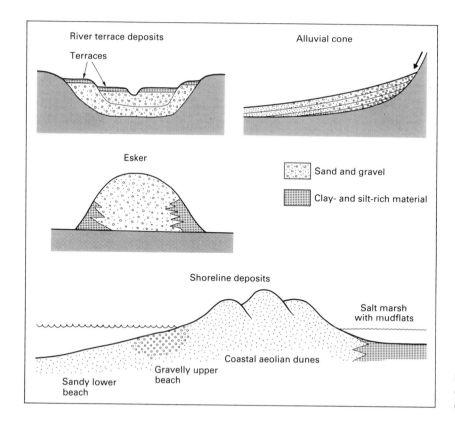

Fig. 21.1 Sketch sections (at different scales) of selected types of sand and gravel deposits.

21.3.5 Cement and concrete

Portland cement is manufactured by calcining a mixture of limestone and clay or shale. About 18–25% of clay is added, less if the limestone is argillaceous. The limestone does not have to be chemically very pure but pure limestones are preferred, and the magnesia content must be low. After calcining, the resultant clinker is ground to produce cement and about 5% of **gypsum or anhydrite** is added to act as a setting retardant. A number of recent structural failures in concrete dams, flyovers and buildings have apparently been caused by **aggregate–cement reactions**. They are due to the presence of alkalis released by cement setting reactions into the residual fluids. These fluids will react with cryptocrystalline silica (in chert and flint, silicified limestone, greywacke etc.) to form an expansive silica gel which can exert pressures as high as 14 MPa and shatter the concrete. Careful selection of aggregates to exclude these reactive materials is at present the only answer to this problem.

Stone quarrying

This was the most important mining industry in mediaeval Europe, possibly more important than all the others combined. France was the richest source of stone and vast volumes were worked in both open and underground quarries. Many of the latter were under towns and Paris has over 300 km of **known** galleries compared with the 189 km of the Paris *Métro*! Notre-Dame Cathedral was built mainly from stone quarried from below the *faubourgs* Saint-Michel, Saint-Jacques and Saint-Marcel!

21.3.6 Building stones

Stone is used in a variety of ways in the building industry. Roughly broken stone may be used for rock-fill and large blocks for coastal defences. Better shaped blocks are used as **armour stone** on breakwaters and other harbour works.

Building stone that is quarried, cut and dressed into regularly shaped blocks is termed **dimension stone** (ashlar). For use as a building stone a rock should possess mineral purity — that is, be free from decomposition products resulting from secondary alteration of the rock. It should have the necessary mechanical strength to sustain loads and stresses in service (i.e. a high crushing strength). The **durability** (resistance to weathering) should be high if the stone is to be used in the polluted atmospheres of cities, sulphur dioxide being particularly deleterious. The **hardness, workability and directional properties** are likewise important. The hardness ranges from that of soft coquina or limestone to that of granite. Workability depends partially on hardness. Limestone is easy and cheap to dress but granite is expensive. Workability also depends on the presence or absence of planes or incipient planes of splitting (joints). Porosity and permeability determine the water content and therefore the **susceptibility to frost action**. The texture affects the workability: fine-grained rocks split and dress more readily than coarse ones. Lastly, the **colour** should be attractive and permanent and not affected by weathering, nor should there be accessory minerals such as pyrite, that will oxidize and produce unsightly stains.

21.3.7 Glass

The Crystal Palace, built for the Great Exhibition in London in 1851, was the first building in which glass was a major feature. Glass is now an important material in buildings and large sheets are produced by rolling or drawing, or by floating on molten metal. Over 90% of manufactured glass is of the **soda–lime–silica type** for which the raw materials are high purity quartz sand (glass sand), limestone and soda ash (sodium carbonate).

Some dolomite and/or feldspar (or aplite or nepheline syenite) is also added to improve **chemical durability** and **resistance to devitrification**. To obtain certain characteristics (e.g. heat resistance) other mineral products, such as borates, fluorite, arsenic and selenium, may be added. Loosely consolidated sands and sandstones may make suitable **glass sands**. In these the silica content ranges from 95 to 99.8%. Alumina up to 4% is acceptable for common glass production but it must be <0.1% for optical glass, although a small amount does help to prevent devitrification. Iron oxides can be tolerated only for green and brown **bottle glass**; otherwise iron as Fe_2O_3 must be <0.05% or even <0.015% for some purposes. The **sand grains** should be even-sized (i.e. **well sorted**) and less than 20 but more than 100 mesh (A.S.T.M.). Few sands are sufficiently pure that no processing is necessary. Refractory grains, such as chromite and zircon, must be removed as these will not dissolve in the melting furnace but will persist to form black and colourless specks in the glass.

21.3.8 Gypsum

This mineral has been long used by man. It was known and prized by the Assyrians and Egyptians and the latter used it to make **plaster** for the pyramids. In the form of **alabaster** it has been long used as an ornamental stone and in mediaeval times alabaster ornaments, sculptures and altarpieces were exported form the Nottingham area of England to many other parts of Europe.

When heated to 107°C gypsum becomes the hemihydrate $CaSO_4 \cdot 1/_2 H_2O$—**plaster of Paris**. It then can be mixed with water, and sometimes with fine aggregate, for rendering walls and ceilings and the manufacture of **plasterboard**. This accounts for more than 70% of world consumption. Other uses are in cement and fertilizers and as a filler in paper, paint and toothpaste and in the manufacture of gypsum muds for oil well drilling.

Gypsum is an evaporite and as such accumulated in basins or on supratidal salt flats under arid conditions (sabkhas); see Section 21.6. It is now quarried or mined from seams in rocks or the caprocks of salt domes, where ages range from Cambrian to Tertiary, but the largest accumulations in Europe and North America are in the Permo-Triassic. Gypsum is a low cost product, about £6–7t⁻¹ C.I.F. in the U.K., and only in favourable circumstances can it be transported over long distances. Nevertheless, Spain exports nearly 1 Mt p.a. to the eastern U.S.A. for plasterboard manufacture. The future of gypsum mining lies under a cloud as considerable substitution of natural gypsum by **desulphogypsum** from coal-fired power stations can be expected in the future. Japan is now a leading country in this trend.

21.3.9 Insulators and lightweight aggregates

A number of rocks and minerals are used for noise and heat insulation in buildings. **Pumice** is mixed with cement and moulded to form lightweight insulating blocks. **Perlite** (considered later) has a similar use, as does **vermiculite** [see Section 10.3(b)]. Pumice and perlite — being rhyolitic, volcanic glass — are obtained only from Tertiary and younger rocks. Other minerals used for this purpose are **diatomite** and **expanded micas**.

21.4 Asbestos

This material is now considered to be the cause of **asbestosis** (scarring of the lung tissue) and many types of cancer. Certain forms, e.g. **crocidolite** (blue asbestiform riebeckite), have been banned in a number of European countries and there is a virtual ban on all forms of asbestos for most uses in the U.S.A. Natural and artificial substitutes are now used whenever possible. About 95% of world production is of **chrysotile** (asbestiform serpentine) which is less dangerous than crocidolite. It occurs in veinlets and veins forming stockworks in serpentinites. The asbestos fibres grew perpendicular to the vein walls and hence their **fibre length** and value is governed by vein thickness. Deposits are mined in bulk and the asbestos separated in the mill.

21.5 Clays

Clay minerals are **layered silicates** with sheets of SiO_4 tetrahedral combined in various ways with sheets containing $Al(O,OH)_6$ and/or $Mg(O,OH)_6$ octahedra. Disorder in the stacking of the layers results in the growth of only very small grains (usually <0.002 mm diameter) with unusual surface properties. Water and other molecules are attracted to their surfaces resulting in the development of plasticity. In some clay minerals exchangeable cations (Na^+, K^+, H^+, Ca^{2+}, or Mg^{2+}) are essential to their structure and give them important chemical properties with industrial applications. Clay deposits normally consist of one or more clay minerals with varying proportions of non-clay minerals, the most common being quartz, feldspar, carbonate minerals, gypsum and organic material. The types of clay raw materials and their uses are given in Table 21.1.

21.6 Evaporites

The principal products of these widespread deposits are **gypsum**, **common salt** and **potash**. Concentrations of **borates**, **nitrates** and other salts are relatively uncommon.

Table 21.1 Clay raw materials, their mineralogy and uses.

Type	Mineralogy	Uses
Kaolin (sometimes known as china clay)	Kaolinite of well ordered crystal structure plus very minor amounts of quartz, mica and sometimes anatase. Highest quality material will be almost pure kaolinite and contain >90% of particles <0.002 mm. Low to moderate plasticity	Paper filler and coating; porcelain and other ceramics; refractories; pigment/extender in paint; filler in rubber and plastics; cosmetics; inks; insecticides; filter aids
Ball clay (plastic kaolin)	Kaolinite of poorly ordered crystal structure, with varying amounts of quartz, feldspar, mica (or illite) and sometimes organic matter. Kaolinite particles usually finer grained than in kaolin. Moderate to high plasticity	Pottery and other ceramics; filler in plastics and rubber; refractories; insecticide and fungicide carrier
Halloysitic clay	Halloysite with varying amounts of quartz, feldspar, mica, carbonate minerals and others. Similar particle size distribution to ball clay. Moderately plastic	Pottery and other ceramics; filler in plastics and rubber
Refractory clay or fireclay	Structurally disordered kaolinite with some mica or illite and some quartz. Very plastic	Refractories (firebricks); pottery, vitrified clay pipes
Flint clay	Kaolinite and diaspore. Plastic only after extended grinding	Refractories
Common clay and shale	Kaolinite, and/or illite, and/or chlorite, sometimes with minor montmorillonite, quartz, feldspar, calcite, dolomite and anatase. Moderately to very plastic	Bricks, tiles and sewer pipes
Bentonite and Fullers' earth	Montmorillonite (smectite) with either Ca, Na or Mg as dominant exchangeable cation. Minor amounts of quartz, feldspar and other clay minerals. Forms thixotropic suspension in water. Forms a very plastic sticky mass	Iron ore pelletizing; foundry sand binder; clarification of oils; oil well drilling fluids; suspending agent for paints; adhesives etc.; absorbent
Sepiolite and attapulgite	Sepiolite or attapulgite with very minor amounts of montmorillonite, quartz, mica, feldspar. Thixotropic and plastic as montmorillonite clays	Absorbent; clarification of oils; some oil well drilling fluids; special papers

21.6.1 Nature, occurrence and genesis

Wherever evaporation exceeds rainfall, evaporites may form. This may occur in the supratidal zone (sabkhas) or within a restricted body of water, which may occupy a small or large basin. Modern examples of these evaporite-forming environments are well known. There are also thick and extensive Phanerozoic evaporitic deposits, the so-called **saline giants**, of which **no modern equivalents** have been found. Some of these may have formed in deep restricted waters, others have clear shoreline and shallow-water features.

Quaternary and recent evaporites occur mostly in subtropical zones between 15° and 35° latitude, in equatorial elevated plateaux, arctic deserts and rain shadows of high mountain ranges and continental regions, as in Central Asia. **Ancient evaporites** may have been formed in any of these environments. The oldest date back well into the Archaean but their best development was during the Phanerozoic in which their distribution is quite irregular in both time and space. **Most evaporites** are of **marine** origin but terrestrially formed deposits can be of economic importance.

Gypsum, anhydrite and halite are the most common evaporitic minerals but thirty-nine other minerals — too many to discuss here — are listed as common by Schreiber. Although many of these minerals may occur in the same deposit, many evaporites consist only of gypsum–anhydrite, whilst others consist of halite with little calcium sulphate. Potassium and magnesium salts are much less common but important reserves are present in the Devonian of the Williston and Elk Point Basins of western North America and in Belarus, in the Carboniferous of the Paradox Basin, Utah, the Permo-Carboniferous of the Upper Kama Basin along the western edge of the Urals, R.F., the Permian Zechstein of north-western Europe and the Permian of New Mexico and Texas, the Tertiary of Italy and Spain and other regions.

The evaporation of sea water leads first to the **precipitation** of calcium carbonate, followed by calcium sulphate, halite, potassium and magnesium salts. The natural deposition of minerals from concentrated sea water in brine pools and other restricted bodies of water follows this pattern. For the **formation of deposits of any size** a great volume of sea water must evaporate, e.g. if a basin of sea water 1000 m deep was evaporated completely, only about 14 m of evaporite would be deposited, of which 12.9 m would be halite, 1.5 m $MgCl_2$, 1 m $MgSO_4$, 0.7 m $CaSO_4$ + $CaCO_3$ + $CaMg(CO_3)_2$ and 0.4 m KCl reflecting the ratio of these salts in normal sea water. Clearly for the development of thick evaporite deposits considerable **replenishment of the body of evaporating water** by new sea water must take place. **Relative humidity** is an important control on the precipitation of evaporitic minerals in arid climates. For the development of brines from which halite can be precipitated the mean relative humidity must be <76% and for potassium salts <67%; but most low latitude coastal regions have relative humidities of 70–80%; however, low values do occur over land masses. Thus, along many arid coastlines $CaSO_4$ will be the only precipitate and the best situation for halite and potassium salt precipitation will be a marine basin nearly surrounded by land.

Evaporites may suffer a number of **post-depositional changes**. They may be replaced by other minerals, dissolved away to produce collapsed strata and dissolution residues or deformed during or after burial to produce contorted and brecciated beds and diapiric structures, including **salt domes**.

Sabkha deposits. The relative importance of these compared with basinal deposits is a moot point, particularly as they may flank evaporitic basins and one deposit type may pass laterally into the other. In addition, if sabkhas sink to form a deeper water lagoon, a vertical transition may develop. We therefore must not expect that individual evaporite deposits fall simply into one or another depositional model. **Ancient basinal evaporites** are believed to have developed in shallow and deep water environments. **Shallow water evaporites** tend to occur as massive beds with considerable lateral continuity (tens of kilometres) and the associated sediments yield abundant evidence of their shallow water origin. **Deep water evaporites** also have great horizontal continuity but now great vertical continuity as well. They are thin-bedded to laminar and may contain sulphate and halite sections tens to hundreds of metres in thickness. The majority of these **saline giants** were formed within interior basins, but others occur in rift valley settings, e.g. the Gulf of Mexico, Gabon and Red Sea deposits.

21.6.2 The Zechstein evaporites

Some workers have suggested that during the formation of certain saline giants both models operated at different times. This idea has been put forward to explain the complications seen in the Permian Zechestein evaporites of north-western Europe. The water level in the Zechstein Sea (see Fig. 16.1) is believed to have been controlled by worldwide eustatic changes, such that at low sea level stand it became a completely enclosed basin leading to a high degree of desiccation.

The Zechstein succession is up to 3000 m thick and has been divided into **five depositional cycles**. The ideal cycle starts with a thin clastic unit and passes upwards through limestone, dolomite, anhydrite, halite to potassium and magnesium salts as evaporitive concentration increased the salinity. These evaporites are exploited right across the site of the former sea from the U.K., through the Netherlands and Germany to Poland. The Zechstein evaporites are also important as **cap rocks** in the gas fields of the Netherlands and the southern North Sea. The **reservoir rocks** here include the Rotliegendes sandstone (Fig. 16.2) and the **gas originated in the underlying Carboniferous coals**.

Sodium carbonate

This was originally prepared from the ashes of seaweed and was called **soda ash**, but it was not used on a large scale until the French chemist Leblanc devised a process for its production from sodium chloride. Leblanc's process was superseded by the less expensive, currently used Solvay process.

21.6.3 Salt

About 50–60% of world production goes to the chemical industry for the manufacture of chlorine and sodium chemicals, and salt is used to produce soda ash (sodium carbonate) in the Solvay process. Large amounts are used in northern countries as a de-icing agent on roads to the detriment of the environment and concrete structures in which it may set off, or accelerate, aggregate–cement reactions.

21.6.4 Potash

About 95% of potash production goes into the manufacture of fertilizers; the remainder into the manufacture of potassium hydroxide, which is used in soaps, detergents, glass, ceramics, dyes and drugs. About 93% of world production of potash comes from only six countries: R.F., Canada, Germany, U.S.A., France and Israel.

21.6.5 Borates

The main uses of borates in the U.S.A. in 1989 were **insulation and textile fibreglass**, 42%; borosilicate glass, 9% and detergents, 9%. In western Europe the detergent use is about 3% and fibreglass only 18%. The other principal uses are in agriculture, fire retardants and enamel glazes. The U.S.A. and Turkey produce about 85% of world output. The American production is from Boron, Death Valley and Searles Lake in the desert region of California and the Turkish from three districts about 200 km south of Istanbul. World reserves have been estimated at 322 Mt.

21.6.6 Nitrates

Although **the world acquires most of its nitrogen from the atmosphere**, natural nitrates have been produced in Chile since about 1830. Approximately 750 000 t of sodium and potassium nitrates are produced each year forming, less than 0.3% of the world's nitrogen needs. The ores occur in a belt about 30 km wide and extending for about 700 km through the Atacama Desert. They are largely in the form of surface caliche (calcrete) and the **grade is around 7–8% nitrate**.

21.7 Fluorspar

Some writers use this name for the mineral fluorite, others for the raw ore containing fluorite and gangue and yet others, particularly in industry, for the finished product, which comes in three grades: acid, ceramic and metallurgical. About half the fluorite produced is consumed in iron and steel making, but this demand is decreasing as new technology is adopted by smelters. **Acid grade** (minimum of 97% CaF_2) is used for the manufacture of hydrofluoric acid (HF), much of which is consumed in the production of fluorocarbons and of artificial cryolite (Na_3AlF_6), the latter being essential for the smelting of aluminium. **Ceramic grade**, No. 1 (95–96% CaF_2) and No. 2 (80–92% CaF_2), is used in the manufacture of special glasses, enamels and glass fibres as well as in other industrial applications. **Metallurgical grade** must run 60% CaF_2 and have less than 0.3% sulphide and less than 0.5% Pb.

China is the leading producer followed in order by Mexico, Mongolia, the R.F. and R.S.A. Much fluorite is won from Mississippi Valley-type deposits (Chapter 18) although the U.S.A. is now a minor producer. It also occurs in pipes (mineralized diatremes), cryptovolcanic structures, alkaline igneous complexes, carbonatites and other deposit types.

21.8 Graphite

21.8.1 Price and production

This is a valuable mineral whose price in October 1996 varied according to its grade and purity from $U.S.650–850 t^{-1}$ for crystalline 92–95% pure to $U.S.220–300 t^{-1}$ for amorphous powder running 80–85% graphite. World production is difficult to assess for a number of reasons but is ~700 000 t. Top producers are China, South Korea, R.F., Brazil, India, Mexico, Austria, Czech Republic and North Korea. Two new mines have recently opened in Canada, three more are under development and a number of possible ore-bodies are being investigated. It appears at the time of writing that the market may be moving into an oversupply situation.

Graphite is marketed in a number of forms. The two basic divisions are into **amorphous** and **crystalline**. Amorphous is a misuse of the term as this graphite is really microcrystalline. It has a black earthy appearance, is graded primarily on its carbon content and commercial ores of this type, which come mainly from China, Mexico and South Korea, contain 50–90% C. Its main geological source is **metamorphosed beds of coal or carbonaceous material**. Crystalline graphite may be subdivided into **flake**, **vein** and **powder**. Flake graphite consists of lustrous, mica-like grains disseminated through metamorphosed carbonaceous rocks (e.g. black shales, marbles) that have suffered at least garnet grade metamorphism. These rocks may grade from 1 to 90% C. Many countries supply it but the major producers are China and the Malagasy Republic (Madagascar). Within a year or two Canada will be an important producer too. Vein graphite (sold in lump, chip and dust forms) is relatively rare, the best deposits being

---

Synthetic graphite

This is manufactured by graphitizing petroleum, coke or anthracite in an electric furnace. In general it has different properties, and therefore uses, from natural graphite, so there is little competition between the two, except in the process of recarburizing steel where it competes with amorphous graphite. The world annual production of synthetic graphite is about 1.5 Mt.

---

found in Sri Lanka. Some of these veins may consist almost entirely of graphite; others many run 80–90%. The graphite is often coarse-grained; fine-grained material is sold in powder form as '**chip**' and '**dust**' graphite.

21.8.2 Uses

Graphite is important for one or a combination of the following properties:
1   high lubricity,
2   low coefficient of thermal expansion,
3   good electrical and heat conductivity,
4   flexibility and sectility over a wide temperature range,
5   chemically inert and non-toxic,
6   generally not wetted by metals.

In 1987 the major uses to which graphite was put in the U.S.A. were refractories, 30.5%; foundries, 15%; lubricants, 14%; brake linings, 13%; pencils, 7%; steel making, 5% and batteries, 3%.

21.9   Limestone and dolomite

Limestone is probably the most important industrial rock or mineral used by man and, together with dolomite, it is quarried at thousands of localities throughout the world. Limestone contains at least 50% of the minerals calcite and dolomite, with calcite predominating. In dolomites (dolostones), the mineral dolomite is the major carbonate. There is a complete gradation from impure to high calcium limestone (>95% $CaCO_3$). High purity dolomites are defined as carrying >87% $CaMg(CO_3)_2$. Common impurities are clay, sand, chert and organic matter. Limestones and dolomites possess a wide range of **colour**, **grain size** and **bed thickness**. All these features can have an influence on their utilization.

The most commonly exploited limestone is compact and lithified, but softer porous varieties may be used for certain applications (e.g. Cretaceous chalk in the U.K. and France). **Other naturally occurring forms** of calcium carbonate, e.g. aragonite and shell sands, marble (much commer-

Table 21.2 Uses of limestone in terms of product size (excluding cement production). Important assessment requirements are given in parentheses for each size.*

Size	Use
>1 m	Cut and polished stone ('marble'). (Mineable as large blocks containing no planes of weakness. Consistent white or attractive patterns of coloration. Low porosity and frost resistant)
>30 cm	Building stone. (Wide spacing of bedding and jointing. Low porosity and frost resistant. Consistent appearance. High compressive strength)
>30 cm	Rip-rap or armour stone. (High compressive and impact strength. High density. Low porosity and frost resistant. Wide spacing of bedding and jointing)
1–30 cm	Kiln feed stone for lime kilns. (Chemical purity. Degree of decrepitation during calcination and subsequent handling. Strength in relation to design of kiln and crushing plant. Burning characteristics of stone)
1–20 cm	Aggregate, including concrete (cement and bituminous), roadstone, railway ballast, roofing, granules, terrazzo and stucco. (Impact strength. Abrasion resistance. Resistance to polishing. Soluble salts. Alkali reactivity with cement. Tendency to form particles of particular shape)
0.2–5 cm	Chemicals and glass. (Chemical purity. Organic matter. Abrasion resistance)
3–8 cm	Filter bed stone. (Compressive strength. Chemical purity. Moisture absorption. Abrasion resistance. Crust formation)
3–8 mm	Poultry grit. (Chemical purity. Shape of grains)
<4 mm	Agriculture. (Chemical purity. Organic matter)
<3 mm	Iron ore sinter, foundry fluxstone and non-ferrous metal fluxstone, self-fluxing iron ore pellets. (Chemical purity)
<0.2 mm	Filler and extender in plastics, rubber, paint, paper, putty. (Chemical purity. Whiteness and reflectance. Oil, ink and pigment absorption. pH. Nature of crushing and grinding circuits in relation to compressive strength)
<0.2 mm	Asphalt filler. (Most pulverized limestone including dust collector discharge can be used)
<0.2 mm	Mild abrasive. (Near white colour. Low quartz content)
<0.2 mm	Glazes and enamels. Mine dust. Fungicide and insecticide carrier. (Chemical purity. Near white colour. Organic matter)
<0.1 mm	Flue gas desulphurization. (Chemical purity. Surface area. Microporosity)
Various	Bulk fill. (Depends on customer requirements. Size gradation important)

* Elements usually determined in chemical analysis are Si, Al, Fe, Mn, Mg, Ca, Na, K, S, P and loss on ignition. Other elements may be important for some uses, e.g. toxic elements in filler for plastic toys.

cial 'marble' is polished limestone), carbonatite, calcareous tufa, vein calcite and cave onyx, may compete with, or be an alternative source of, $CaCO_3$ where limestone is absent. Vein calcite may come from veins composed almost entirely of that mineral or may be a by-product of fluorspar and baryte operations. Table 21.2 gives some of the more important end uses of limestone according to product size.

21.10 Magnesite and magnesia

Magnesia is one of the world's much used and versatile chemicals. Its uses range from refractories through fertilizers to lining upset stomachs and it is obtained from three sources: magnesite, sea water and natural brines. A further magnesite product is **magnesium metal**. Magnesia for the refractory industry is prepared from magnesite by calcining at 1 600–2 000°C to produce dead-burned magnesia:

$$MgCO_3 \rightarrow MgO + CO_2 \uparrow.$$

For other industrial users who require an active magnesia, calcination is at 800–1 000°C to produce caustic calcined or light-burned magnesia. In the sea water process a more complicated procedure must be followed. Magnesium is the lightest structural metal (density 1 740 versus 2 700 kg m^{-3} for aluminium) and its alloys have very high strength to mass ratios and are used in the transport, particularly aerospace, industries. Non-structural uses of the metal include desulphurization of iron and steel, metallothermic reduction of titanium and zirconium, explosives, dry batteries and anodic corrosion rods. The demand for magnesium metal in the late 1980s was increasing steadily at about 5% p.a.

21.10.1 Uses of magnesia

About four fifths of world production goes into the manufacture of **magnesian refractories** which, because of their inertness and high melting points, are used in lining steel and non-ferrous metal furnaces, cement kilns, etc. The refractories industry has rigid specifications: MgO $\nleqslant$ 95%, Fe$_2$O$_3$ < 1%, a lime to silica ratio of 2:1, bulk density of 3 400 kg m^{-3} and a low boron content. The manufacture of magnesia–carbon refractories calls for such pure magnesia that only a few magnesite mines can satisfy this market and much of the magnesia used comes from sea water processing plants.

The remaining one-fifth or so of magnesia production goes into such diverse uses as animal feedstuffs, fertilizers, special cements, gas scrubbing equipment, paper, pulp, rubber, plastics, fire-proof boards and Milk of Magnesia!

21.10.2 Magnesite deposits

The mineral magnesite is rarely found pure as it is the end member of a complete solid solution series to siderite (FeCO$_3$) and also Mn and Ca can substitute for Mg to a limited extent. Thus the purity of dead-burned magnesia produced from different deposits is very variable and variations in composition within a single orebody may create production difficulties. Magnesite of industrial interest exists in two principal natural forms: **macrocrystalline (sparry) magnesite** and **cryptocrystalline magnesite**. The former is coarse-grained and largely confined to deposits of metasomatic or sedimentary origin hosted by ancient marine platform carbonate

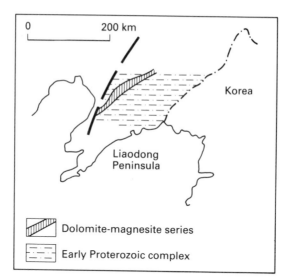

Fig. 21.2 Sketch map showing the location of magnesite deposits in the Liaodong Peninsula, China.

suites; the latter has a porcelain- or bone-like appearance and is extremely fine-grained, and occurs as lenticular masses, veins and stockworks in serpentinites. These are usually smaller deposits than the carbonate-hosted ones. However, very large sedimentary deposits of cryptocrystalline magnesite have been discovered recently in Queensland, Australia and their exploitation will put that country into the forefront of magnesia production. The world's largest magnesite deposits occur in the Liaodong Peninsula, China (Fig. 21.2). They and their host rocks are of early Proterozoic age. The magnesite-bearing zone has been outlined along approximately 60 km and has an outcrop width of 1–2 km, but is known to continue further to the north-east. There are seventeen moderate to large deposits and many smaller ones. Reserves of magnesite for this district have been put at 2.5 Gt!

21.11 Nepheline syenite

This is an alkaline, plutonic igneous rock with high sodium and aluminium contents and low silica — a chemical make-up that enables it to compete with feldspar and aplite as a source of sodium and aluminium in the manufacture of container and sheet glass. Its lower fusion temperature cuts energy costs, but feldspar has the edge on it in price. The sodium acts as a flux and the aluminium reduces the tendency of glass to devitrify and improves its chemical durability. Nepheline syenite is also used in the manufacture of some vitreous whiteware, glazes and enamels and as a filler in paint, rubber and plastics.

21.12 Olivine

Olivine is a solid solution series having two end members — forsterite (Mg_2SiO_4) and fayalite (Fe_2SiO_4). Fayalite melts at 1205°C, forsterite at 1890°C; thus, magnesian-rich olivine is a highly refractory material. Substitution by other cations, including Ca^{2+}, is very low. Most olivine in workable deposits runs 48–50% MgO, 41–43% SiO_2, 6.1–6.6% FeO and 0.2–0.8% CaO.

21.12.1 Production and price

World production increased dramatically in the last decade and in 1989 was about 4 Mt; but as a number of firms do not release production figures this is very much an estimate based on limited information. Norway is the leading producer with an output of about 2.1 Mt (but has a production capacity of 4 Mt). Spain also has a high production capacity, about 1.5 Mt p.a., and enormous reserves (>100 Mt), but inferior transport facilities mean that Spanish output is mainly marketed within the E.U. Specimen olivine prices are given in Table 21.3.

21.12.2 Uses

The large increase in production in recent years has been in response to olivine's use as a **slag conditioner in iron and steel making**. The usual slag forming agents in iron smelting are dolomite or limestone with fluorspar to lower the melting point. Silica may be added too, according to the iron ore composition, its presence being an aid in lowering the melting point of the charge. Because olivine contains both magnesia and silica, it is an excellent substitute for dolomite and silica when treating low silica ores and, as it has a higher magnesia content than dolomite, fewer tonnes are required and less slag produced.

A traditional use of olivine has been as a **foundry sand**, thanks to its high melting point, good thermal conductivity, high heat capacity (giving it resistance to thermal shock), low thermal coefficient of expansion and high green strength. It has the additional advantage that it is not, like silica

	$£ t^{-1}$
Bulk, crushed for blast furnace	£9–13 C.I.F.
Foundry sand, bagged, delivered U.K.	£52–56 C.I.F.
Blast cleaning agent	£70–76 F.O.B.

Table 21.3 Some olivine prices in October 1996.

† C.I.F., carriage, insurance and freight; F.O.B., freight on board.

sand, a health hazard and there is no risk of workers developing silicosis. Another traditional use is as a **blast cleaning agent** — again there are no health risks involved. This use is likely to increase as more and more countries prohibit the use of quartz sand for sand blasting. Many European countries have already done so.

Among the other uses to which olivine is put are the manufacture of refractory bricks, ladles, torpedo tubes and heat storage units (night storage heaters). In this last use olivine is facing strong competition from magnetite ('FeOlite'), which has better storage properties. Olivine's high density ($3300\,kg\,m^{-3}$) has led to its being used in the North Sea oilfields as ballast for oil platforms and for covering undersea pipelines. Specifications for some of the above uses are given in Table 21.4.

21.12.3 Geology

Olivine is an essential mineral of peridotite and rocks composed almost exclusively of this mineral are known as **dunite**; this is the only rock of economic interest as a source of olivine. Most dunites of commercial interest are probably of alpine type (see Section 12.1.3), but some deposits are not sufficiently well described for one to be dogmatic on this point. **Norway's** production comes largely from the Åheim district. This is in the Proterozoic North-western Gneiss Complex of southern Norway. The Åheim Mine is situated close to a deep water harbour about 300 km south-west of Trondheim. The dunite runs 90–95% olivine with minor quantities of pyroxene, chlorite and other minerals, and the quality of the rock is very

Use	Specifications

Steel slag conditioner	MgO/SiO_2 molar ratio $>1.75:1$
	MgO/SiO_2 wt % ratio approximately $1:1$
	CaO, Na_2O, K_2O all low
	Size of particles
	    Lump olivine: 1–5 cm diameter
	    Sinter feed: all <6 mm, 65% >0.85 mm
	    Pellet feed: all <2 mm

Foundry sand	High refractoriness: i.e. low alkalis and CaO, high MgO content 97% $>90\,\mu m$ and $<325\,\mu m$
Refractory bricks	Mole ratio of MgO/SiO_2 as high as possible. Ratio made up to $>2:1$ by the addition of MgO Cr_2O_3 and Al_2O_3 both useful Low Na_2O, K_2O and CaO Size range: 1–8 mm
Abrasive	No silica minerals present Normal grade in range 1–1.5 mm, special types <0.25 mm

Table 21.4 Specifications for olivine for various uses.

consistent. The deposit occupies about $6.5 \, km^2$ and is surrounded by gneiss of the Fjordane Complex. Gravity anomalies over the Åheim body indicate an eastward extension at depth and the total volume is estimated to be $25–30 \, km^3$. It is no wonder that the reserves have been described as limitless!

A/S Olivin, which is wholly government owned, began operations in 1948 and slowly increased production to meet the demands of the traditional blast cleaning and foundry sand markets. However, at the low output of about 100 000 t p.a., the operation was barely economic and it was realized that a really profitable operation could only be achieved by increasing production and lowering unit costs in order to **expand existing markets and create new ones**. A unique example of the latter is the use of olivine in North Sea oilfield operations (mentioned above under 'Uses'), which are on A/S Olivin's doorstep. To cut mining costs, ripping, rather than drilling and blasting, is employed to break up much of the dunite. The six open pits worked separately before 1980 were consolidated into a single operation. A new processing plant was installed and the capacity is 4 Mt p.a., although present output is around 2 Mt p.a. Loading facilities at the harbour can accommodate vessels of up to 80 000 t. The processing plant produces three blast furnace grades, two refractory grades and five foundry sand grades. A refractories plant produces bricks and other refractories. Sixty per cent of sales are in Europe but bulk carriers take foundry grade sands to the U.S.A., South America, New Zealand and Iran. Another Norwegian company, Franzefoss Bruk A/S, is shipping crude ore to the Eastern Seaboard of the U.S.A. for processing at Aurora, Indiana, where the finished product can compete successfully with that from Washington State, which is handicapped by the high overland transport costs.

Spain has dunite in abundance in Galicia. The Landoy Mine has reserves of about 100 Mt and a production capacity of 1–1.5 Mt p.a. However, its prices are dependent on a number of factors, including the port used. Puerto de Carino, 12 km from the mine, can only handle vessels of up to 6 000 t. The second port used is 52 km away, involving higher overland transport costs.

In the **U.S.A.**, olivine production is largely concentrated in Washington State where, although there are large reserves, production is only around 150 000 t p.a. At Twin Sisters Mountain, reserves of approximately 200 000 Mt of good quality olivine have been outlined. This is worked by the Olivine Corporation, which is **carving out a market** for itself by developing and producing refractory incinerators for waste disposal. Waste disposal is a large problem in the U.S.A. and, as each incinerator requires 20–100 t of olivine per unit, the company is establishing a useful new market.

Other countries producing olivine include Austria, Italy, Japan and Pakistan.

21.13 Perlite

The important property of perlites is that they are **hydrated rocks** carrying 2–5% water and on heating by flash roasting to temperatures close to their melting point, the contained water is converted to steam and the grains swell into light, fluffy, cellular particles. This results in a volume increase of 10–20 times and produces a material with low thermal conductivity, considerable heat resistance and high sound absorption.

21.13.1 Uses

Over half the world's production goes into the construction industry as aggregate for insulation boards, plaster and concrete in which weight reduction and special acoustic or thermal insulation properties are required. It is used for loose-fill insulation of cavity walls and for the thermal insulation of storage tanks for liquified gases. Horticultural applications include use as a rooting medium and soil conditioner and as a carrier for herbicides, insecticides and chemical fertilizers. It is used for filtering water and other liquids, in food processing, in pharmaceuticals and as a filler in paints, plastics and other products. Its uses for animal feedstuffs, poultry litter and crop farming is of growing importance. For some of these uses, pumice is a competitor; it has been expanded for us by nature and therefore does not require furnace treatment. The crushing strength of pumice is higher than that of perlite so that its use makes for a stronger concrete. Pumice, however, has the economic disadvantage that it must be shipped to processing plants in the **bulky expanded form**, whereas perlite can be shipped as the crude, unexpanded rock. Vermiculite is a much more important competitor with many properties in common with perlite. Its occurrence and uses have been discussed in Section 10.3(b).

21.13.2 Production and price

World production in 1995 was 2.079 Mt and the leading producers were the U.S.A., Greece, China, Turkey, Japan and Hungary. They price of ex-mine, crushed, graded perlite in the U.S.A. is about $30 t^{-1}$.

21.13.3 Geology

Perlite, like other glasses, devitrifies with time so that commercial deposits

> **Variable nature of a perlite deposit**
>
> In the Palhaza Perlite Quarry, Hungary 4 t of waste are mined for each tonne of usable perlite because the deposit is so heterogeneous. The variation in volatile content results in the production of three grades each giving different expansion results: $100\,\mathrm{g\,l^{-1}}$, $70\text{--}100\,\mathrm{g\,l^{-1}}$ and $70\,\mathrm{g\,l^{-1}}$. Eight size grades are made of each of these three qualities of perlite.

are mainly **restricted to areas of Tertiary and Quaternary volcanism**. Perlite occurs as lava flows, dykes, sills and circular or elongate domes. The **domes** are **the largest and most commercially important bodies** and they can be as much as 8 km across and 270 m in vertical extent. Many of these lava domes cooled quickly in their outer parts to obsidian but the interiors remained hot and formed fine-grained, crystalline rock. In certain instances, the obsidian has been hydrated as a result of penetration by ground water forming perlite. Remnants of unaltered obsidian may remain in the perlite, which may also contain phenocrysts of quartz, feldspar and other minerals.

As a result of this mode of formation a particular perlite body may be spherulitic, pumiceous and obsidian-carrying; it may contain much breccia and show a great variation in volatile content. A sound knowledge of the distribution of these variables within a working quarry is essential in order to preserve a consistent finished product and this calls for much advance drilling and bench sampling.

21.14 Phosphate rock

Phosphorus is a fundamental element in life. The bones of mammals consist of phosphate; phosphorus is an important constituent of the genetic material D.N.A. and a primary nutrient in the growth of crops. For modern man to feed himself he must have access to huge quantities of artificial fertilizer. Consequently, a large tonnage of phosphate is mined each year to form, with potash, nitrates and sulphur, the basis of a vast fertilizer industry which consumes over 90% of the phosphate that is mined annually.

Phosphorus is present in most rocks in minor to trace amounts, but it is only in **phosphate rock** that the P_2O_5 content is high enough for it to constitute a phosphate ore, where values as high as 40% P_2O_5 may be attained. With rocks of a suitable type and up-to-date processing plants, ores as low as 4% P_2O_5 can be worked. It should be noted that the phosphate content of phosphate rock is generally quoted as % P_2O_5. Other forms quoted include % B.P.L. (bone phosphate of lime — bones were once the main source of phosphate) and % P. Conversion factors are: % $P_2O_5 \times 2.1853 =$ % B.P.L. and % $P_2O_5 \times 0.4364 =$ % P.

The phosphate in commercial deposits is almost invariably in the form of apatite—commonly **fluorapatite** $Ca_5(PO_4)_3F$ or **carbonate fluorapatite**, which has the approximate formula $Ca_{10}(PO_4)_{6-x}(CO_3)_x(F,OH)_{2+x}$. Phosphate deposits are of two main types—**igneous** and **marine sedimentary**. The former deposits, which supply about 16% of the world's production, are discussed in Chapter 10 and it is the sedimentary deposits that will be discussed in this section.

21.14.1 Uses

Most phosphate rock needs beneficiation to increase the B.P.L. content to between 60 and 80%. This is done by crushing, sizing and flotation to reduce the content of impurities such as quartz, chert, clay or shale, mica and carbonates. A certain amount of crushed rock is applied directly to soils, if they are acidic, as a fertilizer and to increase the pH. Electric furnace treatment is used on some lower grade ores (24–30% P_2O_5) to produce elemental phosphorus, which may be used for fertilizer manufacture, or for the preparation of pure phosphoric acid or food-grade acid. However, the bulk of crushed ore is treated with sulphuric acid to produce phosphoric acid which can be shipped economically by tanker to fertilizer plants around the world. This phosphoric acid is also used for the manufacture of calcium phosphate animal feedstuffs and for certain chemical processes. Various grades of superphosphate are also prepared from ores by reaction with sulphuric acid. About 90% of world phosphate production goes for fertilizer manufacture, the remaining 10% for the manufacture of animal feedstuffs, detergents, food and drink products, fire extinguishers, dental products and the surface treatment of metals.

For every tonne of phosphoric acid produced by sulphuric acid treatment, about 3 t of waste **phosphogypsum** result and vast stockpiles of this material have been built up. Some has been used as a soil conditioner and minor amounts in plasterboard and cement manufacture. A pilot plant was set up in Florida recently to test the possibility of producing sulphuric acid and aggregate from this waste material. Another by-product of the mining of some deposits is **uranium** and the I.M.C. Fertilizer Company produces about 500 t p.a. of yellow cake at its New Wales plant in Florida. In addition, some fertilizer plants already recover **fluorine** and others are installing the necessary equipment to do so. This fluorine is being used to make artificial cryolite.

21.14.2 Production and price

World production of phosphate climbed steadily from 22 Mt in 1950 to

142 Mt in 1980 in response to world demand, but since then there has been **little growth in demand**, and production in 1995 was only about 138 Mt. Demand in many regions has decreased in recent years because of the high level of grain stocks in North America and western Europe. This has resulted in low grain prices and the compulsory set-aside of land, two factors that tend to depress fertilizer usage. For example, phosphate rock usage peaked in western Europe at 23 Mt in 1980 but was down to 17.1 Mt by 1989. This declining demand and an increase in the world's production capacity has led to an **oversupply** of phosphate and the price for phosphate rock has declined from about $U.S.50 t^{-1} in 1980 to $47 t^{-1} for ~73% B.P.L. ore. The top four producers in 1995 were the U.S.A. (45.5 Mt), China (26.8 Mt), Morocco (20.2 Mt), and the R.F. (10.5 Mt). These countries have for some years produced about 75% of world output. The remaining 25% came from twenty-six other countries.

21.14.3 Geology of phosphates

Sedimentary phosphate deposits are known as phosphorites and world reserves of this material are thought to be in excess of 200 000 Mt. Phosphorites occur on every continent and range in age from Precambrian to Recent, but **almost all the commercially exploited deposits are Phanerozoic**. However, some extensive continental areas have no commercial phosphorites, e.g. western Europe, Canada, Greenland and the northeastern R.F.

Phosphorites generally form beds a few centimeters to tens of metres thick that are composed of grains (frequently termed pellets) of cryptocrystalline carbonate fluorapatite, which is often conveniently referred to, particularly in the field, as collophane. Most grains are well rounded and up to about 2 mm across. They may contain grains of quartz, clay, framboidal pyrite and carbonaceous material. In some deposits it is clear that many of these phosphatic grains were originally carbonate that has been phosphatized diagenetically. Oolites occur in many deposits and some may be primary but others clearly are **phosphatized calcite oolites**. Other forms of collophane include nodules, which may be up to tens of centimeters across, and mudstone phosphorites. Phosphorites often show **much evidence of reworking**, perhaps with the addition of new phosphatic material and later *in situ* leaching, which removes carbonate to form enriched residual deposits. Phosphate deposits commonly have interbeds of shale and chert.

At the present day, phosphorites are forming in low latitudes off the coasts of continents where there is known to be **marked upwelling** bringing deep oceanic waters on to continental shelves, e.g. off the coasts of Chile, Namibia and eastern Australia. Major upwelling sites develop

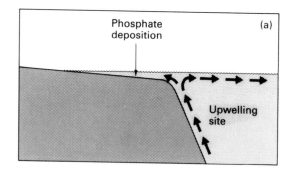

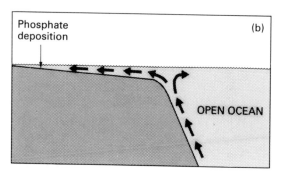

Fig. 21.3 Possible model for phosphorite formation beside an upwelling site initiated in a deep ocean. (a) Nutrient rich, deep ocean water floods on to the continental shelf and low grade phosphorite deposits form. (b) A sea level rise leads to a major marine transgression resulting in the reworking of phosphatic shelf sediments and the shoreward transport of phosphatic grains to form major deposits in the coastal zone and in marginal embayments.

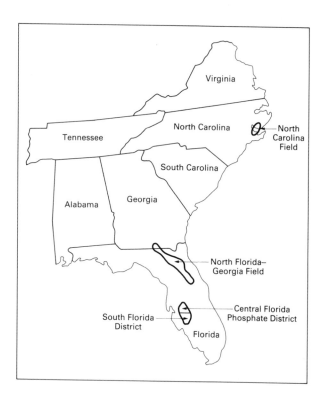

Fig. 21.4 Location of the phosphate districts of economic importance in the south-eastern U.S.A.

where oceanic boundary currents—such as the Gulf Stream off the coast of Florida and the Carolinas (see Fig. 21.5)—steer around bathymetric highs producing gyres within which Ekman pumping occurs (spiral upward movement of water). The deep ocean is a vast reservoir of phosphate, which increases in amount when the rate of deep oceanic circulation slows down. If continental drift gives rise to the development of new seaways or

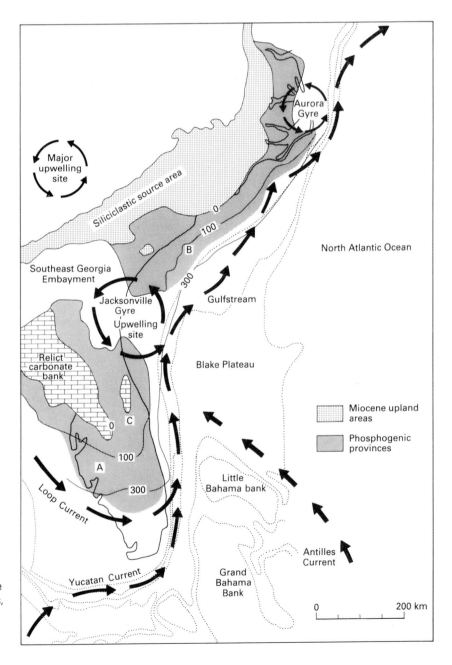

Fig. 21.5 Miocene palaeogeographical reconstruction of the south-eastern continental margin of the U.S.A. showing oceanic currents, major topographically induced upwelling sites and areas of phosphate deposition.

the drift of a continent into low latitudes (see Section 29.2.3), then this phosphate may be recirculated into shallow oceanic areas and on to continental shelves where a great increase (around 500 times) will develop in the biomass, leading to the formation of widespread, low grade phosphate deposits (Fig. 21.3). However, these will be on the deeper parts of shelves and it seems that economic deposits only develop when a rise in sea level results in a **marked marine transgression** that leads to re-working of the shelf deposits, and the shoreward transport of the phosphate grains to form major deposits in coastal traps, such as bays and other marginal embayments, and around structural highs.

The deposits of Florida and North Carolina appear to be a good example of this process. Figure 21.4 shows the distribution of the economic phosphorite fields. These fields are responsible for much of the present day phosphate rock production in the U.S.A. The Miocene phosphatic sediments are usually coarser, with siliceous impurities on the landward side, and they grade seawards to finer grained deposits and carbonate. Sedimentation took place on the shallow water coastal and nearshore platform adjacent to structural highs. The deposits in the Central District have only a moderate overburden thickness, good ore thicknesses, a B.P.L. content averaging 70% and a significant uranium content. A reconstruction of the Miocene palaeogeography with the positions of major upwelling sites which fed phosphate on to the shallow marine platforms is shown in Fig. 21.5.

21.15 Pyrophyllite

This is a soft, white mineral with a pearly lustre that **resembles talc** and is used for many similar purposes. The principal markets for pyrophyllite are **refractories, ceramics** and **steel making**. However, in the last-named market it seems that the advent of continuous casting techniques is leading to a decrease in the use of pyrophyllite in zircon ladle bricks, traditionally one of pyrophllite's important end uses. The Australian prices for bulk purchases ex-store are $U.S.106 t^{-1}$. Unfortunately, pyrophyllite production is lumped statistically with talc, steatite and soapstone. It has been estimated that world production lies in the 0.9–1.1 Mt p.a. range, of which approximately half is in Japan and a quarter in South Korea. China is believed to be a substantial supplier and large new reserves have recently been found in that country.

21.16 Sillimanite minerals

Under this heading, industry groups andalusite, kyanite and sillimanite which are polymorphs of Al_2SiO_5. These are **highly refractory minerals** and this is their major use. Minor uses include abrasives, glazes and non-slip flooring production. Commercial grades must run 56% Al_2O_3 or better and 42% SiO_2; acid soluble Fe_2O_3 must be less than 1%, TiO_2, less than 1.3% and CaO and MgO must not exceed 0.1% each. Sillimanite minerals command a good price: andalusite from the Transvaal \$U.S. 180–240 t^{-1}, according to grain size and Al_2O_3 content—the higher the Al_2O_3 the higher the price; South African sillimanite running 70% Al_2O_3 fetches \$U.S.312. However, as they have different physical properties, their markets are different and they are traded as separate mineral concentrates. World production in 1994 excluding C.I.S. countries, for which no statistics are available, was about 0.38 Mt; of this, R.S.A. produced 54%, mainly in the form of andalusite, the U.S.A., 23.5%, as kyanite and related minerals. France was third with 13% and India, which produces all three polymorphs, came fourth with 7%.

Economic deposits of **kyanite** and **sillimanite** are generally of **regional metamorphic origin** and **andalusite of contact metamorphic origin**. The large South African output comes mainly from the metamorphosed Daspoort Shales, which lie in the contact aureole of the Bushveld Complex (Chapter 12). These shales are over 60 m thick and run 5–20% andalusite. Known reserves will last for thirty years at the present production rate and it has been suggested that the total resources are sufficient to keep the world supplied at present consumption levels for 500 years! The andalusite hornfelses are soft and no blasting is required. The ore is crushed and the andalusite separated in a heavy media plant. France is the other big andalusite producer—about 70 000 t p.a. from the Damrec Mine, Brittany.

The U.S.A. is by far the largest producer of kyanite, much of which comes from **kyanite quartzites** at Willis Mountain, Virginia. India and Sweden are also important kyanite producers, the latter from a mine in Precambrian quartzites running 30% kyanite. Much sillimanite production, about 20 000 t p.a., comes from India; most of this used to come from immense boulders of massive sillimanite (with minor corundum) but **by-product sillimanite** from **placer operations** is now the principal source.

21.17 Slate

Many readers will be familiar with the use of this rock as a roofing material and ornamental stone. Those guilty of a 'misspent youth' will have devoted much time to propelling billiard balls on slate-bedded tables! However, in recent years, slate has been put to other uses. Rapid heating of crushed slate to about 1200°C produces a porous, slag-like material suitable as lightweight concrete aggregate. In North Wales finely powdered slate (fullersite) is produced for use as a filler in bituminous compounds such as automobile underseal, plastics and industrial adhesives. It is used also in insecticides and fertilizers.

21.18 Sulphur

Sulphuric acid is the most important inorganic chemical produced in terms both of volume and use. As has been indicated earlier, it is derived from a number of sources, such as the smelting of sulphide ores and the sulphur recovered from crude oil (approximately 50% of world sulphur production comes from oil refineries). Rather less than 40% of world sulphur output comes from working **beds of native sulphur** and deposits in the caprocks of salt domes. Much of this sulphur is produced by the **Frasch process** in which the sulphur is melted by pumping hot water into it and then pumping the resultant liquid to the surface. Poland is the world's major producer of native sulphur with an output of nearly 5 Mt p.a. using the Frasch process; the U.S.A. is next with about 3 Mt. Other important Frasch producers include Mexico and the R.F. The sulphur in these deposits is believed to have resulted from the **biogenic alteration of gypsum** to sulphur and calcite. In the Polish Miocene deposits this reduction appears to have occurred at places where the gypsum was upfaulted, covered by impervious clay and saturated with saline water and hydrocarbons from lower strata.

Sulphur

Sulphur is a most important and useful element. Early man used sulphur to colour cave drawings. About 1 500 B.C. the Egyptians discovered that SO_2 could be used to bleach cotton and linen. By 500 B.C. the Chinese were using sulphur in gunpowder.

About A.D. 1 800 the commercial manufacture of sulphuric acid started and the increasing uses in industry and agriculture created a large and expanding market for sulphur.

21.19 Trona

The bulk of the world's **soda ash** (sodium carbonate) is produced by the Solvay process, with halite as the starting point. However, there is important production of natural sodium carbonate in the form of **trona** ($Na_2CO_3 \cdot NaHCO_3 \cdot 2H_2O$) in Wyoming, U.S.A., Kenya and Mexico. Botswana and China are expected to become producers in the near future and in due course some Turkish deposits may be exploited. Output in the U.S.A. in 1989 was almost 9 Mt, in Kenya 235 kt and Mexico 145 kt. The remaining 22.34 Mt was produced by Solvay plants scattered across the world. Most soda ash is consumed by the glass industry. Other users include the chemical, textile, paper and fertilizer industries.

The Wyoming deposits belong to the Eocene Green River Formation of **lacustrine origin**. At least forty-two beds of trona are present in the Wilkins Peak Member interbedded with marl, oil shale, halite and clastic sediments. Twenty-five beds range from 1 to 12 m in thickness and extend over 250 km or more. The trona is probably ultimately of **volcanic origin**. Searles Lake in California is another non-marine evaporite deposit from which trona is won. The Kenyan deposit is in Lake Magadi in the Eastern Rift Valley. In the lake, which floods and dries out annually, trona beds up to 35 m thick are present and these are still being formed by brines entering the lake from hot springs. Again the ultimate source is thought to be volcanic. The trona is extracted by a floating dredger, then crushed, slurried and pumped to the treatment plant.

21.20 Wollastonite

This mineral, not mined at all until the 1930s, is now in considerable demand as it can **substitute for asbestos** in some uses of that mineral. This is particularly true of long fibre wollastonite, which is used in fibre boards and panels. Another major use is as a filler in plastics and ceramics. In the U.S.A. the price ex-works in October 1996 was $U.S.170–620 *per short ton* depending on processing and degree of acicularity. Wollastonite has the ideal formula $CaSiO_3$ but Fe, Mg or Mn can substitute for small amounts of Ca and thereby reduce the whiteness. In 1994 recorded world output was of the order of 194 kt; 130 kt were produced in the U.S.A., 36 kt in Mexico and 28 kt in Finland. The Magata Mine, R.S.A. was commissioned in 1990 to produce 2500 t of ore per month and the search is on for more deposits, explorationists being encouraged by the high price and demand. It was reported recently that a 100 Mt deposit has been found in China and a number of encouraging prospects are being investigated in Canada. China, India, Japan, Namibia and Turkey are known to produce some wollastonite but no statistics are available for these countries.

The majority, if not all, of wollastonite production is from **contact metamorphosed impure limestones**. Wollastonite also occurs in a number of alkaline igneous rocks and it is conceivable that economic deposits could exist in some of them. The principal producer in the U.S.A. is the Fox Knoll Mine in the Adirondack Mountains of New York State. Here impure Proterozoic limestone in the contact aureole of an anorthosite has been metamorphosed to a **wollastonite–garnet hornfels**. This contains bands rich in one or the other mineral and the ore averages 60% wollastonite and 40% garnet plus impurities. The ore is crushed and magnetic separation used to free the non-magnetic wollastonite from the feebly magnetic garnet and diopside. Diopside is separated from the garnet electrostatically.

21.21 Further reading

Harben P.W. and Bates R.L. (1990) *Industrial Minerals Geology and World Deposits*. Metal Bulletin, London. A comprehensive review of the subject.

Noetstaller R. (1988) *Industrial Minerals: a Technical Review*. The World Bank, Washington. Full of interesting financial and statistical points that do not find their way into most textbooks. The 'Profile of Industrial Minerals by End-use Classes' will assist geologists in identifying potential uses for any deposits found in their concession areas.

Schreiber B.C. (1986) Arid Shorelines and Evaporites. In Reading H.G. (ed.), *Sedimentary Environments and Facies*, 189–228. Blackwell Scientific Publications, Oxford. An extensive coverage of the subject which contains on pp. 213–228 an excellent summary of the salient features of ancient evaporites.

Scott P.W. (1987) The Exploration and Evaluation of Industrial Rocks and Minerals. *Annual Review 1987*, 19–28. Irish Association for Economic Geology, Dublin. A very valuable little summary of the subject for those without much time to spare!

22: Water—the Mineral of Life

22.1 Introduction

In this chapter we consider:
- the effect of global warming and the global water cycle;
- the availability and use of surface water, building of reservoirs and environmental repercussions;
- the location, use and misuse of ground water;
- ground water contamination;
- other environmental consequences of the abuse of underground water supplies.

Earth has been termed the **water planet** on account of the aspect of blue oceans and white water clouds when viewed from space. Without water, life as we know it would not exist; it is a key substance for humankind. We need to drink it, we need it for our crops and animals, for health and hygiene, for industry and transport. The demands of an increasing world population and its aspirations for higher living standards have brought with them much greater pressure for supplies of fresh water. **Over the last fifty years global water use has quadrupled** (Fig. 22.1). Currently, two-thirds are used for agriculture, a quarter by industry, only about 10% domestically. This rapid growth in demand has considerably increased many nations' vulnerability regarding water supplies, particularly because many of the world's great sources of water are shared. Approximately half the global land area is within **water basins shared by two or more countries**. For example, the Danube passes through nine countries, all of which use its water, the Nile and its tributaries through seven, the Tigris-Euphrates-Shat-al-Arab through four and the Jordan through three. Already, violent disagreements have occurred over water usage in the countries bordering on these river systems.

Global warming will substantially change supplies of fresh water. A temperature increase will stimulate surface evaporation and some parts of the world e.g. southern Europe, are forecast to receive less rainfall, especially in Summer. Higher evaporation and lower rainfall will mean less surface runoff into rivers, lakes and reservoirs.

22.2 The global water (hydrological) cycle

The global water cycle is a fundamental component of the climate system and water is cycled between the oceans, the atmosphere and the land surface (Fig. 22.2). The total volume of water involved is about 1400 million km^3. Of this, over 97% is stored in the ocean and sea floor rocks and is of little use unless desalinated, an energy intensive operation, as would be the melting of ice sheets and glaciers in which 80% of fresh water is

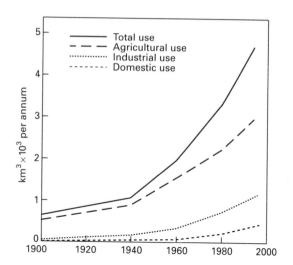

Fig. 22.1 Global use of water for different purposes 1900–2000.

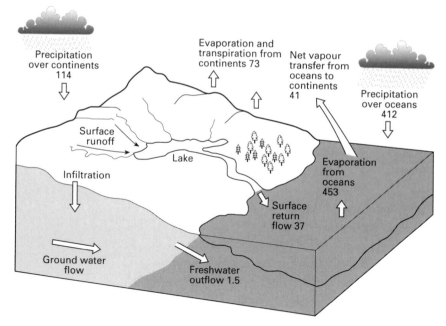

Fig. 22.2 The global water cycle: evaporation, precipitation, vapour transport, transport to the oceans by runoff and ground water flow. Figures are in km³ × 10³ p.a. The missing balance of 2.5 for water returned to the oceans is accounted for by melting ice at sea level.

locked up. The largest reservoir of fresh water is **ground water** mainly present in the saturated rocks beneath the **water table**. Surface water occurs in soils, streams, rivers and lakes.

22.3 Surface water

22.3.1 General points

The **runoff** in streams, rivers and lakes is what is left from the precipitation that falls on the land after some has been taken by evaporation, transpiration and infiltration (ground water seepage) (Fig. 22.2). It must be remembered, however, that some of the last-named may return to the surface as springs or ground water flow into rivers. Surface water for human needs may be supplied by **springs**, **river** and **lake abstraction** and **reservoirs**. Farms and villages may be supplied direct from a spring or stream. Towns may be supplied by pipeline or aqueduct direct from a reservoir *or* from a river whose flow is maintained by the addition of water from an upstream reservoir, one in another drainage basin or an aquifer.

Besides acting directly or indirectly as a water supply, **reservoirs** may be used for generating hydroelectric power, flood control or irrigation. The Tennessee River in the U.S.A. has reservoirs to trap and store water that would otherwise cause floods, the water being released when river levels have fallen. The Aswan High Dam on the River Nile in Egypt is used for all three purposes.

The most straightforward way to construct a reservoir is to build a dam across a river flowing through a narrow, deep, unpopulated, upland valley. This will enable a large volume of water to be stored. Lowland reservoirs have a greater surface area and evaporation rate for a given volume of water and may entail the loss of good agricultural land. On the other hand, being closer to urban areas they may have much greater recreational value. Many factors are important in **selecting a site** for a water supply reservoir:

1   an adequate and reliable supply of high quality water;
2   an elevation high enough to provide gravity flow to the distribution system or river thus saving the capital and recurrent costs of pumping stations;
3   as watertight a reservoir area as possible to prevent loss of water through joints and other permeable structures, such as a well developed joint set trending at a high angle to the proposed dam;
4   the absence of potential geological hazards—for example, large rock or land slides into the reservoir which might give rise to severe stress on the dam and water overflows flooding the downstream area. Such a disaster occurred at the Vaiont Reservoir, Italy in 1963 when there was great loss of life in villages below the dam even though the dam itself was essentially undamaged;
5   a suitable dam site;
6   minimum detrimental effects on ecology or environment.

22.3.2 Environmental consequences of reservoir construction

1   A new base level results and, upstream from the reservoir, new aggradation by sediment deposition may occur. In strongly eroding areas the reservoir itself may silt up, as in Lake Mead behind the Hoover Dam (built in 1937) on the Colorado River, U.S.A. which is now over half full of silt.

2 Reduced river flow below the dam may result in less sediment being deposited at and near river mouths thus increasing coastal erosion.

3 Land loss—reservoirs may occupy large areas of usable land, e.g. Lake Nasser above the Aswan High Dam covers 6 000 km², but again all may not be lost, e.g. Lake Kariba on the Zambesi River now has an important fishing industry.

4 Changes in farming methods—the annual Nile floods which deposited fertile sediment on the lands down river from the Aswan High Dam are now a thing of the past and artificial fertilizers have to be used. The floods used to wash away the salts deposited by the river water but now with irrigation and no flooding salt buildups are developing, leading to infertility.

5 Seismic activity resulting from crustal adjustment to the load of water. Earthquake generation is associated with Lakes Mead and Kariba.

22.4 Ground water

22.4.1 General points

Most of the upper part of the continental crust is filled with fresh ground water largely supplied by rainfall. The fraction of this precipitation that infiltrates the ground is called the **recharge**. This water percolates downward through the **unsaturated** or **vadose zone** to an extent depending on the soil or rock permeability. In this zone the intergranular pores and fissures are unsaturated with water and contain air at atmospheric pressure, hence another name—**zone of aeration**. The vadose zone is characterized by episodic saturation and seasonal capillarity and by evaporation and transpiration. Intense mineral leaching is often characteristic of the upper part of the vadose zone, whilst in the capillary zone above the **water table**, particularly in tropical and semi-arid climates, there tends to be reprecipitation of carbonates, sulphates, quartz or iron oxyhydroxides which form **calcretes**, **gypcretes**, **silcretes** or **hard pans**, respectively. Beneath the vadose zone there is the **saturated** or **phreatic zone** in which the intergranular pores and fissures are completely filled with water at hydrostatic pressures greater than atmospheric. The interface between these two zones forms the **water table**. This is an equilibrium surface, at which the hydrostatic pressure in the pores is equal to atmospheric pressure in the pores of the overlying vadose zone. The water table is uneven, varying with topography, and is variable in its height, fluctuating with rainfall and internal drainage in the host sediment. During periods of high rainfall the water table rises whilst during periods of drought it falls. Deserts characteristically have low water tables; river valleys and coastal systems have characteristically high water tables. The lowest level to which the water table falls in any given locality is known as the **permanent water table**. Where the water table coincides with the ground surface, **seepages** and **springs** form. When below ground level, its depth can be measured in a **well**.

Water saturated rock can constitute an **aquifer**. Ground water aquifers vary from regoliths and other materials such as sand and gravel just below the surface to sedimentary beds or fracture zones that extend to many kilometres depth. To justify the term aquifer they must be able to yield significant quantities of water to wells and springs. Figure 22.3 depicts a common type of aquifer. It could be a permeable and highly porous sandstone formation within which the ground water is confined by aquicludes above and below it. **Aquicludes** may be porous and contain water but they are not, geologically speaking, significantly permeable; examples are clays and shales. The aquifer shown in this figure may also be termed an **artesian aquifer** since its water is under sufficient pressure to drive it to the surface when the aquifer is penetrated by a well or borehole. Today such aquifers are frequently called **confined aquifers**. The classic case is illustrated in Fig. 22.3. A dipping aquifer lies between two aquicludes or confining beds and crops out to form a **recharge area**. Clearly the water-saturated volume below the water table will be at a pressure greater than atmospheric. Water will rise up a well through the upper aquiclude until the water column (**head**) balances the aquifer pressure. With a well sited some way down dip, as at A, water will flow out of the well which is then termed artesian. If there were many wells into the aquifer with their water levels joined by an imaginary surface, that surface would indicate the static head of water in the aquifer. This is called the **potentiometric** or **piezometric surface**. If a large number of wells are sunk into the aquifer then discharge may outstrip replenishment and this will lower the potentiometric surface to a level like that of B. At the same time, springs and seepages at C

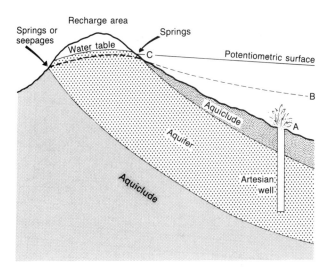

Fig. 22.3 Artesian aquifer with an artesian well.

may cease to flow. If discharge becomes too great the wells may lose their artesian nature and water will then have to be pumped out, which will in turn lower the potentiometric surface still further. A good example of a former **artesian basin** where this has happened is the London Basin, England.

When water is removed from an ordinary aquifer surrounding a well by pumping, then the water table (or the potentiometric surface in an artesian aquifer) is lowered around the well and a **hydraulic gradient** from all directions is established towards the well. The water table in homogeneous aquifers then takes the form of an inverted cone called the **cone of depression** (Fig. 22.4). If pumping is continuous then the cone of depression deepens and spreads out laterally. Continued **drawdown** lowers the water table still further and nearby shallow wells may go dry. This can be very important in **alluvial aquifers** which are composed of unconsolidated sediment. In many parts of the world such aquifers provide the main or only source of groundwater and they form the bulk of the world's developed aquifers. They are generally of recent origin, little compacted or cemented and thus highly permeable. Sands and gravels in large, wide valleys of major rivers can provide a very important groundwater resource, e.g. the Ganges Delta. Aquifers of this type, lacking an upper aquiclude so that their upper saturation limit is the water table, are termed **unconfined aquifers**. Locally important **perched aquifers** occur when water is trapped above aquicludes in the vadose zone.

The **natural flow of ground water** is from recharge areas on high ground to low-lying **discharge areas** where it usually finds it way into river systems. Water from **coastal aquifers** may, however, discharge direct into the sea. Water will flow from an unconfined aquifer wherever the ground surface intersects the water table to form a seepage or spring. Water will flow from a confined aquifer (artesian aquifer) when the potentiometric surface is above the ground surface and where there is some form of permeable channelway through the overlying aquiclude.

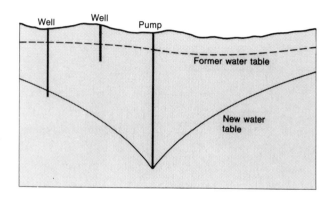

Fig. 22.4 Cone of depression of the water table formed by pumping and its effect upon nearby wells.

When ground water aquifers are undisturbed, an equilibrium prevails between the amount of recharge and discharge. Withdrawal of ground water from wells disrupts this equilibrium, and excessive withdrawal by pumping over a large area can result in regional falls in the water table. There are many parts of the world where ground water is being used faster than it is being replenished. One example is the U.S.A., where, over more than half its land area, in excess of a quarter of the ground water withdrawn is not being renewed, so every year the water has to be extracted from deeper levels. Another example is Beijing in China where the water table is falling by 2 m a year because of the high rate of ground water pumping. Saudi Arabia provides an extreme example — here, withdrawal is now so rapid that complete exhaustion of ground water will occur in about seventy years.

The fact that more than 1 000 million people in Asia and South America alone are estimated to be directly dependent on ground water sources for their water supplies, and that the continuing increase in the world population will result in even higher demands, brings home to us the need to see that our aquifers are not overexploited or that they are replenished by **artificial recharge**. This entails pumping suitably treated water into them. Such water may be taken from surface supplies, e.g. rivers or from other aquifers, when it can be conveyed by pipeline or by making use of a suitable river.

Artificial recharge

This is a technique which will be much used in the future, especially in highly populated countries with wet winters which suffer from summer droughts and water shortages. Vast volumes of treated water can be stored, and good use made of the winter excess of surface water which can then be used to supplement summer supplies. Schemes of this type in the U.S.A. have shown that they can be up to **nineteen times** cheaper than building surface reservoirs and use just a **fraction** of the land surface.

22.4.2 Environmental considerations

The almost worldwide dependence to some degree or other on ground water as a water source for human consumption, not to mention irrigational and industrial use, emphasizes the importance of eliminating or minimizing any pollution. Normally, ground water is of excellent microbiological and, generally, of adequate chemical quality for such use. The three most common causes of unacceptable quality are:

1 natural problems of hydrogeochemical reaction with bedrocks,

2 anthropogenic pollution of inadequately protected aquifers,
3 saline intrusion of mismanaged aquifers.

But first a definition, in our context, of **pollution**, which has been aptly described as 'the presence of abnormal substances or abnormally high concentrations of normal substances in the natural environment'. **And**, before we go further, what is considered to be **adequate water quality**? Waters carrying greater than 500 ppm dissolved salts are considered unsuitable for drinking or cooking purposes and those with more than 2 000 ppm are unsuitable for most other human uses. The dissolved salts are derived from the soils and rocks that the ground water has been in contact with. Many governments and the World Health Organization have published maximum recommended limits for the concentrations of many inorganic, organic and synthetic substances in public water supplies. These are too comprehensive to quote in full here and lists are often complicated by including two values: the first for the highest desirable level and the second for the maximum allowable level. But, as illustrations, here are *some* maximum allowable levels as laid down in the U.S.A. all, except one, in ppm: arsenic, 0.05; cadmium, 0.010; coliform bacteria, 1/100 ml (mean); copper, 1.0; fluoride, 2.4; iron, 0.3; lead, 0.05; lindane, 0.004; methoxychlor, 0.005; nitrate (as N), 10; selenium, 0.01; sulphate, 250; 2,4-D, 0.1. Especially worrying is the presence of any organic chemicals as these often have toxic effects even at low concentrations or they may be carcinogenic. It is important to differentiate between natural pollution and anthropogenic pollution.

(a) *Natural pollution.* Health problems can be caused as much by **deficiencies** as by **excesses** of elements or compounds in ground water, and plants dependent on it. Examples are soils derived from ultrabasic rocks that are so poor in potassium and phosphorus that good crops cannot be grown on them, or the development of the thyroid disease goitre in humans whose drinking water lacked adequate iodine. On the other hand, selenium-enriched plants and surface waters can cause blind staggers in horses and cattle that eat and drink them. We usually look on natural fluoride in drinking water as a welcome preventative against tooth decay. However, fluoride in ground waters of the Awash Valley, Ethiopia causes bone and teeth deformities and other severe health problems in the local population. Investigations showed fluoride concentrations increase from 0.92 ppm in river water to 11.8 in wells in the alluvium to about 20 in the volcanic bedrock — the primary source of the fluoride — and within this area hot springs yield values of up to 600. Experimental treatment by ion exchange to reduce the fluoride concentrations was uneconomic and the problem has been solved by piping water from the Awash River, which has values of 3–4 ppm.

(b) *Anthropogenic pollution* or contamination by humans. This sometimes seems to be as diverse as man's activities and varies from concentrations of animal waste and residues of pesticides and nitrates on farmland, through runoff from urban paved areas and motorways which may also involve toxic spillages and contaminate infiltrating waters with salt, hydrocarbons, pesticides and other chemicals, to mining, quarrying and other industrial activities, the disposal of sewage, organic wastes from intensive livestock farming and waste disposal in landfill sites. Some of the resulting pollution of ground water can be traced to one or more discrete sources, e.g. factories, landfill sites etc., and is termed **point-source pollution**. A more widespread influx of contaminants such as agricultural wastes, polluted rivers or acid rain is called **diffuse pollution**. Unconfined aquifers, having no overlying aquiclude to limit the downward infiltration of pollutants, are particularly susceptible to contamination. In these aquifers contaminated water from a point source commonly forms a **plume** (Fig. 22.5) which gradually enlarges down the regional flow path of the ground water. The plume's ultimate form and behaviour is largely governed by the pollutant's miscibility with water and its density. Light hydrocarbons remain close to the surface, sometimes forming pools at the water table whence volatile components may then invade sewers and basements, causing explosions such as the 1992 disaster in Guadalajara, Mexico, which killed several hundred people. Pollutants miscible with and having about the same density as water will develop more dispersed plumes, but those of higher density will sink through the aquifer, with their more soluble constituents dissolving into the ground water, to form a plume flowing along a basal aquiclude. **Examples of such point sources** are chemical and engineering works and landfill sites lacking a clay liner. In developed countries, stringent anti-pollution legislation has reduced such pollution to a manageable level but there is still much contamination of ground

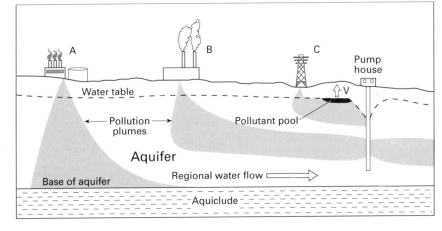

Fig. 22.5 Schematic illustration of pollution plumes in an aquifer. Point source A is leaking higher density contaminants, e.g. chlorinated solvents. B is spilling same density pollutant, e.g. aqueous solutions of toxic metals. C is losing lower density organic liquids rich in hydrocarbons. V indicates volatile components from a pollutant pool rising through the vadose zone.

water from abandoned mines and mine tips, former chemical and tar works and old landfill sites.

With **diffuse pollution** we are dealing with situations such as a large area having many industrial sites, a heavily polluted river draining through an industrial region and contaminating large volumes of an aquifer that it recharges, or the infiltration of soil water over wide areas of arable land where there has been overuse of pesticides, herbicides and nitrate fertilizers.

(c) *Saline intrusion of freshwater aquifers.* When no abstraction has taken place, then coastal freshwater aquifers are in equilibrium with the sea water into which their ground water flows and saline waters are unable to seep inland into the aquifers. Unfortunately, from the 1950s onwards, extraction of ground water in numerous coastal areas in many parts of the world has changed the natural hydrological balance, the freshwater–saltwater interface has moved landwards, resulting in **saltwater intrusion**. The saline waters, being denser, tend to form an intrusive plume beneath the freshwater (Fig. 22.6). Intrusions of this sort have forced the abandonment of dozens of wells all down the eastern seaboard of the U.S.A. from Boston to Miami. A similar situation occurs along much of the Californian coast. Saltwater problems may also occur inland; for, example an extensive body of saline water lies below the freshwater aquifers of the Central Valley of California and there is much concern that excessive extraction of ground water will allow this saline water to move up into the aquifers. Often during oil production saline waste water is produced, and in Mississippi this was reinjected to avoid contaminating surface waters; unfortunately, this led to ground water contamination and the abandonment of many water wells.

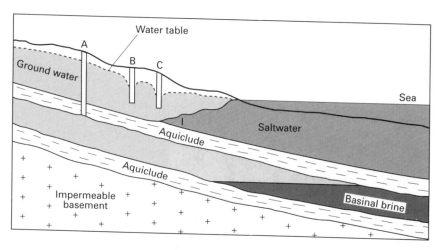

Fig. 22.6 Actual and potential effects of overpumping. Too high a rate of extraction from well A will cause the saline basinal brine to rise up the confined aquifer and contaminate it. Overpumping in wells B and C from the unconfined coastal aquifer has given rise to a saltwater intrusion I, which is in danger of engulfing both wells.

(d) *Subsidence and related problems.* In some confined or semi-confined aquifers, the pore fluid pressure is partially supporting the overlying rocks. Fluid withdrawal results in the rock grains bearing more of the load consequently undergoing further compaction, this being most marked in fine-grained rocks. Naturally the result is a slow, and often irreversible, subsidence of the land surface. This resulted in a lowering of about 25 cm affecting London, England over several centuries, quite a minor effect because of the presence of fairly strong subsurface sediments. An extreme case affecting a large conurbation is that of Mexico City where about 9 m of subsidence has occurred as a result of ground water abstraction. Similar overabstraction has created a potentially catastrophic situation at New Orleans. This large city now lies 2–3 m below sea level and the coincidence of a hurricane and a strong tidal surge could cause tremendous damage and much loss of life. Even lower degrees of subsidence can cause considerable damage including:

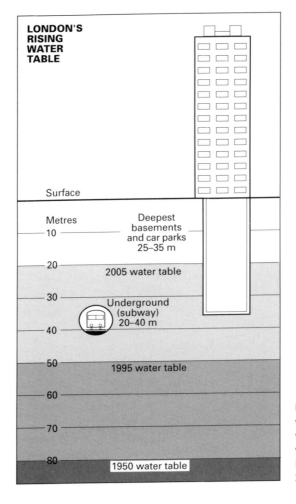

Fig. 22.7 An example of a rising water table resulting from the widespread cessation of ground water extraction, London, England. The estimated level for 2005 A.D. is shown.

1 structural damage to buildings, roads, bridges;
2 damage to sewers, water pipes, buried cables and well casings;
3 changes in the gradients of canals and irrigation systems;
4 flooding of low-lying inland and coastal areas.

Examples of more devastating effects are the rapid **sinkhole developments** (sometimes in a matter of hours) resulting from dewatering of aquifers in limestone regions, e.g. Alabama and Georgia, U.S.A. and the Transvaal, R.S.A.

A related problem has now arisen in a number of developed countries, e.g. in Europe where various combinations of circumstances—for instance, the decline of heavy industries, the development of surface water supplies, leaking water mains—has led to **rising water tables under cities**. This has caused re-swelling of clays, with resultant structural damage to buildings, tunnels, sewers etc., and the flooding of subways, cellars and other subsurface structures. These have either had to be waterproofed, rebuilt or have permanent pumps installed.

In parts of London the water table is rising by more than 2 m a year (Fig. 22.7) and is now close to some subway (underground) tunnels and deep basements. Similar rises are occurring elsewhere in the U.K. where the demand for ground water has decreased significantly; the cities concerned once contained much more heavy industry than is now the case and include Birmingham, Glasgow, Liverpool and Nottingham. These and numerous other formerly extensively industrialized towns and regions of the U.K. have vast areas of soil, subsoil and bedrock heavily polluted with chemicals following a century or more of mismanagement. Rising water tables may result in these chemicals being soaked out of their host sediments and conveyed into rivers. This risk is thought to be particularly high in the north-west and the West Midlands of England. A plan to restart large scale ground water abstraction from beneath London has had to be abandoned after a £1M study demonstrated that the water is too contaminated to be of any beneficial use.

22.5 Further reading

Brassington R. (1988) *Field Hydrology*. Open University Press, Milton Keynes. A useful introduction to practical techniques and hydrological theory.

Craig J.R., Vaughan D.J. and Skinner B.J. (1996) *Resources of the Earth Origin, Use and Environmental Impact*. Prentice-Hall, Upper Saddle River. Chapter 11.

Price M. (1985) *Introducing Groundwater*. Allen & Unwin, London. A very readable introduction to the fundamentals of hydrogeology.

23: The Changing World of Energy

23.1 Energy sources

■ Energy sources and consumption are the subjects of this short chapter.
■ How much coal, gas, oil etc. is left?

At first, human beings required only **wood** or, failing that, **dung**, to meet their needs to keep warm, cook food and, in time, to perform simple metallurgical and ceramic tasks. For land cultivation, **human and animal muscle power** was employed. Later, **water power** (river and tidal) and **wind power** were harnessed, but the amount and form of energy these sources could deliver were minuscule compared with the great demands created by the Industrial Revolution.

As recounted in Section 1.3, **charcoal shortages** in England in the eighteenth century led to the use of **coal** which became the first of the fossil fuels to be exploited on a large scale. **Oil** was next on the energy scene. Although oil and tar seepages had been worked in the Middle East for thousands of years, commercially significant production of oil only commenced in the U.S.A. in the latter half of the nineteenth century. Despite this, oil, as an energy source, only began to rival coal in post-W.W.1 times (Fig. 23.1). **Natural gas** soon followed and became important from about 1940 onwards, to be joined by nuclear power in the 1960s. **Hydroelectric power** generation has played an important role in some countries for many years but is unlikely ever to supply more than 10% of world energy requirements.

23.2 Energy consumption

In immediate post-W.W.2 decades, energy demands were climbing exponentially (Fig. 23.1) but the **oil crises** of 1973 and 1979 brought home to the developed world the need for **energy efficiency** and conservation and this is reflected by the downturns in Fig. 23.2 following these years. However, the rise in energy consumption has continued in response to an upturn in world trade and the rapid industrialization of some Asian countries. It is a sobering thought to note that world energy consumption has increased by nearly **threefold** between 1960 and 1990 (Fig. 23.1).

Many commentators feel that the time has come when the developed world should mend its spendthrift policies. For example, in Europe we

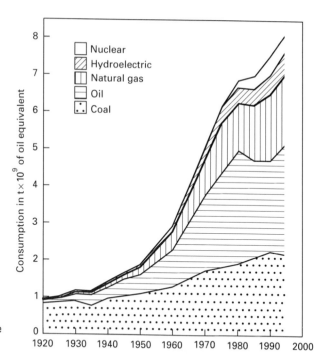

Fig. 23.1 Total world energy consumption since 1920.

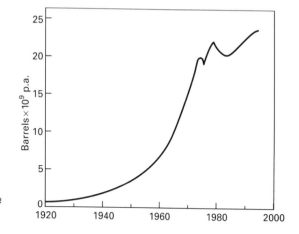

Fig. 23.2 World production of oil since 1920. Note the sharp rise in the 1950s and 1960s and the effects of the oil embargoes in the 1970s.

actually throw away more than 90% of the potential energy in our fossil fuel, and North Americans are even more wasteful. Since A.D. 1800, world annual primary energy consumption has risen from 50 Mt of coal equivalent to more than 12 Gt today. The developing world is watching our consumption of energy resources with envy and is beginning to emulate it. More than half the world has no electricity supply and it has been estimated that if every inhabitant of China switched on one extra light bulb, the resulting demand would require the building of thirty large generating stations, which in China would probably be coal fired. What would be

One barrel of oil weighs 136.4 kg. In energy terms it is equivalent to approximately 0.22 t of coal.

the climatic consequences of **such a huge increase in greenhouse gas emissions**?

The Chinese economy has grown at 10% *per annum* over the past ten years, but it would have grown much faster but for electricity shortages. Another developing country with extensive coal reserves is India; clearly with such enormous populations involved, the growth in global energy consumption will be huge. What will our descendants make of today's spendthrift energy policies? What sort of prospect are we handing down to them?

23.3 Energy supplies in the future

Will future generations face an energy crisis?

(a) *Oil.* Of our three important fossil fuels, the outlook for oil is the most critical. Figure 23.2 shows that despite the effects of drastically increased prices in the 1970s, world consumption and hence production is once more on an upward path. **How much oil have we left?** Omitting possible contributions from oil shale, heavy oil and tar sands which are discussed later, there appears to be a general consensus of opinion among government officers, oil company geologists and university researchers that we have **about 1500 billion barrels**. This consists of proven reserves of 1000 billion barrels and 500 billion of possible resources. (Over 60% is in Middle Eastern and about 75% in O.P.E.C. countries!) At our present rate of use, this means our oil stocks will last, assuming unrestricted production, until about A.D. 2060. With some restrictions and some totally unexpected finds, another fifteen years might be added to this time span. As a postscript to the oil situation, it must be mentioned that in due course, as the price of oil rises, then the processing of oil shale, heavy oils and tar sands to produce oil products will become economically viable.

(b) *Natural gas.* The curve showing our demands for natural gas (Fig. 23.3) is at present even steeper than that for oil. Known reserves are put at about $96.28 \times 10^{12} \, m^3$ and possible resources at $450 \times 10^{12} \, m^3$. **At the present rate of consumption** these reserves will last **about forty-five years**. The resources, provided this figure does not prove to be excessive, will add another 210 years. The tremendous demand for natural gas means that it can be piped great distances, e.g. from Siberia to Western Europe and Algeria to Portugal, or shipped in liquified natural gas tankers all over the world, as from Alaska to Japan. The rate of consumption will continue to increase dramatically. Like oil, most of the major gas fields have probably been found and future discoveries will be smaller and more costly to develop. However, the exploitation of coal bed methane (see Section 25.7.1) will undoubtedly provide substantially more gas.

(c) *Coal.* Consumption of this fuel has also risen considerably in post-W.W.2 decades but at a more measured rate than the two relative newcomers oil and gas (Fig. 23.4). **Known reserves are about 895 $\times 10^9$ t**, with further possible resources estimated to be $2\,000 \times 10^9$ t. Our **reserves will last about 190 years** at the present consumption rate, so clearly we have ample reserves of coal to tide the world over until alternative energy sources (biomass, geothermal, solar, tidal, wave, wind) begin to produce a significant proportion of the world's energy needs. Moreover, these coal reserves are in quantity far more evenly distributed around the world than oil — a point which might not appear to be immediately obvious from a comparison of Figs 24.2 and 25.1 but which can be concisely stated as

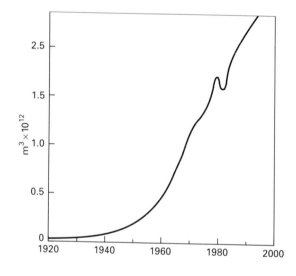

Fig. 23.3 World consumption of natural gas since 1920. Note the sharp rise since 1950 when pipelines and marine transport became available.

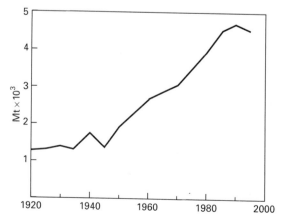

Fig. 23.4 World consumption of coal since 1920. Note the steep rise in use after W.W.2.

follows. About 42% of the world's crude oil reserves lie in six neighbouring Middle Eastern territories: Iran, Iraq, Kuwait, the Neutral Zone, Saudi Arabia and the United Arab Emirates — of which Saudi Arabia has the lion's share, about 43% of these territories' total. No other region can approach this concentration of oil reserves. The whole of North America has only 14%, Latin America America about 11%, Western Europe (mainly the North Sea Province) 2% and the rest of the world put together — all Africa, Russia, China, south-eastern Asia, Australasia — but 31%!

(d) *Hydroelectric, geothermal and nuclear energy.* As can be seen from Fig. 23.1, **hydroelectric power** production has increased very little over recent decades, indicating that most of the potential sites for generating stations have been exploited and thus there will be little further growth of this energy source.

Geothermal energy is still in its infancy and experts are disagreed as to its importance from a world viewpoint. Locally, where the earth's thermal gradient is markedly above average, this energy source is, or could be, playing an important role; the extent of its future role is less obvious.

Nuclear power generation has in theory great attractions, particularly as it does not produce polluting greenhouse gases. Enthusiasm for its use has, however, been dampened by the partial meltdown at **Three Mile Island**, Pennsylvania in 1979 and complete meltdown at **Chernobyl**, Ukraine in 1986. Nevertheless, some countries with little or no fossil fuel resources, e.g. France, Japan, are still constructing nuclear power stations. Table 23.1 gives an idea of the present importance of nuclear power generation in selected countries. The global picture is of 420 stations in twenty-

Country	Production (%)
Lithuania	87
France	78
Belgium	59
Taiwan	52
Sweden	42
South Korea	40
Czech Republic	38
Switzerland	38
Spain	36
Japan	31
Finland	30
Germany	30
U.K.	26
U.S.A.	21

Table 23.1 Total electricity production generated by nuclear power for selected countries.

five countries generating over 300 000 MW of electricity. **Present uranium reserves** would last until **about A.D. 2050** but **resources are truly extensive**. Furthermore, the commercial development of fast breeder reactors would greatly extend the lifetime of uranium derived nuclear energy. Will nuclear fusion ever become technically and financial viable?!

24: Coal—the Fuel of the Industrial Revolution

24.1 What is coal?

24.1.1 The scientific viewpoint

Coal is a readily combustible organoclastic sedimentary rock composed mainly of lithified plant remains and containing more than 50% by weight of carbonaceous material and inherent moisture. **Coalification** progresses from moist, spongy, partially decomposed vegetation such as **peat**, through **brown coal (lignite)** and **sub-bituminous coal**, **bituminous coal**, **semi-anthracite** to **anthracite**. This transition is a response to **diagenesis** associated with burial and tectonic activity. These are the **banded** or **humic coals**. The other main group contains the non-banded or **sapropelic coals** derived from algae, spores and finely divided plant material. During coalification, the percentage of carbon increases, volatiles and moisture are gradually eliminated, the calorific value increases and so does the reflectance of the vitrinite content.

Microscopic examination indicates that coal consists of particles and bands of different kinds of carbonaceous material. These are the coalified remains of plant material that existed at the time of peat formation. They differ from each other in their morphology, hardness, optical properties and chemical characteristics. They are analogous to the minerals that form other rocks, but they are organic materials characterized by botanical structure rather than crystallographic properties and are called **macerals**. They are studied in polished sections using the techniques of ore microscopy.

Vitrinite is the principal maceral in most coals. It is mainly formed of the remains of trunks, branches, stems, roots and leaves of land plants. In reflected light, vitrinite is a medium grey in contrast with the lighter inertites and the darker exinites. The Carboniferous coals of the Northern Hemisphere contain at least 30–40% vitrinite whilst Gondwanaland coals rarely contain >20–30%. The **exinite** group is composed of a diverse assemblage of small organic particles such as algae, spores, cuticles and resin bodies. This group is characterized, especially in **low rank coals**, by high

---

Banded coals

Banded coals are stratified with centimetre or so thick layers of different maceral composition having varying appearance.

Non-banded coals, on the other hand, are homogenous, tough and often have a marked conchoidal fracture.

---

hydrogen and volatile contents. The most common member of this group in **banded coal** is **sporinite**—coalified spores and pollen.

Alginite consists of coalified algal remains and is characteristic of **boghead** and **cannel coals** which are themselves composed dominantly of algal material. It is also abundant in some **oil shales**, but rare in **banded coals**. Boghead and cannel coals are thought to reflect depositional environments in which clear, aerated surface waters, free of humic matter, permitted algal colonies to flourish. On dying, the remains of these colonies accumulated in **euxinic** bottom oozes, to form deposits of alginite or **bituminite**. The compositions of these coals are similar to the kerogen precursors of oil and when raised to higher temperatures and pressures they yield oil and gas rather than the black vitreous macerals of humic coals.

Intertite group macerals have higher carbon and lower hydrogen contents than macerals in coals of equivalent rank. They are harder than the other macerals and therefore tend to have a high relief in polished sections. The name of this group implies their relative inertness during **coke** manufacture and other industrial processes.

In addition to the assemblage of macerals, the general appearance, chemical composition and petrographical properties of coal are affected by the effects of post-depositional increases in pressure and temperature. A coal little affected by burial or tectonism such as **brown coal** is called a soft or **low rank coal**, whilst one much modified by these processes is called a hard or **high rank coal**.

24.1.2 The economic viewpoint

(a) *Coal classification.* Various properties are used for the classification of coal. These include the percentages of carbon, hydrogen and volatiles, the specific energy or **calorific value** and the coking and agglomerating properties. These properties may be measured and reported on in various ways. Most analyses are performed on air-dried samples. This method excludes the surface moisture always present in coal that is mined, shipped and delivered to the customer. Analyses on this material are reported as 'as received' or 'as sampled'. Other analyses may be presented as 'dry' or 'moisture free', representing coal after removal of both surface and inherent moisture. For data described as 'dry, ash-free' (d.a.f.), the analysis is recalculated after subtraction of the **ash** and moisture content, whereas 'dry mineral matter-free' (d.m.m.f.) excludes volatile mineral matter (e.g. CO_2 from carbonates, SO_2 from sulphides and H_2O from clays) as well as ash.

Many different schemes have been drawn up for the classification of coals and most require reference to tables that are too extensive to reproduce here. Seyler's classification dating from 1933 is still much used (Fig. 24.1). It is based on plots of hydrogen versus carbon or, alternatively, of volatile matter versus calorific value. Most coals plot within the curved band. Low rank coals (lignites) have lower carbon and higher hydrogen contents than high rank coals and fall on the right of the diagram. **Carbonaceous coals** (anthracites) with very low hydrogen contents plot at the other end of the band. The meaning of other terms can be read off the diagram.

The American Society for Testing and Materials (A.S.T.M.) classification is widely used in North America and other parts of the world. It is based on variations in the **fixed carbon** content and the **calorific value**. The

---

Ash

The non-combustible inorganic residue remaining when coal is burnt. It represents the bulk of the mineral matter in the coal. High ash coals are generally less valuable than low ash coals as they have a lower calorific value and produce greater amounts of waste for removal after combustion.

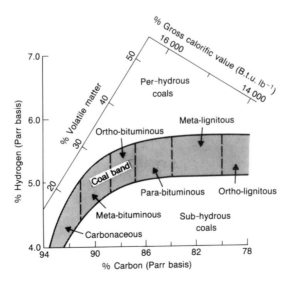

Fig. 24.1 Seyler's classification of coal. All results are given on a dry, mineral matter-free basis.

International Coal Classification drawn up by the United Nations Economic Commission for Europe uses a series of numbers to represent the chemical and physical properties of the coal. In the U.K. the scheme devised by the former National Coal Board (now privatized) is used. This is again a series of numbers expressed on a d.m.m.f. basis to represent the chemical and physical properties that represent the usefulness of the coal for **coke** production and steam-raising applications. The Australian Standard Coal Classification is similar to the International Coal Classification.

(b) *World production*. Coal still represents the world's biggest mined mineral tonnage (see Table 2.1). Over sixty countries produced 10 000 t or more in 1994 and about 90% of coal production is still consumed in its country of origin. The geographical distribution of coalfields is shown in Fig. 24.2. The twenty leading producers are listed in Table 24.1 and the leading exporters in Table 24.2. China and the U.S.A. dominate production, having 27.2 and 21.0%, respectively, with the next most important producers being responsible for much smaller percentages: R.F., 6.1%; India, 6.0%; Germany, 5.8%; Australia, 5.2%; Poland, 4.5%; R.S.A., 4.4%. Following the recession of the early 1990s, demand and prices are rising again as industrial output increases. Demand is now increasingly determined by the rapidly expanding economies, particularly those of the Far East. According to the International Energy Agency, by A.D. 2010 the developed countries' share of total energy consumed will drop to below 50% for the first time since the Industrial Revolution.

It is estimated by the U.N. that global employment in the coal industry is about 9.4 M, of whom 5.4 M work in China. Approximately 11 000

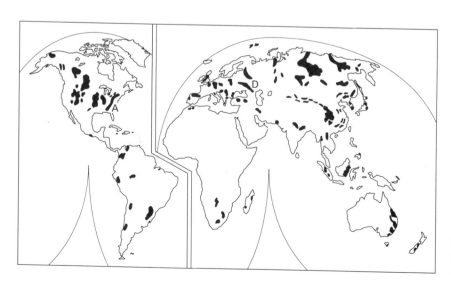

Fig. 24.2 World distribution of coalfields. A, Appalachian; D, Donets.

Country	Production (Mt)
China	1 210
U.S.A.	935
R.F.	271
India	267
Germany	259
Australia	230
Poland	200
R.S.A.	195.3
Kazakhstan	104
Ukraine	94.9
Czech Republic	74.1
Canada	70.3
Greece	56.3
Turkey	50.0
U.K.	47.8
Romania	40.5
Indonesia	30.5
Bulgaria	29.6
Spain	29.5
Colombia	23.5

Table 24.1 World coal production in 1994.

Country	Exports (Mt)
Australia	131.2
U.S.A.	64.3
R.S.A.	51.2
Canada	31.6
Poland	25.0
China	24.3
Indonesia	24.0
R.F.	19.6
Colombia	17.7
Venezuela	4.2

Table 24.2 World coal exports in 1994.

workers in the industry are killed each year in about a million serious accidents. The best safety record is Australian—0.1 fatalities Mt^{-1}—followed by the U.S.A. (0.11), U.K. (0.17) and Canada (0.18).

24.2 Peat and coal formation

The process of coal formation requires the **terrestial accumulation** of plant material in swamps, its **conversion to peat** and its **preservation by burial**. The word terrestial must be emphasized because it is these organic deposits that may be converted to coal or gas whereas those in the marine *milieu* are usually altered into oil and gas.

24.2.1 Plant material

Although land plants developed in the late Silurian, which puts a lower stratigraphical limit on significant coal formation, extensive coal swamps did not form until Upper Carboniferous times. During this time, the great coalfields of the eastern U.S.A. and Europe were formed. Another great coal forming period stretches from the beginning of the Jurassic to the mid Tertiary when, *inter alia*, the very important coals of the western U.S.A. were formed. Indeed, over half the world's coalfields are Tertiary.

24.2.2 Accumulation in peat swamps and subsequent burial

The development of humic coals started with the accumulation of organic débris in peat swamps which were sufficiently anaerobic to prevent its oxidation, decomposition and dispersion. The water table must have been almost coincident with the sediment surface for this requirement to be fulfilled.

An important region of Carboniferous coal formation developed along the equatorial southern margin of Laurasia from the Donets Basin in the Ukraine (Fig. 24.3) through western Europe and the British Isles to the central U.S.A. Along this region Lower Carboniferous (Mississippian) carbonates were covered by Upper Carboniferous (Pennsylvanian) deltaic and delta swamp deposits in continually sinking sedimentary basins (**coal basins**). The deltaic deposits show pronounced **cyclic sedimentation** of the type: (limestone) shale, sandstone and coal (**coalfield cyclothems**) repeated many times. The development of a cyclothem is shown in Fig. 24.4. This cyclic deposition results from a widespread marine transgression over the lower delta plain followed by regression caused by delta progradation. This progradation takes the form of elongate lobes of

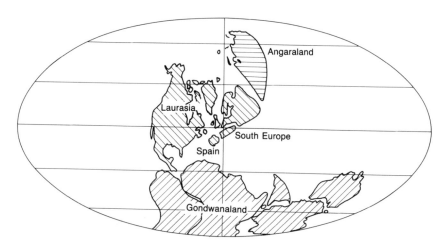

Fig. 24.3 A possible distribution of continents in the Upper Carboniferous.

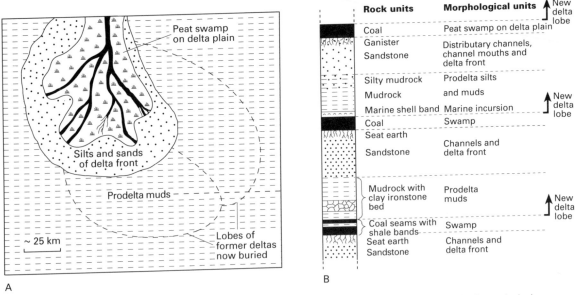

Fig. 24.4 (A) Three elongate delta lobes superposed upon each other with the three principal morphological units of the latest delta indicated. (B) A typical sequence of cyclothems. The positions of the approximate boundaries between the deposits of the three main morphological units of the delta are indicated with pecked lines.

sediment as **birdsfoot deltas** (Fig. 24.4A), which may form where rivers debouch on to a marine shelf or lake shore whose coastal waves or tides are too weak to carry away much of the sediment. In an active lobe, coarse sediment is deposited in and at the mouths of the distributaries and carried on to the delta front; finer sediment, silt and mud, finds its way on to the prodelta unit to be covered in time by the prograding sands of the delta front which in turn are covered by the advancing peat swamp.

As the lobe advances, the gradients of the distributaries decrease, sediment deposition increases and less water can pass. At a time of extensive flooding, the main channel may burst its banks and develop a new channel to the sea so initiating the formation of a new lobe. The weight of the extensive area of sediment causes crustal subsidence and the new lobe may then cover large parts of the older lobe. The buried rocks are lithified, the peat becoming coal, and thus stratigraphical sequences such as are shown in Fig. 24.4B may arise with many coal seams in one vertical sequence. In such cyclic sequences, three other rock-types of present or past economic importance may be present. **Ganister** is a sandstone or siltstone underlying a coal seam. These rocks have usually been highly leached by plant roots and, being very pure silica, can be used for making refractory bricks for lining furnaces. **Seat earth**, also called **fireclay**, is a strongly root-leached underclay rich in kaolinite and low in mica and iron. Fireclays now face many competing materials for the manufacture of refractory bricks, so they are now increasingly used for non-refractory products such as facing bricks, pipes and ceramic stoneware. **Clay ironstones** consist largely of siderite and occur in thin bands or nodules in some argillaceous rocks. Their iron content is low by present day iron ore standards, but those of the British Coal Measures were an important source of iron during the Industrial Revolution. In the British Isles cyclothems are well developed towards the top of the Mississippian (in the Namurian) and, in the eastern and central U.S.A. and the Donets Basin, they extend to the top of the Carboniferous.

Another widespread facies in this southern Laurasian region is that of the upper delta plain, fluvial and delta swamp (**sheet deltas**). This facies, which in many regions replaced birdsfoot deltas, is particularly associated with emergence and the falling worldwide sea levels of the Pennsylvanian. The peat swamps grew on extensive delta plains fed by distributary channels that often terminated in small lacustrine deltas. Along the tropical southern shoreline of Laurasia, thick peat deposits accumulated in extensive swamps on these delta plains which, after burial and compaction, produced the celebrated Carboniferous coal beds (seams) of high quality coal in coalfields from the Donets Basin to the central U.S.A. Scores of individual coal seams, separated by limestones, shales and sandstones, range from a few centimetres to tens of metres in thickness. The individual coal seams often split into two or more where deposition in the peat swamp was interrupted by incursions of clastic deposition. On the other hand, rapid subsidence of part of the swamp area may have created a topographical low into which a new distributary flowed, eroded down into the peat swamp and produced a **washout** in the coal seam—that is, a break in the seam occupied by a sinuous deposit of clastic sediment.

24.3 Coal mining and its environmental impact

Coal is mined from both open pit and underground mines. The choice of mining method is largely governed by the depth of overburden and rule of thumb has it that if the ratio of the thickness of overburden to be removed to the coal thickness does not exceed 20:1 then the operation could be financially viable. Otherwise the coal must be exploited by underground mining or by auguring. Augurs are drills a metre or so in diameter that can be driven into horizontal or gently dipping coal seams. As it revolves the augur cuts the coal and feeds the broken fragments out. Only about 50% of the coal is recovered by this mining method, but augurs are used in situations where no other mining operation would be financially possible.

24.3.1 Surface mining

When buried, flat lying or gently dipping beds of coal, iron ore or other wanted minerals are extracted then **strip mining** is employed. The first operation is the removal and stockpiling of the topsoil from the area where mining is to commence. Overburden is then removed and stockpiled separately to permit mining of the coal or ore. After this initial procedure, mining proceeds across country in parallel strips with the overburden from the newly opened strip being dumped in the previously worked strip, its surface graded and recontoured and topsoil spread on it (Fig. 24.5). If necessary, fertilizers are added, and this action, plus the greatly improved drainage resulting from the fragmentation of the overburden, often renders the ground more productive than before mining took place. Mining companies prefer open pit mining to underground mining, where there is a choice, because it is more profitable (since less labour intensive), safer for the miners and it involves fewer or no complications with power supplies, water, ventilation and rock handling; however, the surface disruption, although temporary, is much greater.

Contour mining is a form of strip mining used around the sides of hills when economic considerations rule out the removal of overburden from the entire hill (Fig. 24.6). A bench or **berm** is cut into the hill at seam level and the wanted material removed, the bench being excavated to the economic limit of overburden removal. An augur may then be used to drill out parts of the seam under deeper cover. Mining then moves laterally parallel to the contours with the mined ground being made good with waste material and topsoil.

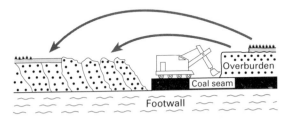

Fig. 24.5 Strip mining.

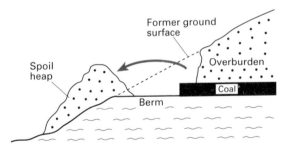

Fig. 24.6 Contour mining.

Strip and contour mining acquired a bad name in the past because overburden was not graded, leaving an ugly, hummocky topography, topsoil was not preserved and little or no replanting of trees or grass was carried out. Legislation in many countries has now revolutionized such environmental devastation, and in many mined areas where hedges, trees, buildings, roads and even rivers have been restored, the casual visitor is often unaware that the ground has been so drastically disturbed. Old strip and contour mines for coal that were left as open cuts led to the development of acid mine drainage [see Section 3.2.1(c)] which polluted both surface and underground waters. In modern mines, ground water quality is still an important concern and lime or limestone are often added to worked areas to reduce the ground water acidity. This action may be unnecessary where limestone is present in the overburden.

24.3.2 Underground mining

After visiting one of the coalfields in northern England in the 1930s, George Orwell wrote,

> Each man is shifting coal at a speed approaching two tons an hour. I have just enough experience of pick and shovel to be able to grasp what this means . . . in my garden, if I shift two tons of earth during the afternoon, I feel that I have earned my tea. But earth is tractable stuff compared with coal, and I don't have to work kneeling down, a thousand feet underneath the ground, in suffocating heat and swallowing coal dust with every breath I take . . . all of us *really* owe the comparative decency of our lives to poor drudges underground, blackened to the eyes, with their throats full of coal dust, driving their shovels forward with arms and belly muscles of steel. [*The Road to Wigan Pier*, George Orwell, 1937]

The remnants of horrifying conditions such as these still persist in some small mines in more developed countries and in other parts of the world they are still regrettably widespread. They are practised today in **room and pillar** (also called pillar and stall) **mining** operations (Fig. 24.7). Using this mining method, the coal or ore is only partially removed and the method can be illustrated by 'lifting off' the hanging wall rocks to look at the coal seam and the workings in it (Fig. 24.7). The purpose of the pillars is to support the roof for an indefinite time. The pillars are composed of good coal or ore that can never be mined without causing roof collapse and surface subsidence. The percentage that must be left for support varies with the depth of the seam and the strength of the overlying rock. Normally 40–60% of the coal can be mined. In time many coal pillars fail, particularly where mine managers have been greedy and left insubstantial pillars, or they may have all been deliberately removed, as in parts of the

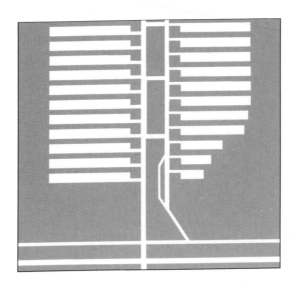

Fig. 24.7 In this simplified plan view of part of a room and pillar mining operation, the mined out areas are left white.

U.S.A. This leads to unpredictable and haphazard surface subsidence which, if it occurs beneath populated districts, can make homes unusable and disrupt transport services.

On the other hand, with **longwall mining** the roof is deliberately allowed to collapse. Coal is cut by a continuous mining machine moving along a face several hundred metres long. The operation is carried out beneath a moveable protective canopy supported by hydraulic jacks that can be slid forwards as the face recedes. As the canopy moves forward, the hanging wall behind it slowly collapses on to the former footwall of the seam in a continuous and controlled manner. Seams a little thinner than 1 m and upwards can be mined in this way with nearly 100% recovery. As the pattern and timing of subsidence over longwall faces is predictable, surface structures can be strengthened, before mining begins, to minimize damage. Unexpected and differential subsidence is rare and the subsidence is always less than the thickness of the mined coal, because a volume increase occurs in the subsiding rocks of the hanging wall as its rocks fracture. Being largely over in a few months, coal mining companies and landowners can agree on the compensation to be paid.

As noted in Section 24.1.1(b), coal mining is still a dangerous occupation due to coal gas (methane) explosions, underground fires, roof falls, water floods etc. Dust is another problem and earlier this century as many as 4000 retired miners in the U.S.A. used to die each year from pneumoconiosis caused by dust inhalation, but strict legislation concerning dust levels in many coal mining countries has greatly improved this situation.

The burning of fossil fuels adds CO_2 to the atmosphere and augments the greenhouse effect. In the case of coal this deleterious effect is exacer-

---

A cheering postscript: Surface coal mining cleans up the environment! The Smotherfly story

The Smotherfly opencast coal site, where it is planned to mine 1.6 Mt of coal, occupies 26.5 km^2 in the North Midlands of England (location shown on p. 34). The site included an area in which old coal mining, clay pipe works and toxic dumps from acid and tar works had left a great deal of industrial dereliction and pollution. When mining is completed, the toxic dumps and other contaminated ground together with unreclaimable scrap will have been carefully buried, the toxic wastes in specially engineered containment cells, and the waste land will have become a woodland area for recreation. Smotherfly is an economic coal mine but elsewhere in the U.K. and many other parts of the world, valuable mineral resources, especially coal, lie beneath derelict land where their mining feasibility is financially doubtful. Small government subsidies might well make such undertakings profitable and lead to low cost land restoration.

---

bated by the accompanying production of SO_2 which is considered to be a significant contributor to acid rain. Sulphur in coal varies from 0.2% to 7%. Usually about half is bound within macerals; the rest occurs as pyrite or marcasite. Many coal burning power stations are now equipped with desulphurization equipment which combines the SO_2 with limestone to form desulphogypsum, now a significant substitute for natural gypsum.

24.4 Further reading

Thomas L. (1992) *Handbook of Practical Coal Geology.* John Wiley, Chichester. Covers the subject very well, especially the exploration side, and takes a global view of all the matters discussed.

Ward C.R. (ed.) (1984) *Coal Geology and Coal Technology.* Deals with all aspects of coal deposits and their assessment.

Whateley M.K.G. and Spears D.A. (eds) (1995) *European Coal Geology.* Spec. Publ. **82**, Geological Society of London. Excellent descriptions of all aspects of European deposits.

25: Petroleum—Oil and Gas

25.1 Introduction

After a quick look at the history of the use of oil and gas this chapter considers:
- some economic matters,
- the nature and formation of petroleum,
- its migration, accumulation and extraction,
- little exploited and potential sources of gas and oil,
- environmental concerns.

First a few words about this chapter's title because there is some confusion between American and British usage of these words and even within these national usages! A definition that is very close to that of the *American Heritage Dictionary of the English Language* is given by Stoneley: 'Petroleum is useful, not because we like long words, but because it is a blanket term to cover all the naturally occurring *hydrocarbons* (a word which itself is sometimes used interchangeably with "petroleum"): gas, crude oil, and also certain solid substances are embraced. Thus gas is gas, oil is liquid, pitch is pretty solid; "petroleum" covers the lot.'

I must add that in this book and in much common usage **crude oil** is often shortened to oil and gas lengthened to **natural gas**. The other main terminological confusion arises from the British use of the word **petrol**. This is **not** an abbreviation for petroleum and should not be used as such. The British run most of their cars on petrol and they would soon be gummed up by petroleum, whereas the Americans run theirs on gasoline which they often refer to ambiguously or confusingly (depending on which side of the Atlantic Ocean one is) as gas, despite its liquid nature! So let's hang on like grim death to Stoneley's definition!

25.2 A brief history of the exploitation of oil and gas

Although the major exploitation of petroleum has only taken place during the last 140 years or so, **bitumen** (pitch) was used as a mortar between the bricks of the walls of Jericho nearly three millennia ago, the ancient Egyptians used it to caulk their boats and mummify their dead and other ancient Middle Eastern peoples used it to glue arrow heads to their shafts, for setting inlays in tile designs and even in oil lamps. This tacky bitumen came from the numerous oil seeps of that petroleum-rich region where there were also gas seeps, often ignited by natural causes, such as the eternal fires of Baku on the western shores of the Caspian Sea where the fire worshipping Zoroastrian religion developed. Greek Fire, the secret weapon of the Byzantines, used successfully to defend Constantinople in

A.D. 674–8 and in many naval battles, may well have been largely crude oil plus a spontaneously inflammable additive. About A.D. 1000 Arab chemists discovered distillation and were soon producing tonnes of kerosene (= British paraffin) — an invention forgotten until the nineteenth century when Abraham Gesner, a Canadian geologist, rediscovered the technique in 1852. Soon entrepreneurs were digging wells to recover oil for kerosene production, but the modern history of oil really started in 1859 when 'Colonel' Edwin L. Drake drilled the first American oil well near Titusville, Pennsylvania, U.S.A. for the Seneca Oil Company of which Professor Benjamin Silliman, of Yale University and sillimanite fame, was a leading investor. Within a few years successful wells had been drilled in many parts of the world although not in the Middle East. Here discoveries came in 1908 in Iran and 1927 in Iraq. The great potential of the region only became apparent in 1938 when the first of Saudi Arabia's large fields was found. However, until the start of W.W.2, oil production was still largely confined to the U.S.A., Venezuela, the Caribbean, Iran–Iraq, Romania and the U.S.S.R. Since then fields have been discovered all over the world from the North Slope of Alaska to south-eastern Australia (Fig. 25.1).

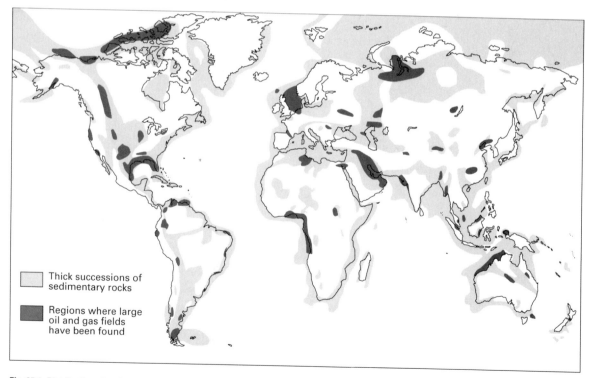

Thick successions of sedimentary rocks

Regions where large oil and gas fields have been found

Fig. 25.1 Distribution of sedimentary basins that contain known or potential oil or gas accumulations and the main known oil and gas bearing regions of the world.

The exploitation of natural gas in modern times preceded that of oil by some three decades or so when gas from an 8 m deep well at Fredonia, New York was passed through wooden pipes to fuel the lights of a local inn. But, as so often, the ancient Chinese were there first! They piped natural gas through bamboo poles to boilers where saline brines were evaporated to obtain the salt. However, exploitation of natural gas in recent times remained for many decades a local industry because of distribution problems—a handicap which did not affect its then competitor, manufactured gas.

From the early 1660s onwards **manufactured gas** (coal gas), made principally from bituminous coal but also from wood or peat, in Belgium and England was used for lighting purposes and by 1802 the first factory in the U.K. was gas lit. In 1812 a gaslight company opened in London, England and in 1816 one in Baltimore, Maryland. It was relatively easy to transport coal to gasworks, which could be very small to very large according to local needs, for the local production and distribution of coal gas. This manufactured gas industry eventually disappeared in most parts of the world when the natural gas distribution problem was solved. **Many oil wells produce gas** as well but often there was no local use for it and it was **flared** (burnt) off for safety reasons. The development of cheap, seamless, steel piping in the 1930s and the widespread discoveries of natural gas in post-W.W.2 decades meant that it could be piped thousands of kilometres over land and under sea to distant markets or it can be liquified under pressure and shipped by sea as **liquid natural gas** (L.N.G.).

25.3 Some economic matters

World consumption and production rates of both oil and gas, the location of our present reserves and the probable time span during which our reserves and resources will last are discussed in general terms in Chapter 23. Now we must look at some more detailed facts.

25.3.1 Oil prices

Trade in crude oil
Most oil, formerly traded on long term contracts, is now traded at world market prices. There are various traded crudes, such as Arabian Light, North Sea Brent and West Texas Intermediate.

The average annual price of crude oil, corrected for the effects of inflation, is given in Fig. 25.2. From this it can be seen that after the initial excitement attending the birth of a new industry when wild fluctuations in supply and demand were reflected by equally large price variations, the price in real terms was relatively stable from 1878 to October 1973 even through the events of W.W.2 and the economic booms that followed. To understand what has happened since that fateful October we must look at some more history. When the oilfields of the Middle East were being brought into production, largely by foreign companies, during the 1940s and 1950s, the

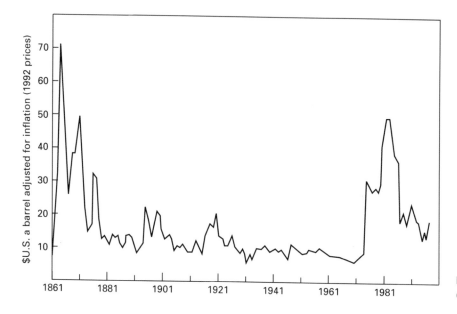

Fig. 25.2 Average annual oil price over the period 1861–1995.

U.S.A. operated **import quotas** to protect its own higher priced domestic oil and this caused the oil companies, now dealing with the massive Middle Eastern production, to seek markets elsewhere, particularly in Europe. The resulting severe competition led to these foreign companies reducing the price of Middle Eastern crude oil without consulting the governments of the producing countries whose revenues from royalties and taxes thereby dropped dramatically; by the mid-1960s the price of Arabian Light was down as low as $U.S.1.30 a barrel! Naturally, these countries were already reacting strongly against this unilateral treatment and in September 1959 the oil ministers of Iran, Iraq, Kuwait and Saudi Arabia were joined by the Venezuelan minister in Baghdad to form the **O.P.E.C.** — the Organization of Petroleum Exporting Countries.

The O.P.E.C. had some initial success in controlling price levels but more in increasing the proportion of its members' share of the profits of oil production, from 50% to over 85%. In the following years more oil producing nations joined the O.P.E.C. to take the membership up to thirteen, but this became twelve when Ecuador resigned in 1994. The membership is given in Table 25.1. Notice the dominance of Arab countries and countries having religious ties with them. For political and economic reasons but particularly because of the vast worldwide overproduction of oil, the O.P.E.C. achieved little in the way of price increases until the start of the 1970s when the world's crude oil output, especially in the U.S.A., began to drop and Arabian Light crept up to $U.S.3.00 a barrel. Then came the Yom Kippur War: Israel was attacked on 6 October 1973 by Egypt and Syria,

Algeria	Iraq	Saudi Arabia
Ecuador*	Kuwait	United Arab Emirates
Gabon	Libya	Venezuela
Indonesia	Nigeria	
Iran	Qatar	

Table 25.1 Membership of the O.P.E.C.

* Resigned 1994.

both nations striving to regain territory lost in the Six Day War of 1967. The O.P.E.C. met in Vienna two days later, heatedly debated the price at which its members would sell oil and decided on a 70% increase. Oil consuming nations were divided into four categories, the U.S.A. being in the embargoed category. The O.P.E.C. communiqué making those announcements stated *inter alia* that the Arab oil embargo would let the U.S.A. know 'the heavy price that the big industrial countries are having to pay as a result of America's blind and unlimited support for Israel'.

This was probably **the greatest intervention in world markets** of all time and its ramifications were endless. The principal energy source of the industrialized nations, long thought to have a low and stable price, rocketed in price and so did the cost of other energy sources, commodity prices in general and of course all industrial costs; a world economic recession was initiated. Worst hit of all was not the U.S.A. but the less developed countries of Africa, Asia and Central and Southern America who did not have the hard currency to pay these high prices. The O.P.E.C. continued to squeeze the oil markets, pushing up the oil price to dizzy heights particularly between 1979 and 1981 and of course the predictable came about. 'You can't buck the markets,' Mrs Margaret Thatcher, U.K. Prime Minister, declared in a different but analogous situation. And indeed you can't. Pushing oil up to such absurd prices stimulated energy conservation (a very good outcome for the health of the environment!), a switch away from oil to coal and gas for energy production, heating etc. and most important of all, from the market point of view, **exploration for new oilfields**. This resulted in a significant increase in the world's oil output. Britain, Norway, Mexico and other countries became net exporters of oil, and techniques of exploring for oil under the sea bed and extracting it were developed at astonishing speed with the result that oil accumulations in oceanic regions, as well as shallow seas such as the North Sea, are now coming into production.

All these factors contributed to the great slide in prices of the early 1980s and the continuing high level of non-O.P.E.C. production has kept crude oil prices at a much lower level (Fig. 25.2). Another factor, however, must be stressed and that is the inability of the O.P.E.C. members to agree

O.P.E.C. and the Iraqi factor

Iraq, one of the O.P.E.C.'s great oil producers, has now been exiled from world markets for six years. This exile has acted as a safety valve for the market. Saudi Arabia, which still has much idle output capacity, has taken over many of Iraq's markets and the other O.P.E.C. nations have also benefited. For the last six years the O.P.E.C. has largely ignored the Iraqi factor and with prices going up as they have in 1995–6 the gamble has paid off. But for how long? A change of regime in Iraq could mean that the world raises the embargo on Iraqi oil exports and a flood of oil would instantly depress world oil prices. Will the other O.P.E.C. members quietly and unprotestingly and honestly carry out significant cutbacks in their production quotas to support the oil price? Their previous record does not point to an affirmative answer.

on national quotas and abide by any agreements they may make. In the short and medium term this has eased the market situation considerably, but in the long term, as the O.P.E.C. moves towards being the dominant supplier of oil, so it will be in a very commanding position. Let's hope that it learns from history and does not turn the screw too tightly as it did in the 1970s, although once again this could be good for the environment!

25.3.2 Gas prices

Gas prices are more variable across the world than oil prices due to the ease with which it can be distributed over vast distances overland and under the sea by pipeline and the fact that **sales are largely by contract at negotiated prices** for delivery over a number of years. Thus, a country might be producing some of the gas it consumes and importing some from neighbours. Contracts to buy this gas might have been signed with different companies in these adjacent countries at different times and for different time periods. Some European contracts now range up to A.D. 2005 To complicate a complex picture further, distributors in our model country might be tempted to buy some L.N.G. from overseas suppliers if the price was right! As an example, we can cite Germany which now has some long term contracts lasting until 2005, produces 22% of its domestic consumption but imports 30% from the Netherlands, 33% from the R.F., 14% from Norway and 1% from Denmark. It has harbours which can accommodate L.N.G. tankers and so could purchase overseas supplies.

L.N.G. prices vary from region to region depending largely on the competition from pipeline delivered gas. Thus European c.i.f. prices for 1993 were $U.S.2.60–3.00 per million B.T.u. but in 1994 the price range dipped to $2.25–2.60. However, on the Japanese L.N.G. market the 1993 range was $3.60–3.80 and this showed a much smaller dip to $3.40–3.70 in 1994. At the time of writing (November 1996), world gas prices are

B.T.u.

B.T.u. stands for British thermal unit which is one of the commonly used units for quoting gas prices, normally the currency is $U.S. However, the reader may encounter $ per cubic foot in North America (presumably the gas volume is measured at S.T.P. although I have never seen this stated). In the U.K. we have the ludicrous situation of the Government's Department of Trade and Industry quoting gas prices in pence per therm but householders' consumption being measured in metres cubed and their bills, after a complicated calculation, are presented in kilowatt hours! (1 B.T.u = 1.055 06 kJ or 251 997 cal.)

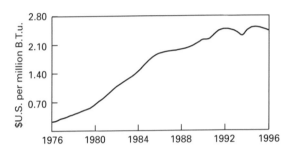

Fig. 25.3 Average annual prices received by producers from sales of North Sea gas during the period 1976–96.

dropping. In North America this has been amplified by a mild Winter in 1994–5 accompanied by a surge in production. As illustrations of these price decreases, we can look at the Alberta wellhead prices: 1993, $U.S.1.64 per B.T.u.; 1994, $1.76; 1995, $1.32; 1996, $1.37; and the North Sea gas prices (Fig. 25.3) which, after about five years of stability, are weakening—in part in response to an oversupply situation.

25.4 Petroleum—its nature and generation

25.4.1 The chemical and physical nature of petroleum

Here we will concentrate on crude oil and gas.

(a) *Crude oil* is a mixture of the organic chemical compounds known as **hydrocarbons**. The principal components belong to the **alkane** or **paraffin** and **naphthene series**. The alkanes consist of straight chain hydrocarbons of the general formula C_nH_{2n+2}. These compounds range from **methane** (CH_4) upwards, with each heavier molecule having a further CH_2 (Fig. 25.4). The first four are gases at standard temperature and pressure (S.T.P.). The first liquid is *n*-pentane (C_5H_{12}) and the first solid is *n*-hexadecane ($C_{16}H_{34}$); *n*-hexadecane and higher members are solid paraffin waxes.

The **naphthene series** (Fig. 25.4) has the general formula C_nH_{2n} and consists of saturated carbon ring compounds. The lowest two members are

(a) Alkanes (paraffins)

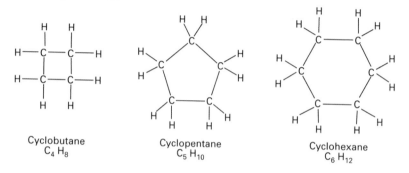

Methane
CH$_4$

Ethane
C$_2$H$_6$

Propane
C$_3$H$_8$

(b) Naphthenes (cyclo-alkanes)

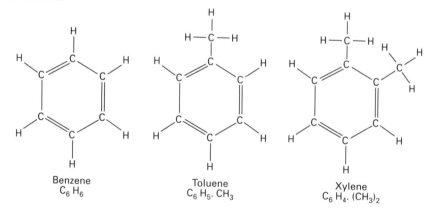

Cyclobutane
C$_4$H$_8$

Cyclopentane
C$_5$H$_{10}$

Cyclohexane
C$_6$H$_{12}$

(c) Aromatics

Benzene
C$_6$H$_6$

Toluene
C$_6$H$_5$. CH$_3$

Xylene
C$_6$H$_4$. (CH$_3$)$_2$

Fig. 25.4 Structural formulae of simple representatives of the main groups of hydrocarbons. Note that these compounds do not have planar structures. The four carbon bonds repel each other and tend to spread out tetrahedrally as in diamonds. (a) Alkanes—note how these develop ever lengthening chains by adding CH$_2$. (b) The three main naphthenes. (c) The three aromatics most commonly found in petroleum.

gases, cyclopentane (C$_5$H$_{10}$) is the first liquid and it and cyclohexane are the commonest in most crude oils.

A third series of hydrocarbons — the **aromatics** (Fig. 25.4) which are benzene-based compounds — occur in many crude oils commonly in amounts less than 1% and only occasionally exceeding 10%. Paraffin-base crudes, the most valued of all oils, only form about 2% of modern world supplies. Naphthenic dominated crudes (**asphalt-based crude oils**) provide about 15%. Most modern crude oils are of mixed base including nearly all Middle Eastern, Mid-Continental (of the U.S.A.) and North Sea oils. The non-hydrocarbon constituents include sulphur, nitrogen, helium and oxygen together with organic compounds of some heavy metals, mainly **nickel** and **vanadium**. **Sulphur** is present in nearly all crudes in quantities up to about 3%. It is higher in heavier than in light oils and occurs mainly in polyaromatics. A small amount occurs as elemental sulphur or H_2S. Oils with detectable H_2S are called **sour crudes**. Oil refineries prefer low sulphur **sweet crudes** and will pay a premium price for them. Nitrogen and oxygen are present in much smaller quantities. Vanadium runs 30–300 ppm in crudes (from which it may be recovered, Section 12.4) and nickel 20–85 ppm.

The relative densities of oils are given on the **A.P.I. gravity scale** which is the American Petroleum Institute standard for expressing the specific gravity of oils. In relation to the density (ρ), $^\circ$A.P.I. value $= 141.5/\rho - 131.5$. This gives water, under S.T.P. conditions, a value of 10° A.P.I. Oils with A.P.I. values higher than 30° are considered light, 30°–22° medium and less than 22° heavy. The **viscosities** of oils vary with the density and amount of dissolved gas. Light oils have viscosities below 30 mPa s, whereas heavy asphaltic oils have viscosities in the thousands of millipascal-seconds range. The **pour point** is also used as a measure of viscosity. This is the lowest temperature at which the oil will flow under specified conditions. High contents of paraffin wax produce pour points above 40°C in some crudes, whilst other crudes may flow at temperatures as low as -36°C. Refractive indices vary from 1.42 to 1.48 in passing from lighter to heavier oils.

The first function of oil refineries is to separate crude oils into fractions, rather than single pure substances, by **fractional distillation**. These fractions contain compounds with roughly the same number of carbon atoms. Table 25.2 shows what each fraction contains and what it is used for. A second function is to break down some of the heavier hydrocarbons into lighter, more useful and more valuable products — the process known as **cracking**. Because of their compositional variation, crudes from different oilfields may have to be distilled in different refineries designed to process oils of particular specifications. Thus it is imperative to analyse newly discovered oil in detail to determine its composition, where it should best be refined and hence its potential value.

(b) *Natural gas* consists of the gaseous members of petroleum and is almost entirely composed of the one or more gases shown in Table 25.2. It occurs in the ground on its own or with oil. In the latter case it may be dissolved in the oil under subsurface pressures or, if there is enough of it, some may separate out as a **gas cap** above the oils. This gas, though dominantly methane, commonly contains significant quantities of ethane, propane and

Fraction	Boiling range (°C)	Number of carbon atoms in the constituents	Uses
Fuel gas	−160 to 20	1–4, mainly methane, ethane, propane and butane	Fuels for gas ovens, L.P.G., chemicals
Petrol (gasoline)	20–70	5–10, e.g. octane (C_8H_{18})	Fuel for vehicles, chemicals
Naptha	70–120	8–12	Chemicals
Paraffin (kerosene)	120–240	10–16	Fuel for central heating and jet engines, chemicals
Diesel oils and lubricating oils	240–350	15–70	Fuel for diesel engines, trains and central heating, chemicals, lubricants
Bitumen	>350	>70	Road asphalt, roofing, waterproofing

Table 25.2 The constituents and uses of fractions from crude oil.

butane and is termed **wet gas**. Gas found on its own is usually almost entirely methane and is termed **dry gas**. Certain mixtures of light hydrocarbons exist as gas under subsurface pressures but condense to very light oils on reaching the surface – these oils have A.P.I. values >50° and are called **condensates**.

25.4.2 Hydrocarbon genesis

Trace amounts of hydrocarbons, especially methane, emanate from deep within the earth's interior and methane has been identified among gases given off by volcanoes. Much evidence, however, indicates that petroleum hydrocarbons are of organic origin; e.g. their distribution within sedimentary sequences, the survival of chlorophyll-like vegetable porphyrins and other molecules, that only occur in living organisms and are destroyed at temperatures above 200°C, and carbon isotopic evidence. Thus, most workers nowadays believe that petroleum hydrocarbons were generated from disseminated organic matter in sedimentary rocks, the controlling factors being: (a) the nature of the organic matter; (b) its abundance; and (c) the extent to which it has been heated during burial.

(a) *The nature of the organic matter.* The decayed matter from fish and other animals is of little importance because of its small bulk compared with that of the very much more abundant, minute organisms of lakes and seas. This organic matter has been classified into three types which during diagenesis can give rise to three types of **kerogen** – the precursor of petroleum hydrocarbons.

Type I forms from the minute algae that flourish in some freshwater lakes in warm climates, these become **sapropelic kerogens**.

Type II, found in marine sediments, consists of the remains of single-celled plankton, algae and bacteria that live in great abundance in certain tropical marine environments, particularly where there are upwellings of cold, nutrient-rich waters as off the coasts of California, Namibia and Peru. These become **mixed planktonic kerogens** which are the **source of most of the world's oil**.

Type III consists of terrestial plant material — wind blown spores and pollen plus plant fragments. During diagenesis these become **humic kerogens**.

These kerogens and their petroleum derivatives are listed in Table 25.3; but to become kerogens, and later hydrocarbons, the organic material must escape destruction by oxidation or by being consumed by other organisms. Anaerobic sea bottom conditions and rapid burial will favour its preservation and thus it is common in shales and some fine-grained limestones which can become hydrocarbon **source rocks** provided they contain enough kerogen.

(b) *Abundance of organic matter.* A lower limit of about 0.4% organic carbon (in kerogen) is generally considered necessary, but most recognized source beds contain 0.8–2% and the best as much as 10%. Not all this organic matter is converted into petroleum, at the very most about 70%, and not all will escape from the source rock due perhaps to great thickness or very low permeability.

(c) *Petroleum generation.* After burial in a potential source rock, anaerobic bacteria decompose the organic matter to structureless kerogen producing by-product methane — the marsh gas of stagnant waters — scientifically called **biogenic gas** which usually escapes into the atmosphere, but may occasionally be trapped in small but exploitable accumulations such as that which supplies Anchorage in Alaska. Biogenic gas may also be

Table 25.3 The three types of organic matter that can give rise to petroleum showing some details of their chemistry, their precursors, their original environments and their principal petroleum products. Their evolution is shown in Fig. 25.5.

Kerogen type	H/C ratio	O/C ratio	Precursor	Environment	Petroleum product
I Sapropelic	High	Low	Freshwater algae	Lakes	High quality but waxy light oil
II Mixed planktonic	High	Intermediate	Marine algae and plankton	Marine	Low wax, medium and heavy oil, some gas
III Humic	Low	High	Spores, pollen, plant fragments	Swamps, nearshore marine	Mainly gas, some waxy oil

trapped as frozen **gas hydrate** (methane clathrate; see Section 25.7.3) under marine ice sheets or in the deep sea.

This bacterial decay proceeds from shallow depths to about 1 000 m, up to about 50°C, at which bacteria die. With deeper burial and heating (~1 000–3 500 m and 50–145°C), the large molecules in kerogen break down to form smaller, lower molecular weight hydrocarbons. Oxygen is lost rapidly by dehydration and loss of CO_2, so that H_2O and CO_2 are the initial products. Heavy oils appear first, then around the point of peak generation medium oils, then at even higher temperatures the decreasing size of the molecules yields light oil and wet gas. At still higher temperatures (~200 + °C) only dry gas is formed. Finally above about 230°C all hydrocarbons are destroyed, regional metamorphism sets in and all carbon is now graphite. This process is illustrated diagrammatically in Fig. 25.5, from which it can be seen that there is a limited temperature-depth range within the earth

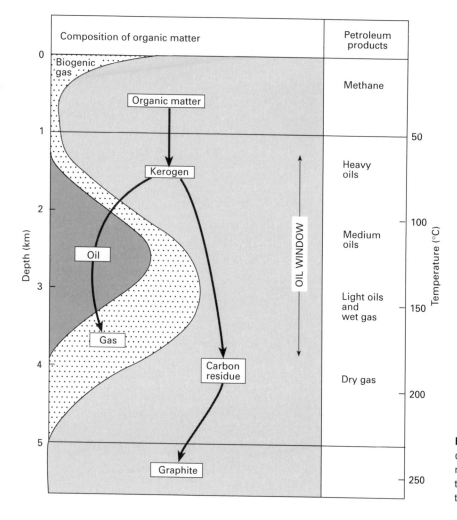

Fig. 25.5 Organic matter diagenesis showing the relationship between temperature, depth of burial and the petroleum products formed.

over which oil is generated **and** preserved. This is sometimes referred to as the **oil window**.

The above story and accompanying diagram can only be approximate because time is also a factor—supply enough of it and the above events can progress at lower temperatures. Nor are temperatures directly linked to depth so oil geologists refer to the **level of maturity** and the whole process as **maturation**.

The maturation of the three kerogen types is depicted in Fig. 25.6 which is a plot of hydrogen to carbon against oxygen to carbon (termed a **van Krevelen diagram** after the first scientist to use it). As noted in Table 25.3, each kerogen type initially has distinctive elemental ratios. During burial diagenesis and hydrocarbon generation, each type changes in terms of its three major elemental components which appears on the figure as a distinct trend of decreasing oxygen and hydrogen. These trends are termed **evolution paths** and they eventually converge during deep burial as the composition of each approaches 100% carbon.

25.4.3 Oil source rocks

Not all shales are potential source rocks. Rich source sediments of different ages are much alike in their gross appearance. They are most commonly brown, often dark brown but may also be black or green. Lamination is generally developed with pale silty bands alternating with bands packed with organic remains and the rocks may be strongly bituminous. In addition, most source rocks are phosphatic and uraniferous; sulphides, especially pyrite, are common. Rich, Albian age, source rock shales occur in the Kazdhumi Formation of the Iraq–Iran Oilfields. On the other hand, two of the most prolific source sediments known are limestones: the Jurassic Hanifa Formation of Saudi Arabia and the Cretaceous La Luna Formation of Venezuela. The latter is a dark brownish-grey to black bituminous limestone, thinly laminated to very well bedded with lenticular bands of dark chert, fish remains and phosphate pellets.

Source sediments are early deposits in most oil basins and are normally older than the main reservoir rocks, from which they may be separated by whole formations devoid of oil accumulation. The combination of high organic matter, undisturbed laminar layering, pelagic flora and fauna, presence of phosphorus and the necessity for anaerobic conditions to preserve the organic matter, indicates a depositional environment below a thermocline, probably in the deepest part of the basin, although the absolute depth need not have been great.

An oil source rock may also produce gas and associated gas is very common. Deep environments may produce non-associated gas and so may other source rocks such as coal. In north-western Europe enormous

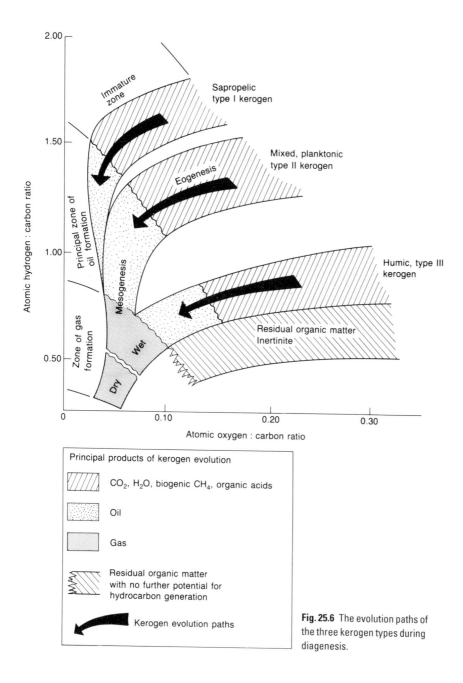

Fig. 25.6 The evolution paths of the three kerogen types during diagenesis.

Legend content:

Principal products of kerogen evolution

- CO_2, H_2O, biogenic CH_4, organic acids
- Oil
- Gas
- Residual organic matter with no further potential for hydrocarbon generation
- Kerogen evolution paths

Diagram labels: Atomic hydrogen : carbon ratio (y-axis, 0.50 to 2.00); Atomic oxygen : carbon ratio (x-axis, 0 to 0.30); Immature zone; Sapropelic type I kerogen; Mixed, planktonic type II kerogen; Humic, type III kerogen; Principal zone of oil formation; Zone of gas formation; Eogenesis; Mesogenesis; Wet; Dry; Residual organic matter Inertinite

gas reserves occur in Lower Permian sandstones extending from the River Elbe westwards into the British sector of the southern North Sea. Here the source rock, Carboniferous coal, lies beneath the porous, desert-deposited, reservoir sandstones of the Rotliegendes Formation which are perfectly sealed by thick Permian Zechstein evaporites.

25.5 Formation of oil and gas accumulations

25.5.1 Introduction

The oil and gas, whose genesis was discussed in the previous section, will be too dispersed within the source rock to be exploitable, so how do accumulations (**pools**) of gas and oil, that we can sink wells into, develop? Well, first of all, the hydrocarbons must move out of the source rock — this is called **primary migration**. Little is known of how this takes place but the buildup of overpressures in the source rock probably plays a leading role. Being driven by pressure gradients, primary migration can be upwards or downwards, but once out of the source rock and into a permeable one the fluids are driven *upwards* (**secondary migration**) by buoyancy forces through the denser interpore water and may eventually reach the surface unless they flow into a suitable **reservoir** and **trap**; the former being, say, a sandstone with high porosity and permeability, **capped** by an impervious rock such as a shale, to form an anticlinal trap (Fig. 25.7).

25.5.2 Migration

We know so little about primary migration that a study of the literature yields only speculation. An attractive one is that it is flushed out by water being squeezed out of the compacting shale — not the compaction waters, these having mostly gone by the time oil is forming, but waters released by **clay mineral reactions** such as the formation of illite from montmorillonite at about 110°C which is well within the oil window.

Secondary migration is better understood. The buoyant hydrocarbons coalesce to form droplets in the brine filled reservoir rock and pass up through it wherever the 'throats' connecting the grain pores are wide enough. If a droplet is held back by a narrow throat, other hydrocarbons will rise up and coalesce with it until the buoyancy forces drive this

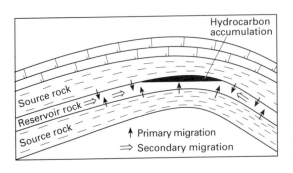

Fig. 25.7 Primary and secondary migration leading to the formation of an oil or gas pool in an anticline. The upper source rock —shale—is acting as a **cap rock**.

through the throat and up the stratigraphy. Grain size does not affect absolute porosity but it does influence throat size and hence migration efficiencies. Packing, degree of sorting, grain shape, clay minerals in the matrix and degree of cementation all affect porosity and permeability in granular rocks. **Carbonate rocks** are rather different and often more permeable due to the development of solution-enlarged pathways along bedding, joints, fossil casts, stylolites and porous reef materials. Dolomitization may be particularly efficacious in developing permeability provided the cavities developed by the associated shrinkage are interconnecting. Most secondary migration only proceeds for a few kilometres but in some oilfields, e.g. western Canada, there is evidence of oil migration over 100 km or more.

25.5.3 Reservoirs

It will already be clear that the best reservoir rocks have both high porosity and high permeability, properties which were discussed in the last section, from which it will also be correctly concluded that the best normal reservoir rocks are sandstones and limestones. It is also necessary for reservoir formations to be of sufficient thickness to contain economic amounts of gas or oil. A further point affecting permeability is the degree of water saturation. The higher this is, the lower the permeability for gas or oil and the lower the rates at which oil can be removed from wells, but this brings us into the realms of the petroleum engineer and out of those of the geologist.

(a) *Sandstone reservoirs*. The points made in Section 25.5.2 indicate that the best reservoir sandstones have well sorted, well rounded grains, are devoid of clay minerals, only lightly cemented and preferably coarse-grained to provide maximum effective porosity. Desert-deposited sandstones come close to meeting these requirements and usually provide excellent reservoirs, e.g. see Section 25.4.3.

(b) *Carbonate reservoirs*. As pointed out in Section 25.5.2, the permeability of limestones is usually due to secondary solution, exceptions being coral reefs or piles of shell débris which may have some primary porosity. This secondary solution may create wide permeable channels providing high storage possibilities and permitting rapid hydrocarbon flow during production. Well cemented limestones are rigid but brittle rocks susceptible to tectonic fracturing which can generate excellent channelways for oil. The fractured limestones of south-western Iran are notable for their high well production rates as are the fractured chalks of the North Sea Ekofisk Field.

Effective porosity

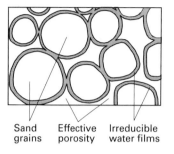

Sand grains   Effective porosity   Irreducible water films

Fig. B. 25.1

The grains of most waterlaid clastic rocks have a very thin film of water sticking to their grain surfaces. This water is held in place by strong interfacial tension and surface bonding and cannot be shifted. It therefore occupies some of the pore space and impedes the flow of hydrocarbons (Fig. B.25.1). The coarser the grain size the less is the surface area to which this water can adhere.

25.5.4 Petroleum traps

Hydrocarbon traps have been divided into three types: **structural, stratigraphical** and **combination traps**, the last-named being a combination of stratigraphical and structural trapping mechanisms. The most common traps are **anticlinal**, if we take the broadest interpretation of that term. Other important structural traps are developed by faulting and piercement structures, e.g. salt domes. **Stratigraphical traps** are created by variations in the stratigraphy that are not due to structural deformation.

Anticlinal traps are the easiest to locate because they are large and extensive structures which can often be located by surface mapping, or at depth by geophysical investigations. Because tectonic anticlines have considerable vertical continuity of shape, there may be many separate gas or oil pools stacked one above the other. An elongate, open anticline is an ideal petroleum trap since its extended limbs facilitate migration, its hinge zone permits accumulation, provided an impervious **cap rock** is present, and its crest generally contains periclinal areas in which oil pools can form. In asymmetrical folds the gentler-dipping limb often contains the most oil (Fig. 25.8).

Fault traps may form wherever an inclined reservoir is cut by a fault that brings impermeable rocks against that portion of the dislocated reservoir which dips downwards away from the fault (Fig. 25.8).

Salt dome traps may be present in those **oil basins** that are underlain by evaporite beds rich in halite as in the Gulf Coast of the U.S.A. The salt diapirs are quite impermeable and where the upward-moving salt plug has domed and pierced the overlying and surrounding strata, then many different oil traps may be formed (Fig. 25.8).

Stratigraphical traps result from lateral and vertical variations in thickness, texture, lithology and porosity of reservoir rocks. An inclined sand-

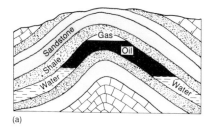

(a)

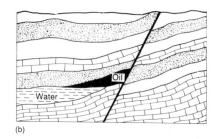

(b)

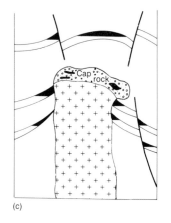

(c)

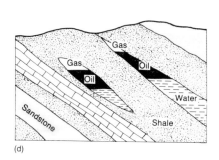

(d)

Fig. 25.8 Structural and stratigraphical oil traps. (a) Anticlinal trap developed in a sandstone reservoir in an open, asymmetrical fold. (b) Oil trapped by a fault seal. (c) Schematic diagram of salt dome traps, in supercap, cap rock and flank sandstones (abutting, fault sealed and pinch out). (d) Two types of stratigraphical traps. Right, sandstone wedge out; left, sandstone lens.

stone reservoir intercalated in shale may wedge out up dip, as beach and bar sands do, thus forming a wedge-out trap. A variation of this is the sandstone lens trap which may or may not be inclined (Fig. 25.8). Shoe-string sands are long narrow bodies of sandstone enclosed by shales; they may have been formed as offshore bars or by meandering rivers. Small oil pools generally characterize these sandstones but some of the Texan occurrences have very large oil accumulations. Porous limestone reefs when buried by impermeable sediments form potential reef traps, but these are generally small by comparison with other traps [Fig. 25.9(b)]. Unconformities can also produce important oil traps. Underlying tilted reservoir rocks

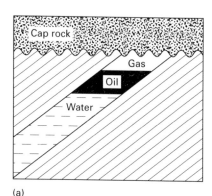

(a)

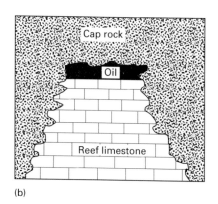

(b)

Fig. 25.9 (a) Unconformity trap. (b) Reef trap.

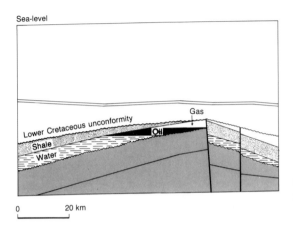

Sea-level

Gas

Lower Cretaceous unconformity

Oil

Shale

Water

0   20 km

Fig. 25.10 Combination trap. A north–south section through the Prudhoe Bay Oilfield, Alaska.

Type of trap	Number of fields	Proportion of reserves (%)
Structural	132	78
Stratigraphical	44	13
Combination	24	9

Table 25.4 Classification of two hundred giant oilfield traps and percentage of oil contained in each class.

may be sealed at the unconformity to form an **unconformity trap** [Fig. 25.9(a)].

The Prudhoe Bay Field in Alaska (Fig. 25.10) is a striking example of a **combination trap**. Here the overall structure is a huge anticline but the actual trap for the principal oil pool occurs where the homoclinally dipping reservoir rock is truncated up dip by a normal fault, which brings down impermeable shales against the reservoir rock, and where it is cut across by a major unconformity.

25.5.5 Relative importance of traps

An analysis of two hundred giant oilfields made in 1975 yielded the results given in Table 25.4 which show the dominance of structural traps not only in number but in size and hence in contained oil reserves.

25.6 Production techniques

The presence, nature and extent of hydrocarbons in a potential trap is established by drilling. This is anything but a simple process and various terms are used to describe the findings. **Oil in place** is the total volume of oil present in an accumulation and usually refers to what was there before any extraction took place. It is impossible to recover all this oil which

Giant gas fields

These are fields with very large natural gas reserves. Nearly one hundred fields each contain more than $140 \times 10^9 \, m^3$ but so far only fifteen have each produced more than $100 \times 10^9 \, m^3$. Some of these giant fields are also **giant oilfields**, e.g. Prudhoe Bay in Alaska and A.J. Bermudez in Mexico. The Urengoi Field in Siberia is capable of supplying 15% of the present world demand for natural gas.

means that the **recoverable reserves** must be determined and these are divided into **primary reserves** which can be produced without artificial assistance other than pumping, **secondary reserves** which would require enhanced recovery techniques (such as driving out the oil by injecting water below it) and **tertiary reserves** using more exotic techniques. Each phase of recovery will clearly be more expensive and so economic considerations enter the calculations governed by many factors such as geographical location, nature and value of the particular oil and its current price, transport arrangements and costs etc. Reserves may also be classified as **proven** or **probable**, but oil companies differ on how to calculate these classes. The tendency today is to designate as proven those reserves that they believe have an 85% probability of being present and as probable those that have only a 15% or less probability of existing.

25.6.1 Oil extraction

To keep the number of production wells to a minimum, they must be carefully located to give maximum production for as long as possible. A simple situation illustrating this point is drawn in Fig. 25.11. Oil flow must be regulated, otherwise a drawdown situation, as with water wells (see Fig. 22.4), could develop, resulting in early gas production. Oil is forced into a well by the downward pressure of the compressed gas cap and the upward pressure from the bottom water (Fig. 25.11). With sufficiently high pressures the oil is forced up to the surface where, at the wellhead, there are valves to control the flow and separate the gas which bubbles out of solution, otherwise pumping is required. With natural flow we have what are called **drive mechanisms**. If the bottom water is pushing the oil upwards to the well, this is known as **water drive**; in a **solution gas drive** or **gas-cap drive**, gas bubbling out of the reservoir oil as the pressure decreases actually serves to keep that pressure high, thus forcing oil into the well.

In **enhanced oil recovery**, **secondary recovery techniques** are employed to maintain reservoir pressure. If water drive is the operative drive mechanism, then water is injected beneath the oil pool; with gas-

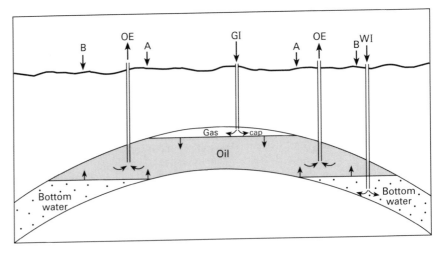

Fig. 25.11 An anticlinal trap from which oil is being drawn. The gas–oil and oil–water contacts will move during production so the producing wells must be drilled accordingly at sites OE. Wells at A would quickly go over to producing gas, whereas those at B would soon produce only water. The wells at OE would be expected to recover the last of the producible oil! OE, oil extraction wells; GI, a gas injection well; WI, a water injection well.

cap drive, gas is pumped into the gas cap (Fig. 25.11). This saves the gas from being flared and wasted, and conserves it for future use. Some more exotic, generally uneconomic techniques are employed in **tertiary recovery** such as steam or detergent injection to recover the remaining viscous heavy oils.

25.7 Unconventional sources of gas and oil

25.7.1 Coal bed methane

Coal bed methane is the natural gas found within coal and was greatly feared by coal miners until modern detection devices and efficient mine ventilation methods kept its concentration below explosive mixture levels — less than 5% methane in air is inert, 5–15% is violently explosive. Methane in coal occurs primarily as a layer adsorbed on to the internal surface of the coal. **Coal is full of microscopic fractures** and can adsorb as much as 30 ml of methane per gramme; it also contains a variable amount of free methane in the **cleat**. In the ground, the adsorbed methane is held in place by the hydrostatic pressure of the ground water. **When the pressure is reduced**, e.g. by mining, **it desorbs** and enters the mine whence it is removed by the ventilation system and either vented to the atmosphere, where it adds to the greenhouse gases, or used locally by industry as a fuel. The annual value of this production in the U.K. in 1993 was £12M.

Cleat

This is the closely spaced jointing found in coal seams. There are usually two approximately orthogonal sets of joints roughly normal to the bedding. The fracture planes are much more closely spaced than the joints in the adjacent non-coal strata. Cleat planes are important in the exploitation of coal, in regard to both the cutting of the coal and because they affect the size of the fragments produced. The latter point is important in the design of coal processing plants and in coal utilization and marketing.

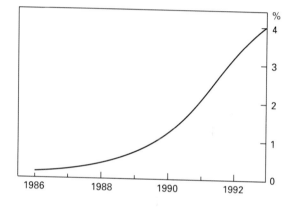

Fig. 25.12 Coal bed methane production in the U.S.A. as a percentage of that country's total natural gas production.

In the U.S.A. (Fig. 25.12) and other countries **this gas is being tapped in unmined areas**. Production is achieved by drilling a borehole into one or more seams, casing off the surrounding strata, stimulating increased permeability by hydrofracturing (see Section 6.4) near the borehole and pumping out the ground water from the seam to lower the hydrostatic pressure and allow the methane to desorb from the coal, thus enabling it to diffuse along the pressure gradient through the cleat to the borehole. It is necessary to reduce the hydrostatic pressure in the seam by 90% to remove 50% of the contained gas. The methane then travels primarily in solution in the water, but also as free gas, to the borehole where it is pumped out and separated out at the surface. Single boreholes extract the gas over roughly circular areas of about 32 ha.

Unlike coal, which is mainly carbon, **methane is a high hydrogen, low carbon fuel** and on burning releases far less CO_2, the worst greenhouse gas. Furthermore, coal bed methane production leaves the coal intact for future generations. An added bonus is that coal too deep for economic underground mining can be exploited for methane production and, in any case, coals at <200 m depth have usually lost their methane to the surface. This means that heavily mined regions, such as the U.K., still have **a high coal bed methane potential at depth**. In areas where no mining has

occurred, the potential will be even higher, e.g. the best wells in the Cretaceous coals of the San Juan Basin in Texas, U.S.A. are expected to yield over
$28\,000\,m^3$ each day. Present research suggests that the maximum depth for
viable recovery is $2\,000\,m$. The main environmental issue will probably be
the disposal of the ground water which will be saline and contaminated
requiring treatment before surface disposal or re-injection.

Well, there's plenty of good news in the above review and here's some
more: although the recoverable amounts of coal bed methane are not well
known, the estimates of world *resources* are very high — higher than our
reserves of natural gas!

25.7.2 Geopressure gas

This is methane dissolved in water in the deeper parts (5–15 km) of the
world's sedimentary basins. Although the gas concentration is low, the
volume of water is high and hence the gas potential is huge. Drilling costs,
however, will be high and the solvent waters highly saline and environ-
mentally unwelcome if pumped to the surface.

25.7.3 Gas hydrate

This is **methane clathrate**, a frozen form of methane in water. It is stable at
above zero temperatures if under the pressure of a few hundred metres of
water and forms beneath ice sheets or permafrost and under the low temperature, high pressure conditions that prevail in the shallow parts of some
marine sedimentary basins. In both environments, the trapped methane is
of biogenic or deeper thermogenic origin. Seismic surveys of these basins
suggest that these accumulations could represent our most important
natural gas reserves. They could be exploited by heating or depressuriza-
tion methods. Perhaps they should be our **number one priority** as it has
been estimated that were global warming to increase substantially for
more than a hundred years, releases from hydrates could become the
largest single contributor of methane to the atmosphere. The distribution
of gas hydrates is shown in Fig. 25.13.

25.7.4 Heavy oil and tar sands

Heavy oil has such a high viscosity that tertiary recovery techniques (see
Section 25.6.1) are necessary to recover it. Oil that will not flow is known as
tar or **bitumen**. Both heavy oil and tar are less desirable than crude oil, as
gasoline (petrol) is not easily extracted from them and they yield a larger
fraction of heavy oil products. Most are the result of the reaction of crude
oil with ground water and bacteria at relatively shallow depths, a process

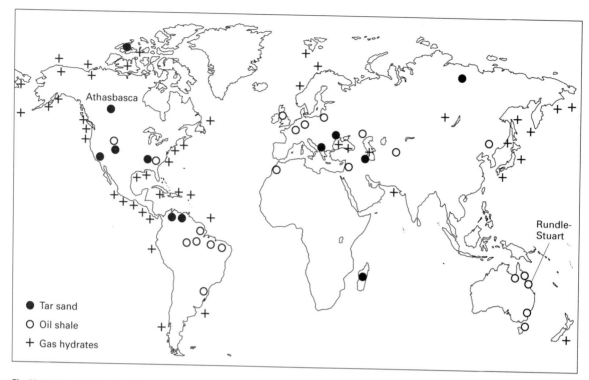

Fig. 25.13 World distribution of important tar sand and oil shale deposits, and the known occurrences of gas hydrates.

known as **degradation**. The Orinoco Oil Belt of Venezuela contains over half the world's reserves and a mixture produced from these deposits—an oil–water emulsion, Orimulsion—is burnt in some electric power stations. However, its obnoxious fumes have caused such an outcry in the U.K. that its use is being phased out.

Tar sands contain even more viscous hydrocarbons and are usually worked by opencast mining and then heated in a retorting plant. Both these materials will probably be worked in the future as oil supplies grow scarce and prices rise, but present estimates suggest that new operations will not break even until oil prices reach $U.S.24 per barrel (at present, November 1996, they are about $U.S.22 per barrel).

25.7.5 Oil shales

These are source rocks that have not given up their oil. They can be mined and the oil distilled by heating the shale to about 450°C. Small oil shale industries have flourished in many countries; one of the best known was in central Scotland from 1850 to 1963. Average yields were about 110–135 l t^{-1} and total production about one hundred million barrels. In the 1970s and

1980s several billion dollars were invested in oil shale projects in Colorado, U.S.A. but all have been financial failures, as has the Rundle Oil Shale Project in Queensland, Australia. The location of world heavy oil, tar sand and oil shale deposits is shown in Fig. 25.13.

25.7.6 Coal

In 1933 a plant near Leipzig in Germany began production of about 100 000 t p.a. of gasoline by the method of direct coal hydrogenation. The plant was enlarged and by 1943 at the height of W.W.2 there were twelve plants which that year produced 2.64 Mt of which 73% was gasoline, 23% diesel oil and 4% mainly fuel oil. Germany could not have fought the war without this production. Nor could the U.K. without its production which provided 0.5 Mt p.a. of very high octane gasoline for aircraft. Much research has been carried out since W.W.2 to produce petroleum more efficiently from coal but the cost has been too high and in June 1995 British Coal shut down its pilot coal liquifaction plant in North Wales into which £14.4M had been invested. The process will not be a financial success until crude oil prices have risen to over $U.S.30–35.

A more successful story comes from the R.S.A. Under threats of trade embargoes and international sanctions, the Sasol Corp. was set up in 1950 to extract oil and chemicals from coal. Coal production costs in the R.S.A. have always been very low (one fifth or less than the cost of U.K. deep mined coal), and Sasol was also helped with a government subsidy and a tariff on imported oil. Oil production commenced in the 1950s and today there are two huge complexes producing over 4 Mt p.a. of high quality liquid fuels that are sold at filling stations throughout the country.

25.7.7 Biofuels

Although hardly in the realm of geology this source of petroleum products must be briefly mentioned for the sake of completeness. Sewage, manure, municipal wastes can be and are processed in various ways to produce methane. Liquid fuels such as alcohols can be made from biomass by **pyrolysis**. Brazil now produces large volumes of ethanol (ethyl alcohol) from sugar cane and beet, cassava and sorghum. This is blended with gasoline to produce **gasohol** which contains up to 20% ethanol. Many Brazilian car engines have been designed to run entirely on ethanol and Brazil now has a production of about 100M barrels p.a. This petroleum alternative is attractive to Brazil as: (i) it imports much oil which has to be paid for with foreign currency; and (ii) it has vast sugar cane producing areas in the depressed north-eastern part of the country.

25.8 Environmental matters

25.8.1 Atmospheric pollution

The burning of any fossil fuel is bad news for the atmosphere at this time when global temperatures appear to be rising, due at least in part to man's contribution to the greenhouse effect. Complete combustion of pure fossil fuels yields CO_2 and H_2O but there are always impurities and these give rise to various unwanted consequences such as acid rain [see Section 3.2.1(h)]. Internal combustion engines produce substantial amounts of the highly toxic gas carbon monoxide (CO) from the incomplete combustion of gasoline which is also responsible for fine carbon particles (soot) that are very damaging to those suffering from respiratory problems, as are the similar particulate products of diesel engines. Both engines are also guilty of producing substantial amounts of toxic nitrogen oxide gases; these are also produced by burning fossil fuels.

25.8.2 Extraction worries

Forests of oil drilling derricks are very much a thing of the past and although oil or gas drilling rigs are unsightly, they are only in place for a matter of months and then, if the well is a producer, they are replaced by a system of valves and/or a nodding donkey pump or two. This apparatus is often easily screened from sight by a little careful shrub and tree planting. Transport of oil or gas to a refinery or port is then by unobtrusive pipeline which can, if desired, be buried from sight.

The worst hazard that may be met whilst drilling is a **blowout** when the gas or oil pressure is so high that it blows the column of heavy drilling mud out of the drill hole. Fire hazard is then high and the environment, especially if marine, may be polluted very rapidly. Improved wellhead and monitoring equipment have resulted in blowouts being now very rare. The other major hazard is that of oil spills from tankers, one of the largest being when the Exxon *Valdez* ran aground in Prince William Sound, Alaska in 1989 releasing more than 260 000 barrels of crude oil which spread along nearly 800 km of coastline. Flora and fauna were devastated, e.g. an estimated 400 000 sea birds including 900 bald eagles died. The best way of preventing recurrence of such disasters is making the polluter pay for the clean up and fining him. The Exxon *Valdez* affair resulted in the payment of over $U.S.1 000M in fines!

25.8.3 As the wells run dry

Companies operating offshore oil and gas production platforms are now

Protecting the environment with directional long reach drilling

Drilling straight holes is not easy and there is the apocryphal story from the early days of drilling in California of the drilling crew who were astonished to see their drill bit emerge from the ground a few metres from the rig! Drillers are now very skilled operators and, using various tricks such as special wedges, bent ends to the drillstring, down the hole motors, etc., can now keep drills on target or turn them to reach locations that are impossible to reach from above. In oil prospecting and extraction this development is invaluable; oil accumulations under lakes, rivers or bays can be tapped, as can accumulations in vertical fractures or steeply dipping reservoir beds.

This technique is of special value in broadening the area that can be tested from one site, e.g. a single offshore platform. Beneath the waters of Poole Harbour the B.P. Co. discovered an extension to its Wytch Farm Oilfield, which makes it the largest onshore field in Europe. Poole Harbour is a magnificent natural harbour nearly 100 km round and little affected by modern developments. To exploit its newly found oil the obvious answer seemed to be to construct an artificial, and intrinsically ugly, drilling island in Poole Bay. B.P. thought again and decided to use directional long reach drilling as indicated in Fig. B.25.2. To accomplish this the drilling head had to be turned through 74° to keep it within the reservoir sandstone, this extreme angle being necessary because the reservoir is only 1 600 m below sea level. The term long reach is very apt in this example as the far end of the reservoir is almost 7 km from the drilling rig.

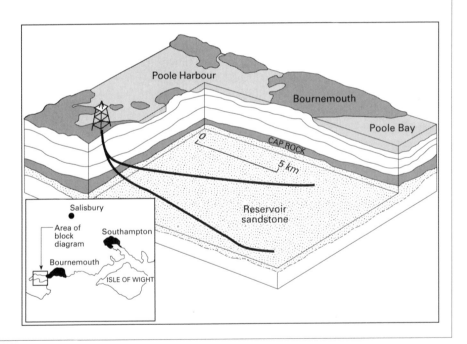

Fig. B.25.2

facing the problem of what to do with them when the wells run dry. In the North Sea the first platform was towed out in 1967; since then more than two hundred have followed and thousands of kilometres of pipeline have been laid along the sea bed. The International Maritime Organization set the legal framework for abandonment. Structures must be removed to provide a water clearance of 55 m to ensure there is no risk to navigation. Any platform in water depth of less than 75 m must be removed entirely. About 130 platforms in shallow water will have to go completely. For those in deeper water the U.K. government may permit removal of the upper section to the required depth or their complete toppling. The biggest headache is a few deep water concrete platforms weighing over 200 000 t. Oil companies have of course been budgeting for abandonment costs but have they put enough aside? And where would they stand if the oil price fell sharply one or two years before abandonment? A recent survey of platform operators came up with total abandonment costs of £7 000M in 1994 values!

25.9 Further reading

Allen P.A. and Allen J.R. (1990) *Basin Analysis*. Blackwell Scientific Publications, Oxford. An advanced but very readable and well illustrated text which, after a thorough discussion of basin analysis, applies this analytic approach to oil exploration.

North F.K. (1985) *Petroleum Geology*. Allen & Unwin, Boston. A comprehensive discussion of the occurrence of oil and gas and how to find accumulations of these desirable fuels.

Stoneley R. (1995) *An Introduction to Petroleum Exploration for Non-Geologists*. Oxford University Press, Oxford. Simply written as the title suggests, but nevertheless it covers a lot of ground in good detail—a rewarding read.

Yergin D. (1991) *The Prize*. Simon & Schuster, New York. An entertaining history of the worldwide oil industry.

26: Energy from the Atom—Nuclear Power

26.1 Introduction

■ This chapter is largely concerned with radioactive waste disposal.
■ The reader is reminded of where to look for discussions of nuclear power and uranium deposits.

This subject was touched on in Section 23.3(d) where it was pointed out that, despite the Three Mile Island and Chernobyl disasters, countries with few or no fossil fuel resources are still constructing nuclear power stations. At the present day about 125 nuclear power stations are in the planning or construction stage. This will result in a steady increase in uranium demand over the coming decades (Fig. 23.1). Research continues apace to develop commercial **nuclear fusion** of hydrogen isotopes which can be harnessed for power generation, but this dream now appears to be decades away, leaving us to use our present technology of **nuclear fission** to produce electricity. Nuclear fission, no matter the type of reactor (pressurized water, advanced gas-cooled etc.), requires supplies of the radioactive isotope ^{235}U which makes up but 0.7% of natural uranium. The supply picture is briefly covered in Section 23.3(d) and it is clear that we have the reserves and resources to sustain many more nuclear power stations over many decades, particularly if we bring fast breeder reactors on stream, install safety procedures that will restore public confidence in nuclear reactors and develop safe solutions for nuclear waste storage and disposal.

26.2 Uranium deposits

These are of various types which appear in a number of previous chapters covering metalliferous deposits. The reader wishing to seek further information on the relevant deposit types should select those listed in Table 8.1 as important uranium producers.

26.3 Radioactive waste disposal

This is at present the subject of a heated national and international debate and it remains a largely unresolved problem. This problem arises because radioactive wastes cannot be neutralized relatively rapidly like chemical wastes or disposed of by any of our accustomed methods such as incineration. They are greatly feared by the general public because their harmful, indeed sometimes lethal, effects are invisible and cannot be detected by human senses.

26.3.1 Categories of radioactive waste

Most radioactive wastes are mixtures of isotopes having both short and long half lives. They are therefore divided into three categories, depending upon the concentrations of radioactive isotopes present, into **low level**, **intermediate level** and **high level** rather than into categories related to the time over which radioactivity is expected to decay away to background values. Low level wastes have a radioactivity up to 1 000 times that considered acceptable in the environment, intermediate level wastes have values between 1 000 and 1 000 000 and high level wastes have even greater radioactivity. The danger of these wastes lies almost entirely in the **alpha** and **beta particles** and **gamma rays** they give off. These can damage human tissue. Alpha particles are stopped by a few centimetres of air, beta particles by a thin sheet of metal or a few metres of air but gamma rays (being similar to X-rays) require several centimetres of lead shielding to absorb them. The volumes of solid radioactive wastes from fuel manufacturing, power generation and reprocessing resulting from a 1 MW nuclear power plant operating for one year are: low level (untreated), 2 000 m³; intermediate (after treatment), 100 m³; and high level, 2 m³. Thus the bulk of the waste is low level.

(a) *Low level wastes*. The production of these wastes starts at the mining stage. At uranium mines the tailings often contain a greater combined bulk of radioactive elements than is shipped out in the processed ore! These mill

Half life

The time it takes for half the atoms of a radioactive isotope and their radiation to decay away. Half lives vary from milliseconds to several million years.

High level waste disposal

This is expected to take a multibarrier form: the waste to be placed in metal drums, encased in concrete and stacked in a concrete vault constructed at depth in impermeable rocks or rocks where any leakage will only percolate into deep sedimentary basins. The vault would be backfilled with clay. Suggested storage sites are evaporites in Germany, granite in Sweden and deeply buried volcanic rocks by the Sellafield nuclear installation on the Irish Sea coast of northern England, an excellent site for processing domestic and imported wastes but geologically a dubious choice. After much havering the U.S.A. has chosen **Yucca Mountain**, Nevada—a site with a thick succession of poorly permeable tuffs. This is not too badly placed for the seven operating reactors in Arizona and California but is distant by anything from 1 500 to over 3 000 km from the rest, several hundred kilometres from the nearest port and not expected to be in use until 2010! It is proposed that the violently radioactive spent fuel rods, now building up in increasing numbers at nuclear power plants, will be set in cylindrical storage cavities in the floors of galleries in the repository. How will such rods be transported past the homes of protesting owners—perhaps an even more intractable problem?!

tailings are finely ground and readily leachable. Normally they are accumulated at or near the mine, then stabilized with earth cover and vegetation. These sites must be so chosen and, if necessary, sealed that percolating rain water cannot enter groundwater systems which may be exploited for human or animal consumption. Besides that produced by nuclear power stations and their ancillary activities large quantities of fairly low level waste comes from hospitals, research laboratories and various nuclear industries. These wastes are usually incinerated to reduce their bulk, and buried in specially sited trenches with a metre of soil cover. The surrounding environment is regularly monitored to detect any form of contamination. More radioactive low level wastes have been incorporated into concrete in sealed drums and dumped at internationally agreed sites in deep ocean waters. Other low level wastes at power stations and nuclear fuel processing plants may undergo treatment that then allows them to be discharged into rivers or seas provided their radioactivity can be kept down to acceptable levels.

(b) *Intermediate level wastes* are, in general, at present being stored in special containers at nuclear plants awaiting disposal, some perhaps to go with the high level wastes, others to be treated to remove isotopes with long half lives and then disposed of as low level wastes.

(c) *High level wastes*. It must be emphasized that, although these wastes carry about 95% of the radioactivity of nuclear power station waste, they only form about 0.1% of the volume. To reduce their radioactivity to about one ten thousandth of the original level, they must be isolated for approximately 10000 years. Most are still being stored in stainless steel tanks at the plants where they originated awaiting a solution to their disposal problem. Possible disposal sites are now being explored in a number of countries because the country that solves this problem could find the solution to be a money spinner if it is prepared to deal with other people's wastes!

26.4 Further reading

Craig J.R., Vaughan D.J. and Skinner B.J. (1996) *Resources of the Earth Origin, Use and Environmental Impact.* Prentice-Hall, Upper Saddle River. Chapters 4 and 6. Very readable coverage of the use of nuclear power, different reactor types and the problems of waste disposal.

Patterson W.C. (1986) *Nuclear Power.* Harmondsworth: Penguin. A comprehensive and informative discussion with a good bibliography.

27: Nature's Hearth—Geothermal Energy

27.1 Introduction

The earth is slowly losing heat; the surface heat flow is about $60\,mW\,m^{-2}$, which is a very low figure. This heat is derived from the residual heat of the lower crust, mantle and core and from the decay of radioactive isotopes concentrated in the upper crust. This quantity of heat is just about sufficient for low temperature applications such as **space heating**, but the employment of geothermal heat is much more attractive in areas of high heat flow, i.e. those with high geothermal gradients. These gradients vary from 15°C to 75°C km^{-1} and average 25°C km^{-1}. The energy available in favourable locations is used for **heating purposes** or for **generating electricity**. These applications may be very important in particular places, e.g. Iceland, but currently they contribute less than 0.2% of total world energy consumption. In the coming decades this figure might rise to 1%.

> ■ The earth is steadily losing heat, but harnessing the heat in the upper crust is such a difficult engineering problem that this heat source is not likely to be a significant energy provider in the coming decades.
>
> ■ The three types of geothermal system are considered briefly in this chapter.

27.2 Geothermal systems

These are of three types: geothermal fields, geothermal aquifers and geothermal hot dry rock systems. At the moment the first named is by far and away the most important.

27.2.1 Geothermal fields

In geothermal fields, ground water seeps downward and is heated to boiling point or above due to the presence of hot intrusive magmatic bodies close to the surface [see Fig. 6.4(b)]. The heated water or steam may re-emerge as hot springs, geysers or fumeroles. If significant quantities of fluid collect in structures analogous to oil traps — a permeable structure overlain by an aquiclude — then these pools can be tapped by drill holes and the hot water or steam utilized. Low temperature waters ($\leq 85°C$) are used for space heating in homes, greenhouses and factories, the entire city of Reykjavik, Iceland being heated with geothermal hot water. High temperature waters at, say, 250°C and high pressures, 'flash' to steam as they encounter atmospheric pressures and are piped away to drive turbines for electricity generation. The cold water recharge of these systems [see Fig. 6.4(b)] is usually sufficient to sustain them and keep the 'kettle' boiling! Suitable geothermal fields are limited to regions with plate boundary or intraplate volcanism and geothermal electricity generation plants are mainly concentrated in California, U.S.A.; Iceland; northern Italy; Japan; Kenya; North Island, New Zealand and the Philippines.

27.2.2 Geothermal aquifers

This geothermal system is usually **developed in sedimentary basins** with higher than normal thermal gradient where high porosity and high permeability aquifers underlie thick formations of low thermal conductivity, e.g. mudrocks, which act as insulators to the rising heat flow. Temperatures in the aquifers may then reach about 65–100°C at a depth of 2–3 km or so and the hot, usually saline, water can be pumped to the surface, passed through a heat exchanger and then re-injected into the aquifer about 1 km away to maintain water pressure in the aquifer (Fig. 27.1). The secondary (heated) water is mainly used for **space heating**. The heat flow in many of these systems is too low to sustain the system for more than a few decades and this resource type is thus **often non-renewable**. There are numerous space heating projects in the Paris Basin, France; Hungary; Iceland; Italy and the U.S.A.

27.2.3 Geothermal hot dry rock systems, known as H.D.R.

H.D.R. has been shown to work on the research scale but it appears to be too expensive and risky to be commercially viable at present. The principle

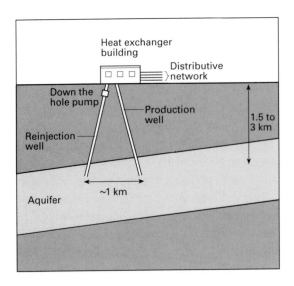

Fig. 27.1 Exploitation of a geothermal aquifer lying between two low thermal conductivity aquicludes.

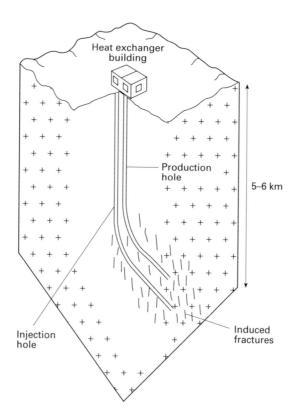

Fig. 27.2 Diagrammatic representation of a hot dry rock (H.D.R.) geothermal energy system.

is simple. Hot dry rocks in areas of high heat flow are drilled to depths of 5–6 km to exploit rock temperatures of 150–250°C. Most experiments so far have been in **high heat flow granites**. These are granites with high concentrations of radioactive elements—K, Th, U. Paired drill holes are employed, one to inject water, the other, about 250 m away, to pump the heated water out (Fig. 27.2). It is important to choose granites with good fracture systems so that the water can travel through the hot rock easily, but the permeability has to be enhanced by hydraulic fracturing (see Section 6.4) or using explosives to increase the density of fractures and so increase circulation rates. Experimental programmes have been carried out in France; Germany; Japan; New Mexico, U.S.A.; Sweden and the U.K.

27.3 Further reading

Fridleifsson I.B. and Freston D.H. (1994) Geothermal Energy Research and Development. *Geothermics*, **23**, 175–214. A comprehensive, up-to-date picture of geothermal energy research and development.

Patterson W.C. (1990) *The Energy Alternative: Changing the Way the World Works*. Boxtree, London. An excellent text for the general reader.

Pine R.J. and Green A.S.P. (1996) Geothermal Energy—a Review of Current and Potential Activity and Research. *Minerals Industry International*. No. 1030, May, 15–22. A short introduction to geothermal power generation followed by a useful summary of the sate of play with regard to H.D.Rs.

28: The Metamorphism of Ore Deposits

28.1 Introduction

It is all too often forgotten that many ore deposits occur in metamorphosed host rocks. Among these deposits, the substantial proportion that are syndepositional or diagenetic in age must have been involved in the metamorphic episode(s) and consequently their texture and mineralogy may be modified considerably, often to the advantage of mineral separation processes. In addition, **parts of the orebody may be mobilized** and epigenetic features, such as cross-cutting veins, may be imposed on a syngenetic deposit. Ores in metamorphic rocks may be **metamorphic** in the sense that their economic minerals have been concentrated largely by metamorphic processes; these include the pyrometasomatic (skarn) deposits and some deposits of lateral segregation origin. On the other hand, they may be **metamorphosed**, in the sense that they predate the metamorphism and have therefore been affected by a considerable change in pressure–temperature conditions, which may have **modified their texture, mineralogy, grade, shape and size** — structural deformation often accompanies metamorphism. This chapter will be concerned entirely with metamorphosed ores.

It is important from the economic as well as the academic viewpoint to be able to recognize when an ore has been metamorphosed. Strongly metamorphosed ores may develop many similar features to high temperature epigenetic ores for which they may be mistaken. This can be very important to the exploration geologist, for in the case of an epigenetic ore, particularly one localized by an obvious structural control, the search for further orebodies may well be concentrated on a search for similar structures anywhere in the stratigraphical column, whereas repetitions of a metamorphosed syngenetic ore should be first sought for along the same or similar stratigraphical horizons. Notable examples of orebodies long thought to be high temperature epigenetic deposits, but now considered to be metamorphosed syndepositional ores, are those of Broken Hill, New South Wales and the Horne Orebody, Noranda, Quebec. In both areas recent exploration having a stratigraphical basis has been successful in locating further mineralization.

In this chapter we learn of:
- the importance of recognizing whether we are exploring epigenetic or metamorphosed syngenetic ores,
- some of the features which metamorphosed ores may display,
- the effects that deformation may have on syngenetic orebodies.

Certain types of deposit, by virtue of their structural level or development in geological environment, are rarely, if ever, seen in a metamorphosed state. Some examples are porphyry copper, molybdenum and tin deposits, ores of the carbonatite association and placer deposits. On the other hand, certain deposits, such as the volcanic-associated massive sulphide class, have generally suffered some degree or other of metamorphism such that their original textures are commonly much modified.

Three principal types of metamorphism are normally recognized on the basis of field occurrence. These are usually referred to as contact, dynamic and regional metamorphism. **Contact metamorphic rocks** crop out at or near the contacts of igneous intrusions and in some cases the degree of metamorphic change can be seen to increase as the contact is approached. This suggests that the main agent of metamorphism in these rocks is the heat supplied by the intrusion. As a result, this type is sometimes referred to as thermal metamorphism. It may take the form of an entirely static heating of the host rocks without the development of any secondary structures, such as schistosity or foliation, though these are developed in some metamorphic aureoles.

Dynamically metamorphosed rocks are typically developed in narrow zones, such as major faults and thrusts, where particularly strong deformation has occurred. This type of metamorphism is accompanied by high strain rates and a range of PT conditions depending on the structural level and other factors, such as proximity of igneous intrusions and penetration by hot hydrothermal solutions of deep seated origin, which may cause what is sometimes described as hydrothermal metamorphism. Epigenetic ores, developed in dilatant zones along faults, often show signs of dynamic effects (brecciation, plastic flowage etc.) owing to fault movements during and after mineralization.

Regionally metamorphosed rocks occur over large tracts of the earth's surface. They are not necessarily associated with either igneous intrusions or thrust belts, but these features may be present. Research has shown that regionally metamorphosed rocks generally suffered metamorphism at about the time they were intensely deformed. Consequently they contain characteristic structures, such as **cleavage, schistosity, foliation** or **lineation**, which can be seen on both the macroscopic and microscopic scales, producing a distinctive fabric in rocks so affected. Whereas regional metamorphism is normally associated with major orogenic activity, thick basinal sequences not involved in orogenesis may show incipient regional metamorphism; this may be referred to as **burial metamorphism**. Metamorphism developed at mid-oceanic ridges, although it may be regional in extent, is referred to as ocean-floor metamorphism.

How can we recognize when ores have been metamorphosed? One way is to compare their general behaviour with that of carbonate and silicate rocks that have undergone metamorphism. In contact and regionally metamorphosed areas these rocks generally show:

1 the development of metamorphic textures;
2 a change of grain size—usually an increase;
3 the progressive development of new minerals.

In addition, as noted above, regionally metamorphosed rocks usually develop certain secondary structures. Let us examine each of these effects in turn.

28.2 Development of metamorphic textures

All three types of deformation, elastic, plastic and brittle, are important— **elastic deformation** largely because it raises the internal free energy of the grains and renders them more susceptible to recrystallization and grain growth. **Plastic deformation** occurs by primary (translation) gliding or by secondary (twin) gliding. **Brittle deformation** takes the from of rupturing or shearing and both generally follow lines of weakness in grains, such as cleavages, twin planes etc. At a given degree of deformation the stronger minerals, such as pyrite and arsenopyrite, may fail by rupturing whilst the softer minerals, such as galena and the sulphosalts, may flow plastically. These processes can give rise to grain elongation and the development of schistose textures.

28.3 Increase in grain size

If the temperature of a grain aggregate rises above a threshold temperature, diffusion along grain boundaries will occur leading to **grain growth**, so that the higher the metamorphic temperature the coarser grained the aggregate tends to be. But grain growth does not go on indefinitely, especially if the ores have reached the grain size of coarse gneisses—of the order of 5–10 mm depending on a variety of factors—and it will often result in a polyminerallic aggregate in which different minerals reach different average grain sizes.

28.4 Development of new minerals

Becuase of the considerable ranges of pressure and temperature over which sulphide minerals are stable, we do not find the progressive development of new minerals with increasing grade of metamorphism that are present in many classes of silicate rocks. Banded iron formations, on the other hand, do show the abundant development of new mineral phases, as do many manganese deposits. Sulphide assemblages by comparison only show minor mineralogical adjustments.

In contact metamorphism, the main effects so far studied are those seen next to basic dykes cutting sulphide orebodies. These include the following reactions and changes:

1 pyrite + chalcocite → chalcopyrite + bornite + S;
2 pyrite + chalcocite + enargite → chalcopyrite + bornite + tennantite + 4S;
3 marginal alteration of pyrite to produce pyrrhotite + magnetite.

The principal mineralogical change reported from regionally metamorphosed sulphide ores is an increase in the pyrrhotite to pyrite ratio with the appearance of magnetite at higher grades.

28.5 Some effects on orebodies and implications for exploitation and exploration

Many massive sulphide deposits occur in regionally metamorphosed terrane of all grades of metamorphism; some are very severely deformed and a massive sulphide deposit after undergoing even mild metamorphism is often radically different in many ways from its original form, such that metamorphic changes can directly influence the exploration for, and the commercial development of, these orebodies. Metamorphism of the host rocks can so change them that it is difficult for the geologist to **recognize and trace favourable rock environments**. Because many of these orebodies are associated with acid volcanics, it is imperative that the distinction be made between metamorphosed silicic volcanics and metaquartzites, meta-arkoses or granite-gneisses. Similarly, the recognition of a metamorphosed cherty tuff layer can be of great assistance as a marker horizon, in structural interpretation and because it may represent a former exhalite with associated massive sulphide deposits [see Section 16.3.4(d)]. The explorationist has to **'see through' metamorphosed suites of rocks** to reconstruct the original stratigraphy and environment in the search for favourable areas and horizons. Metamorphism and deformation of a volcanic-associated massive sulphide orebody may also give rise to a different mineralogy, grain size and shape of the deposit, which can affect its geophysical and even its geochemical response.

Such deposits were probably originally roughly circular or oval in plan and lenticular in cross section. Deformation may change them to blade-like, rod-shaped or amoeba-like orebodies, which may even be wrapped up and stood on end, and then be mistaken for replacement pipes, e.g. the Horne Orebody of Noranda, Quebec! A good example of what can happen to what was possibly originally one single massive sulphide body is shown in Fig. 28.1. When considerable shearing is involved in the deformation, a flattening effect by transposition along shear or schistosity planes may transpose the alteration pipe and the stockwork beneath the massive sulphide lens into a position nearly parallel to the schistosity and the lens (Fig. 28.2). In extreme cases, the sulphide lens may be sheared off the stockwork zone, separating the two.

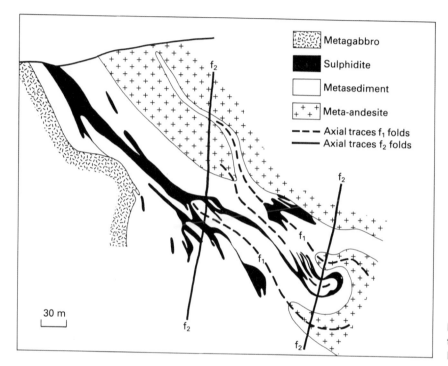

Metagabbro

Sulphidite

Metasediment

Meta-andesite

– – – Axial traces f_1 folds
——— Axial traces f_2 folds

30 m

Fig. 28.1 Cross section through the Chisel Lake ore zone, Manitoba.

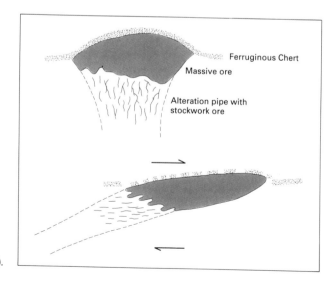

Ferruginous Chert
Massive ore

Alteration pipe with stockwork ore

Fig. 28.2 Schema showing the effect of shearing (below) on an undeformed massive suphide deposit (above).

28.6 Further reading

McClay K.R. (1983) Deformation of Stratiform Lead–Zinc Deposits. In Sangster D.F. (ed.), *Short Course in Sediment-Hosted Stratiform Lead–Zinc Deposits*, 283–309. Mineral. Assoc. Can., Toronto. Contains some very good illustrations of the deformation styles in four stratiform orebodies.

Part 3
Mineralization in Time and Space

And his feet shall stand in that day upon the mount of Olives, which is before Jerusalem on the east, and the mount of Olives shall cleave in the midst thereof toward the east and toward the west, and there shall be a great valley; and half the mountain shall remove toward the north, and half of it toward the south.

The Holy Bible, Authorized King James Version,
Zechariah 14: 4

Surely there is a vein for the silver and a place for gold where they find it. Iron is taken out of the earth, and brass is molten out of the stone.

The Holy Bible, Authorized King James Version,
Job 28, 1–2

There rolls the deep where grew the tree.
   O Earth, what changes hast thou seen!
   There where the long street roars, hath been
The stillness of the central sea.
The hills are shadows, and they flow
   From form to form, and nothing stands;
   They melt like mist, the solid lands,
Like clouds they shape themselves and go.

From *In Memoriam CXXII*. This was the reaction of Alfred, Lord Tennyson (1809–1892, Poet Laureate) to the nineteenth century discoveries of the aeons of geological time and the enormous changes occurring in the Earth's crust as so vividly described in Charles Lyell's *Principles of Geology* (1830–1833).

29: The Global Distribution of Ore Deposits: Metallogenic Provinces and Epochs, Plate Tectonic Controls

29.1 Introduction

It has long been recognized that specific regions of the world possess a notable concentration of deposits of a certain metal or metals and these regions are known as **metallogenic provinces**. Such provinces can be delineated by reference to **a single metal** (Fig. 29.1) **or to several metals or metal associations**. In the latter case, the metallogenic province may show a **zonal distribution** of the various metallic deposits. The recognition of metallogenic provinces has usually been by reference to epigenetic hydrothermal deposits, but there is no reason why the concept should not be used to describe the regional development of other types of deposit provided they show a geochemical similarity. For example, the volcanic-exhalative antimony–tungsten–mercury deposits in the Lower Palaeozoic inliers of the eastern Alps form a metallogenic province stretching from eastern Switzerland through Austria to the Hungarian border.

Within a metallogenic province there may have been periods of time during which the deposition of a metal or a certain group of metals was most pronounced. These periods are called **metallogenic epochs**. Some epochs are close in time to orogenic maxima, others may occur later.

Tin deposits are an excellent example of an element restricted from the economic point of view to a few metallogenic provinces and those around the Atlantic Ocean are shown in Fig. 29.1. Even more striking is the fact that **most tin mineralization is post-Precambrian** and confined to certain well marked epochs. Equally striking is the strong association of these deposits with **post-tectonic granites**. Among tin deposits of the whole world, 63.1% are associated with Mesozoic granites, 18.1% with Hercynian (late Palaeozoic) granites, 6.6% with Caledonian (mid Palaeozoic) granites and 3.3% with Precambrian granites. Further examples of tin belts are shown in Figs 29.2 and 29.3.

The recognition of the development of tin provinces is of fundamental importance to the mineral exploration geologist searching for this metal. It is probable that any further discoveries of important deposits of tin will be made within these provinces or their continuations and those of central

- The study of metallogenic provinces and epochs presents us with regional ore deposit distributions that suggest an underlying tectonic control.
- Plate tectonic theory provides a very compelling mechanism with which to explain these distributions.
- In this chapter we look at the principal plate tectonic domains and their characteristic mineralization.

Fig. 29.1 Tin belts on continents around the Atlantic Ocean. Stippled areas indicate areas that have been worked for tin.

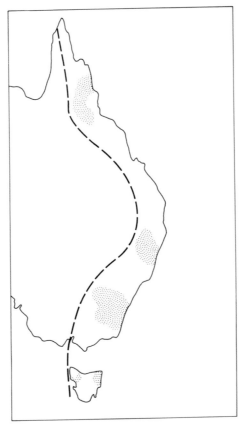

Fig. 29.2 The Palaeozoic tin belt of eastern Australia. The principal fields are shown as stippled areas.

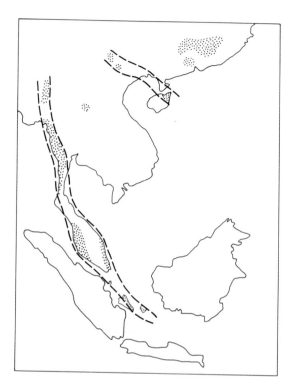

Fig. 29.3 Tin belts and fields of
south-eastern Asia.

Asia. Continental drift reconstructions suggest that a continuation of the
eastern Australian tin belt may be present in Antarctica.

Many other metals and minerals, e.g. nickel sulphides and the titanium–anorthosite association, are concentrated in space and time and this
has obvious implications for mineral exploration programmes.

29.2 Mineral deposits in the principal plate tectonic regimes

A vast deal of information has now been published on this subject
including books by Mitchell and Garson and by Sawkins, but space will
allow only a bare outline to be given here. For this purpose, the six tectonic
settings discussed by Mitchell and Reading will be adopted; these are:
1   interior basins, intracontinental rifts and aulacogens;
2   oceanic basins and rises;
3   passive continental margins;
4   subduction-related settings;
5   strike-slip settings;
6   collision-related settings.

29.2.1 Continental interior basins, intracontinental rifts and aulacogens

There are two types of sedimentary basins within continental interiors: **large basins** often over 1000 km across and relatively narrow, fault-bounded **rift valleys**. The large basins may contain entirely continental sediments, much of which may have been deposited in large lakes like those of the Chad Basin in Africa. Some interior basins, e.g. Hudson Bay, have been inundated by the sea and contain mainly marine sediments. These can carry important mineral deposits, an example being the Permian Zechstein Sea (see Fig. 16.1) of northern Europe. Here the porous and permeable aeolian sandstones of the Lower Permian Rotliegendes, which provide the important **gas reservoirs** of north-western Germany, Holland and the southern North Sea, are succeeded by the marine Kupferschiefer, the world's best known **copper shales**, which were exploited in Germany for about one thousand years and whose mining in Poland makes that country Europe's leading copper producer.

Less than 100 m above the Kupferschiefer, the famous Zechstein cycles, with their valuable **sulphate**, **sodium** and **potassium salts**, commence. Other evaporites in this setting are those of the Devonian Elk Point Basin of western Canada and the Silurian evaporites of the Michigan Basin. These and similar evaporites are important for potash and soda production. In some basins, e.g. the Paris Basin (France) and the Triassic basins of the British Isles, important **gypsum** deposits occur. Similar basins host **sandstone–uranium deposits**, e.g. the Wyoming and Colorado Basins (see Section 18.3.1 and Fig. 18.3).

Interior basins may also be important **petroleum** and **coal** producers. The Williston Basin underlying parts of Montana, the Dakotas and southern Saskatchewan contains valuable oil and gas accumulations located in very gentle anticlines within the basin. Many coalfields are developed in intracontinental basins such as many of the interior coalfields of the U.S.A. and southern Africa.

The first **continental basins** were developed as soon as sufficient craton was present — about 3000 Ma ago in southern Africa, where the Dominion Reef and Witwatersrand Supergroups with important **gold and uranium mineralization** were laid down. Similar Proterozoic basins contain uraniferous conglomerates at Elliot Lake in Ontario and the rich unconformity-associated uranium deposits of Athabasca, Canada and Alligator River, Australia. Following the development of stable lithospheric plates, **banded iron formation** was deposited synchronously over very wide areas; this took place possibly in continental basins and certainly on continental shelves.

Rifts and aulacogens (failed rifts) are initiated by the doming of continental areas over mantle plumes (hot spots). Plumes may cause melting of the continental crust forming **A-type granite intrusions** with important **tin mineralization** as in Nigeria, Rondônia, Brazil and the Sudan. Other mineralization that may be associated with hot spot activity includes the **chromium–P.G.M.** of the Bushveld Complex and the Great Dyke, southern Africa, the Palabora Complex, R.S.A. with its **copper-bearing carbonatites**, **phosphate** rock and **vermiculite** and the alkaline complexes and carbonatites of the Kola Peninsula, R.F. with their important production of phosphate rock, **nepheline**, **iron**, vermiculite and various by-products.

The initial stage of rift valley development is often marked by alkaline igneous activity with the development of carbonate lavas and intrusives and occasionally **kimberlites**. Erosion of the lavas may lead to the formation of **soda deposits**, e.g. Lakes Natron and Magadi in East Africa, and the intrusive carbonatites may carry a number of metals of economic interest (**Fe, Nb, Zr and R.E.E.**) as well as **fluorite**, **baryte** and **strontianite** whilst some provide a **source of lime**. Carbonatites are highly susceptible to weathering and therefore may have residual, eluvial and alluvial placers associated with them. Some of the kimberlites are diamondiferous.

The sediments and volcanics within rift valleys and aulacogens may host important orebodies. The Central **African Copperbelt** rocks (Zambia and Zaïre) were probably formed within a rift or a continental basin as were those hosting the important Cu–Pb–Zn Mount Isa and Hilton orebodies of Queensland and the 2000 Mt Cu–U–Au deposit of Olympic Dam, South Australia. Recently formed **aulacogens may contain** important **oilfields** (Ras Morgan, Egypt, North Sea) and many are important for evaporites. A number of important Proterozoic aulacogens have been identified in North America (Fig. 29.4). Those running perpendicular to the western margin of the continent carry great thicknesses of rocks of the Belt Series. The northernmost trough contains the Coppermine River Group consisting of more than 3 km of basalt flows with native copper mineralization overlain by greater than 4 km of sediments with evaporites near the top. Epigenetic deposits associated with rifting include the Great Bear Lake Uranium Field. To the south is the Athapuscow Aulacogen with alkali syenite and granite and associated **beryllium–R.E.E.** deposits, red beds, **sedimentary uranium deposits** and evidence of the former presence of evaporites. Further south is the Alberta Rift which passes into British Columbia and contains about 11 km of late Precambrian sediments in which important stratiform and epigenetic **lead–zinc** mineralization occurs, including the famous Sullivan Mine at Kimberly, British Columbia. Phosphorite deposits are also present. To the east, the Keweenawan Aulacogen, which is the same age as the Coppermine Aulacogen, also carries a considerable thickness of basalt and clastic sediments, and again **native copper mineralization** is present.

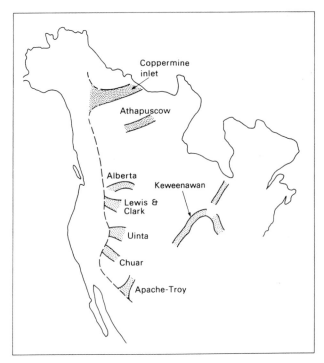

Fig. 29.4 Some of the Proterozoic aulacogens of North America.

29.2.2 Ocean basins and rises

There is strong evidence that a number of mineral deposit types are formed during the birth of new oceanic crust along **mid-ocean ridges**, **back-arc spreading ridges** and the **median zones of embryonic oceans** like the Red Sea. The possible mode of development of oceanic crust along the median zone of the Red Sea is shown in Fig. 29.5. As this new crustal material moves away from the median zone, layer 1 of the oceanic crust, in the form of pelagic sediment, is added to it. There is strong evidence that many **Cyprus-type sulphide deposits** (see Table 16.2) are formed during this process of crustal birth. If slices of the oceanic crust are thrust into **mélanges** at convergent junctures or preserved in some other way, then sections similar to that shown in Fig. 29.5 may be expected. Similar situations to the oceanic spreading ridges are found in back-arc basins and indeed many workers hold that the Troodos Massif of Cyprus was developed in such a milieu. In this massif, non-economic **nickel–copper** mineralization and economic **chromite deposits** occur in the basic plutonic rocks.

Hydrothermal mineralization with the development of epigenetic copper, zinc, silver, tin, mercury, fluorite and baryte has been reported from a number of oceanic ridges but the most striking is the exhalative

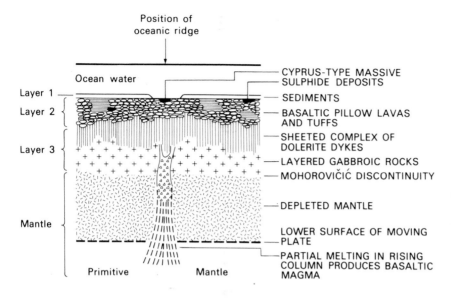

Position of oceanic ridge

Ocean water

Layer 1

Layer 2

Layer 3

Mantle

Primitive   Mantle

CYPRUS-TYPE MASSIVE SULPHIDE DEPOSITS

SEDIMENTS

BASALTIC PILLOW LAVAS AND TUFFS

SHEETED COMPLEX OF DOLERITE DYKES

LAYERED GABBROIC ROCKS

MOHOROVIČIĆ DISCONTINUITY

DEPLETED MANTLE

LOWER SURFACE OF MOVING PLATE

PARTIAL MELTING IN RISING COLUMN PRODUCES BASALTIC MAGMA

Fig. 29.5 Schematic representation of the development of oceanic crust along a spreading axis. The crustal layering and the possible locations of Cyprus-type massive sulphide deposits are shown.

copper and zinc massive sulphide mineralization associated with **black smokers**. Ten deposits estimated to contain 3–5 Mt of sulphide have been found along an 8 km segment of the eastern axial valley at Southern Explorer Ridge, about 150 km west of British Columbia. Gold values up to 1.4 ppm and silver up to 640 ppm have been found in these deposits. Massive sulphide mineralization is also known from modern back-arc spreading centres, e.g. the Lau Basin, east of the Fiji Islands.

29.2.3 Passive continental margins

As ocean spreading gradually forces two continents apart, both sides of the original rift become passive margins with the development of a continental shelf, bounded oceanwards by a slope and landwards by a shoreline or an epicontinental sea. The type of sediment forming on shelves today depends on latitude and climate, on the facing of a shelf relative to the major wind belts and on tidal range. In the past, shelves in low latitudes were often covered with substantial platform carbonate successions, and these can be hosts for **base metal deposits** of both epigenetic and syngenetic nature (so-called Mississippi Valley-type, Irish-type and Alpine-type deposits).

A number of small, stratiform, **sandstone-hosted copper** and **lead–zinc** deposits occur in the Cretaceous along the western edge of Africa from Nigeria to Namibia, and further north continue inland to the margin of the Ahaggar Massif. Similar deposits are known along the northern margin of the High Atlas Mountains on the margin of the Morocco Rift, and older ones, which are related to Infracambrian rifting and the development of a Palaeotethys Ocean, are found along the northern edge of Gondwanaland. The **lead ores** of Laisvall in Sweden and Largentière and Les Malines in France appear to have been developed in a similar tectonic setting.

Many of the world's most important **sedimentary manganese** deposits occur just above an unconformity and were formed under shallow marine conditions on shelf areas, e.g. Nikopol, R.F.; Chiaturi, Georgia and Groote Eylandt in Australia. Passive continental margins that have suffered marine transgressions are also important for **phosphorite** deposits. Palaeomagnetic research has revealed that the majority of phosphorites formed at low latitudes although a few appear to have been deposited 30–50° north or south of the Equator. Faunal evidence suggests that many Cambrian phosphorite deposits were formed in an east–west seaway extending from Australia into Asia and perhaps into Europe, in a similar manner to the later Tethyan seaway. Both these seaways, and the east–west seaway in which the Cretaceous–Eocene phosphorites were laid down, occurred in low to intermediate latitudes. The fact that a coastline is in low latitudes will not necessarily produce the strong upwelling required for phosphogenesis (see Section 21.14.4). A narrow east–west seaway at a low latitude will probably be marked by considerable upwelling because of the

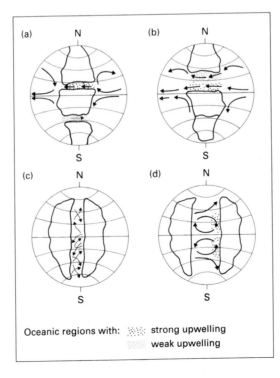

Fig. 29.6 Schematic representation of the probable current patterns and areas of upwelling consequent upon the rifting and drifting apart of continents. (a) A juvenile stage in which phosphogenesis occurs in a low latitude seaway but not in the equivalent high latitude seaway. In the mature stage (b) phosphogenesis is limited to the coast, which has remained in low latitudes. With a longitudinal seaway in the juvenile stage (c) a network of small currents interfere with each other, but in the mature stage (d) large scale oceanic currents can produce upwelling.

Oceanic regions with: ▨ strong upwelling
░ weak upwelling

strong westerly directed flow through it (Fig. 29.6a). In a seaway at higher latitudes, any upwelling will be much weaker and sea water temperatures lower, resulting in a much lower degree of biogenic activity. A newly formed, narrow, north–south seaway will probably have a complex pattern of interfering currents with little or no associated upwelling (Fig. 29.6c). Only when a longitudinally aligned ocean is 2000 km or so wide will ocean-wide currents sweep along continental coasts creating upwelling gyres at points of deflection along the shelf (see Fig. 21.5) and giving rise to extensive and significant **phosphogenesis** (Fig. 29.6d). Thus, as indicated in Fig. 29.6, strong upwelling and associated phosphogenesis can occur in a latitudinally aligned seaway during its juvenile and mature stages but only at a mature stage in a longitudinally directed seaway. The Tertiary phosphorites of the Atlantic Coastal Plain of the U.S.A. appear to have formed under just these conditions. It should be noted that they are a source of **by-product uranium**.

Many workers now favour a continental shelf environment for the deposition of the Proterozoic **Superior-type B.I.F.** and a comprehensive assessment of the evidence from the Hamersley Group, which contains the most extensive accumulation of sedimentary iron deposits known, makes a compelling case for a mid to outer shelf environment.

Along the trailing edges of many continents, particularly those where trade winds blow in obliquely to the shoreline, combining with ocean currents to give rise to marked longshore drift, there are important **beach placer deposits**. Among these we can list the diamond placers of the Namibian coast, the rutile–zircon–monazite–ilmenite deposits of the eastern and western coasts of Australia, and the similar deposits of Florida and the eastern coasts of Africa and South America.

Like beach placers, **petroleum-forming environments** will also depend on local factors such as the amount and nature of sediment reaching the coast, the climate and oceanographical controls. Special situations may develop off large river mouths with huge deltas being built of clastic débris. The Niger and Mississippi deltas are the best examples both containing large numbers of small fields.

29.2.4 Subduction-related settings

In these settings, lithosphere is subducted beneath a volcanic island arc or a continental margin arc on the overriding plate. In continental margin arcs the volcanic arc is situated landward of the oceanic crust–continental crust boundary as in the Andes. Behind volcanic arcs there is the back-arc area usually referred to, when behind island arcs, as the back-arc basin or marginal basin (Fig. 29.7) or, when behind continental arcs, as a foreland basin or molasse trough. It is convenient to divide the mineral deposits in island arcs according to whether they were formed outside the arc-trench environment and transported to it by plate motion (allochthonous deposits), or whether they originated within the arc (autochthonous deposits).

(a) *Allochthonous deposits.* Rocks and mineral deposits formed during the development of oceanic crust may by various trains of circumstances arrive at the trench at the top of a subduction zone (Fig. 29.7). This material is largely subducted, when some of it may be recycled, whilst **some is thrust into mélanges**. There is, however, only one region so far discovered with what appears to be oceanic ridge-type massive sulphide orebodies thrust into a mélange, and that is north-western California, where we find the Island Mountain deposit and some smaller occurrences in the Franciscan mélange. Further north, in the Klamath Mountains of Oregon, the obducted Josephine Ophiolite contains the important Turner Albright deposit.

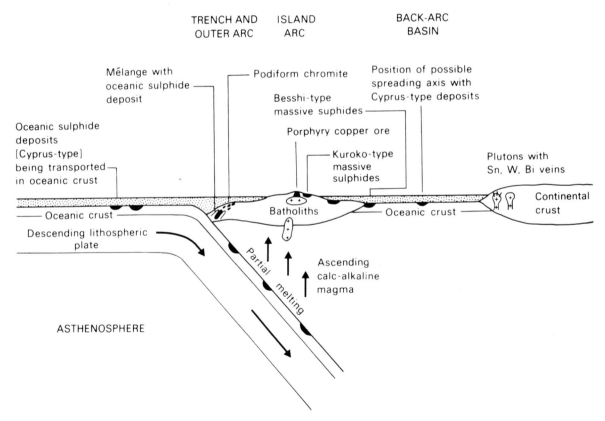

Fig. 29.7 Diagram showing the development and emplacement of some mineral deposits in an island arc and its adjacent regions.

Clearly, any other deposits formed in new oceanic crust at oceanic ridges may also eventually be mechanically incorporated into island arcs and the most likely victims will be the **chromite** deposits of **Alpine-type peridotites** and gabbros. Such podiform chromitites occur in the peridotites of the obducted, ophiolitic Papuan Ultramafic Belt. Small **platinum** metal deposits are known in some of the peridotites; they are only important as the source rocks for **placer deposits** of these metals. Economic deposits of podiform chromitite occur in present day arcs in Cuba, where they are found in dunite pods surrounded by peridotite, and on Luzon in the Philippines, again in dunite in a layered ultramafic complex. Numerous occurrences are present in ancient island arc successions and both recent and ancient settings clearly form prime exploration targets for such deposits. A recent development of great importance in this context, and for exploration programmes, is the conclusion of a number of workers that all major **podiform chromite deposits** were formed in marginal basins and not at mid-ocean ridges.

The plutonic rocks of ophiolite suites have not as yet yielded much in the way of magmatic sulphides. Uneconomic pyrrhotite–pentlandite accumulations are known in the Troodos Massif, Cyprus but the only deposits of this type being exploited today are the nickel–platinum orebodies of the Acoje Mine in the Philippines.

(b) *Autochthonous deposits.* In considering these deposits it is convenient to divide the geosynclinal development into three stages: initial or tholeiitic, main calc-alkaline stage and waning calc-alkaline stage. The main calc-alkaline stage is the principal period of arc development with much plutonic activity and a transition from basaltic to andesitic and dacitic volcanics. Much more rock is exposed to subaerial erosion, thus increasing the volume of sediment which reaches inter-island gaps, the arc-trench gap and the back-arc basin. **Massive sulphide orebodies**, usually yielding copper and zinc and known as **Besshi-type** (after the Japanese occurrences), occur in back-arc turbidite-volcanic (basalt to andesite) associations (Fig. 29.7). Workers investigating the **exhalative tungsten mineralization** in the Austrian Alps (see Section 16.4.2) have concluded that these deposits too were formed in a back-arc basin environment.

Intrusions hosting **porphyry copper deposits** belong to this stage. The deposits are most commonly important for copper, gold and silver; molybdenum is not a usual by-product. **Skarn deposits of tin**, **copper** and other metals may be present at igneous-carbonate country rock contacts. However, much exploration is now in progress in both plutonic and associated volcanic rocks for **disseminated gold deposits**, and the arcs of the south-western Pacific are proving to be well endowed with both porphyry and epithermal gold orebodies of this type.

During the waning stage of island arc formation, the felsic volcanism may be accompanied by the development of **massive sulphide orebodies** containing economic amounts of copper, zinc and lead commonly accompanied by gold–silver values and known as **Kuroko-type**.

The sediments of island arc regimes being rapidly deposited, laterally variable, often mixed with volcanic ash, which adversely affects their porosity and permeability, and complexly folded are not attractive environments for petroleum accumulations and few commercial fields have been found in them. Only in the less compressed zones, e.g. behind the volcanic arc in Sumatra, have significant petroleum reserves been discovered.

Continental margin arcs, like those of the Andes, are frequently regions of high relief exposing regularly arranged, regional metamorphic zones and monzonitic–granodioritic plutons and batholiths, as well as active volcanoes at the higher levels. The crust is of greater than normal thickness and although these are regions of plate convergence with its concomitant compression, many **graben structures** are present running parallel or obliquely to the arcs. Important magmatic activity and mineralization may be related to this rifting e.g. the world's most important **molybdenite deposits**, Climax and Urad-Henderson in Colorado which are related to the Rio Grande Rift System.

The **Andes are immensely rich in orebodies** of various types and metals, but on the whole these show a **plutonic–epigenetic affiliation** rather than a volcanic–syngenetic one which is to some extent a function of erosion. The Andes are characterized by linear belts of mineralization that coincide with the morphotectonic belts which run down the mountain chain. The Coastal Belt, mainly consisting of Precambrian metamorphic rock, is important for its **iron–apatite deposits** of skarn and other types. The Western Cordillera (made up mainly of Andean igneous rocks) is a copper province particularly important for **porphyry-type deposits** carrying **copper, molybdenum and gold**; and the Altiplano, a Cretaceous–Tertiary intermontane basin filled with molasse sediments, is part of a larger province of vein and replacement **Cu–Pb–Zn–Ag deposits** that extends along the Eastern Cordillera. This belt consists of Palaeozoic sedimentary rocks and Palaeozoic and Andean igneous rocks, with the **tin** (plus W–Ag–Bi) belt along its eastern margin.

Metallogenic variations are known from other parts of the continental margin arcs of the Americas. Indeed, the North American Cordillera is a vast storehouse of mineral treasures developed during a complicated tectonic history too involved for discussion here.

29.2.5 Strike-slip settings

Two different types of fault are included here: the transform fault and the long known tear or wrench fault. Strike-slip faults vary in size from plate boundary faults such as the San Andreas Fault of California and the Alpine Fault of New Zealand, through microplate boundaries and intraplate faults, such as those of Asia north of the Himalayas, down to small scale fractures with only a few metres offset. Whether these structures occur as single faults or as fault zones (and where complex patterns with some extension are present in the resulting sedimentary basins), they may be important in locating economic mineral deposits.

Well recognized transform faults in continental crust have little or no associated mineralization, although some potential is perhaps indicated by the location of the **Salton Sea Geothermal System** (which is a potential ore-forming fluid) within the San Andreas Fault Zone, but it must be noted that this system is not underlain by normal continental crust. Other possible examples of transform faults acting as structural controls of mineralization come from a study of the hypothetical continental continuations of transform faults. For example, many of the **diamond-bearing kimberlites** of West Africa lie along such lines and some onshore **base metal deposits** of the Red Sea region show a similar relationship.

Modern strike-slip basins occur in oceanic, continental and continental margin settings. They may also occur in back-arc basins. The classic onshore strike-slip basin is the Dead Sea with its important **salt** production, and the best known offshore basins are those of the Californian Continental Borderland, some of which host **commercial oil pools**. Ancient strike-slip basins are difficult to identify. However, two examples of economic importance are the Tertiary Bovey Basin of England with its economic deposits of **sedimentary kaolin** and minor lignite, and the late Mesozoic, **coal-bearing** lacustrine basins of north-eastern China.

The **38th Parallel Lineament** of the central U.S.A. and its relationship to a number of orefields in the Mississippi Valley and elsewhere is shown in Fig. 29.8. This lineament is a zone marked by wrench faults, lines of alkalic, gabbroic, ultramafic and kimberlitic intrusions, suggesting connexions into the mantle, and changes in stratigraphy across it. Gravity anomaly patterns suggest an 80 km, Precambrian, dextral movement, with a Phanerozoic offset of 8–10 km. **Important orefields occur at the intersections** of this lineament and other structures; for example, the Central Kentucky Orefield occurs where the West Hickman Fault Zone and the Cincinnati Arch cross it and the very important Southeast Missouri Field where it is crossed by the Sainte Genevieve Fault System and the Ozark Dome.

Many workers have commented on the close connexion between **gold deposits** in the Archaean Abitibi Greenstone Belt of Ontario–Quebec and regional fault zones and about 18% of the world's production of **antimony** comes from the Archaean Murchison Schist Belt—a greenstone belt in the Kaapvaal Craton, R.S.A. Here the antimony–gold mines lie along the

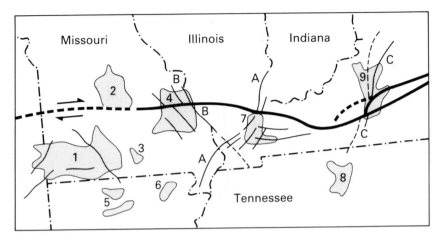

Fig. 29.8 Map showing about half the 38th Parallel Lineament, other faults (narrow lines) and some of the base metal–fluorite–baryte orefields of the Mississippi Valley. Key to orefields: 1, Tri-State; 2, Central Missouri; 3, Seymour; 4, South-east Missouri; 5, Northern Arkansas; 6, North-east Arkansas; 7, Illinois–Kentucky; 8, Central Tennessee; 9, Central Kentucky. Faults: A, New Madrid Fault Zone; B, Ste Genevieve Fault Zone; C, West Hickman Fault Zone.

35–40 km Antimony Line, which is considered to be a zone of shearing hosting most of the mineralization. Finally, we can glance at a small part of the important **lineament analysis** carried out in Australia by O'Driscoll which 'defined the Olympic Dam location as a priority drilling target'. The largest **copper deposits** of the Stuart Shelf–Adelaide Geosyncline region of South Australia lie on two alignments (Fig. 29.9). The giant among these deposits, **Olympic Dam, is close to the intersection of two lineaments** and a sinistral flexure in the N.N.W-trending gravity anomaly (Fig. 29.10).

29.2.6 Collision-related settings

Collision can occur between two active arc systems, between an arc and an oceanic island chain or between an arc and a microcontinent, but the most extreme tectonic effects are produced when a continent on the subducting plate meets either a continental margin or island arc on the overriding plate. Subduction of the formerly intervening oceanic crust can then lead firstly to the generation of **I-type granites** which might carry **porphyry copper, molybdenum** and/or **gold mineralization** (Fig. 29.11a) followed by **melting** in the heated continental slab and the production of a parallel belt of **S-type granites** (Fig. 29.11b). In many continental collision belts, highly differentiated granites of this suite are accompanied by **tin and tungsten** deposits of greisen and vein type. The tin and tungsten deposits of the Hercynides of Europe (Cornwall, Erzgebirge, Portugal etc.) are good

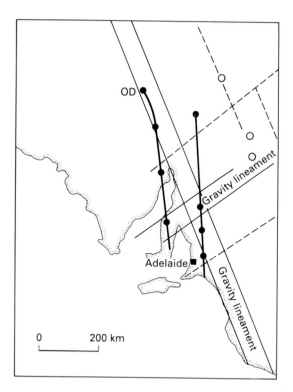

Fig. 29.9 Map showing the two alignments on which the largest of the copper deposits in this part of South Australia lie. The pecked lines represent geological lineaments; open circles are uranium deposits. OD, Olympic Dam.

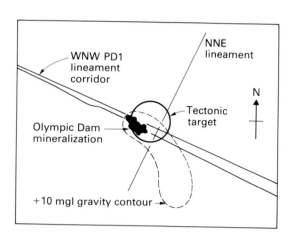

Fig. 29.10 Plan of the Olympic Dam orebody in relation to the west-north-west photolineament corridor PD1 and the north-north-west trending gravity anomaly, which has a sinistral flexure at its northern end, bringing it into line with the photolineament. The location of Olympic Dam is shown on Fig. 29.9.

examples. In south-eastern Asia, parallel belts of porphyry copper and tin mineralization have arisen in this manner.

Uranium deposits, particularly of vein type, may be associated with S-type granites as in the Hercynides and Damarides (Rössing deposit). Unfortunately, not all S-type granites have associated mineralization—for example, those found in a belt stretching from Idaho to Baja California. Sawkins has suggested that the reason that some have associated tin

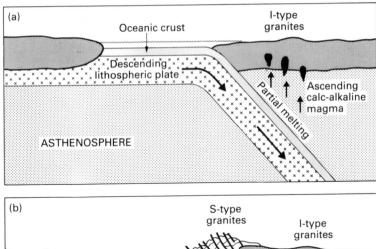

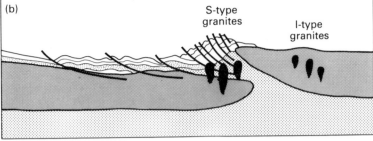

Fig. 29.11 (a) Two continents on a collision course as oceanic crust between them is subducted and I-type granites are generated by melting along the Benioff Zone. Some of these granites may host porphyry copper deposits. (b) Collision has taken place and the leading section of the underriding plate has been blanketed with thick thrust slices of sediment and mélange. Temperatures rise sufficiently in this continental crust to permit partial melting and the formation of S-type granites. If the crustal material contained above average amounts of tin, tungsten etc., then these granites may have associated, epigenetic deposits of these metals. The I-type and S-type granites, with their associated mineralization, may occur in parallel belts, as in Thailand and Malaysia.

mineralization, some uranium and others neither, presumably reflects the **geochemistry of the protolith** from which these granites were derived, and their subsequent magmatic history.

Of course, the various collision possibilities described above can result in deposits developed in arcs of various types being incorporated into collision belts. The other features, such as intermontane basins and foreland basins, which are present in arc orogens will have similar mineralization potential.

It might be thought that when oil-bearing continental shelves collide then **oilfields** are destroyed but quite the opposite was the case 20 Ma years ago when a wide continental shelf on the edge of Arabia collided with another continent, now central Iran, with the ocean between them disappearing in the process. The outer part of the Arabian shelf was compressed and folded to form huge anticlines in Iran which contain **vast oil pools** and redistributed oil from older shelf sediments. The inner parts of the shelf were preserved largely undeformed beneath what is now the Persian Gulf and eastern Arabia. The copious oil accumulations here are still in the original continental shelf structures: supratenuous folds, salt dome related structures etc.

29.3 Further reading

Evans A.M. (1993) *Ore Geology and Industrial Minerals*. Blackwell Scientific Publications, Oxford. Chapter 23. This reading will amplify the above topics.

Mitchell A.H.G. and Reading H.G. (1986) Sedimentation and Tectonics. In Reading H.G. (ed.), *Sedimentary Environments and Facies*. 471–551. Blackwell Scientific Publications, Oxford. Excellent background reading for those who wish to understand more clearly the sedimentary-tectonic setting of the mineralization described in this chapter.

O'Driscoll E.H.T. (1986) Observations of the Lineament–Ore Relations. *Phil. Trans. R. Soc. London*, **A317**, 195–218. A fascinating account of the fracture analysis which led to the discovery of one of the world's largest orebodies.

Sawkins F.J. (1990) *Metal Deposits in Relation to Plate Tectonics*. Springer-Verlag, Berlin. An extremely good coverage of this fascinating subject.

30: Ore Mineralization through Geological Time

30.1 Introduction

■ This chapter is concerned with the changes in mineralization type and style that have occurred as the earth has evolved and tectonic regimes and igneous processes have responded to this evolution.
■ Such a study cannot fail to impress the reader with the absolute necessity for geologists to be aware of these temporal changes when drawing up exploration programmes.

It is now well known to geologists that the earth, and its crust in particular, has passed through an evolutionary sequence of changes throughout geological time. These changes have been so considerable that we must expect them to have had some influence on the nature and extent of mineralization. Reference has already been made (Section 29.1) to the association of most of the world's **tin mineralization** with **Mesozoic** and **late Palaeozoic granites**, to the virtual restriction of **banded iron formation** to the **Precambrian** (Section 19.2.1) and the bulk of it to the interval 2500–1900 Ma ago, and to the importance of the **Precambrian for nickel** (Section 13.5). We will now examine changes in the type and style of mineralization in a little more detail. These changes can be discussed conveniently in terms of the Archaean, Proterozoic and Phanerozoic intervals and the environments that prevailed during them.

First, however, it must be remarked that although the Precambrian occupies about eight-ninths of geological time, very little is known as to whether plate tectonic processes acted then and, if they did, whether their mode of operation differed from those known to us through Phanerozoic studies. Precambrian research suggests that the Phanerozoic plate tectonic model, somewhat modified, is applicable for developments over the last 2500 Ma. As yet there is no general consensus about the type of plate regime in the Archaean; even those indications we have are biased towards the late Archaean, and little is known about large scale structures prior to about 3000 Ma ago.

30.2 The Archaean

This interval, 3800–2500 Ma ago, is notable both for the abundance of certain metals and the absence of others. **Metals and metal associations** developed **in significant amounts** include Au, Ag, Sb, Fe, Mn, Cr, Ni–Cu and Cu–Zn–Fe. Notable absentees are Pb, U, Th, Hg, Nb, Zr, R.E.E. and diamonds. Two principal tectonic environments are found in the Archaean: the **high grade regions** and the **greenstone belts**. The former are not in general important for their mineral deposits, which include Ni–Cu in amphibolites, e.g. Pikwe, Botswana, and chromite in layered anorthositic complexes, e.g. Fiskenæsset (west Greenland) as well as chromitite seams in dunite lenses known to be at least 3800 Ma old. The **greenstone belts** on the other hand are very **rich in mineral deposits** and their principal deposits and those of the high grade regions are related to the major rock groups of the greenstone belts and their adjoining granitic terranes as follows:

1 ultramafic flows and intrusions: Cr, Ni–Cu;
2 mafic to felsic volcanics: Au, Ag, Cu–Zn;
3 sediments: Fe, Mn and baryte;
4 granites and pegmatites: Li, Ta, Be, Sn, Mo, Bi.

30.2.1 Chromite

This is not common in greenstone belts but a very notable exception is present at Sherugwel (formerly Selukwe, c. 3500 Ma) in Zimbabwe. This is a very important occurrence of high grade chromite in serpentinites and talc–carbonate rocks intruded into schists which lie close to the Great Dyke. It resembles the podiform class of deposit, though it is in a very different tectonic environment from that in which Phanerozoic examples of this class are normally found.

30.2.2 Nickel–copper

These deposits are composed mainly of massive and disseminated ores developed in or near the base of komatiitic and tholeiitic lava flows and sills as described in Chapter 12. Only four important fields occur, those of south-western Australia (Kalgoorlie Belt), southern Canada (Abitibi Belt), Zimbabwe and the Baltic Shield of northern R.F. As these metals and their host rocks are probably mantle derived, this suggests the existence of **metallogenic provinces controlled by inhomogeneities in the mantle**, though the anomaly may not consist of an excess of nickel but rather of a concentration of sulphur which led to the extraction of nickel from silicate minerals.

30.2.3 Gold

Gold has been won in smaller or larger amounts from every greenstone belt of any size and its occurrence is the principal reason for the early prospecting and mapping of these belts. The gold is principally in vein deposits cutting basic or intermediate igneous rocks—both intrusions and lava flows—but the more competent intrusives such as the Golden Mile

Dolerite of Kalgoorlie, are more important (see Fig. 17.1). Some gold deposits show an association with banded iron formation (B.I.F.) and these appear to have been deposited from subaqueous brines to form exhalites. The greatest concentration of gold mineralization occurs in the marginal zones of the greenstone belts near the bordering granite plutons and it decreases towards the centre of the belts. This may suggest that it has been concentrated from the ultra-basic volcanics by the action of thermal gradients set up by the intrusive plutons. On the other hand, or as an additional factor, the spatial relationship with regional fault zones that have controlled the distribution of the volcanics, the synvolcanic sediments and the mineralization is clearly significant, and many gold camps in the Abitibi Greenstone Belt of Ontario and Quebec appear to be positioned where cross-lineaments intersect these regional faults (see Section 29.2.5).

Silver is usually present with gold in the greenstone belts. In the Abitibi Belt it is found in an Au–Ag–Cu–Zn association at granite–mafic volcanic contacts. These granites also have porphyry-style Cu–Mo mineralization.

30.2.4 Copper–zinc

Volcanic-associated massive sulphide deposits are very common in the Archaean, especially in the Abitibi Belt. These deposits are principally sources of copper, zinc and gold, but they are of **Primitive-type** (Table 16.1) and their **lead** content is normally **very low**. The virtual absence of lead mineralization from these greenstone belts may to some extent reflect the fact that during the Archaean there had been insufficient time for much lead to be generated by the decay of uranium and thorium in the mantle.

30.2.5 Iron and manganese

Banded iron formation is common throughout Archaean time but not in the quantities in which it appears in the Proterozoic. It is generally the Algoma type (see Section 19.2.1) that is present. There is some production from Archaean iron ore in Western Australia and from the Michipicoten Greenstone Belt, Canada. However, in contrast to all other Precambrian areas, China has large and significant B.I.F. deposits in the Archaean. This B.I.F. development reached a peak in the late Archaean and the B.I.F. occurs both in high grade gneiss terranes and greenstone belts, being more abundant **in the former. This gneiss-hosted occurrence is unique to China**. Another intriguing point concerning these Chinese Archaean occurrences is that nearly all are of **Superior-type** (oxide facies; see Section 19.2.1).

Manganese deposits of any significance, but **still small**, did not develop until 3000 Ma ago, post-dating by at least 800 Ma the oldest known geological sequence containing B.I.F. and base metal sulphide ores.

30.3 The early to mid-Proterozoic

The late Archaean and early Proterozoic (the transition date is arbitrarily put at 2 500 Ma) were marked by a great change in tectonic conditions. The first **stable lithospheric plates** developed, although these seem to have been of small size. Their appearance permitted the formation of **sedimentary basins**, the deposition of **platform sediments** and the development of **continental margin troughs**.

30.3.1 Gold–uranium conglomerates

The establishment of sedimentary basins allowed the formation of these deposits. The best known example is that of the Witwatersrand Basin with its widespread gold–uranium conglomerates, but other examples are known along the north shore of Lake Huron in Canada (Blind River area), at Serra de Jacobina in Brazil and at localities in Australia and Ghana. The Witwatersrand Supergroup, which contains the gold–uranium conglomerates, has been dated at about 3 060–2 708 Ma ago and the Pongola Supergroup of Swaziland, which also carries gold placers, developed on a stabilized craton during the period 3 100–2 900 Ma ago. These deposits represent **a unique metallogenic event** which many feel has not been repeated because **a reducing atmosphere** was a *sine qua non* for the preservation of the detrital uranium minerals and pyrite.

Uranium is a relatively mobile, lithophile element having an average crustal abundance of 2–4 ppm. As with many other metals, economic concentrations show a distinct time-bound nature in the Precambrian that may well reflect **an oxygenation event** in the earth's atmosphere at about 2 400–1 800 Ma ago (Fig. 30.1). Before this time, given the right conditions, large placer deposits with detrital uranium minerals were able to form. After this oxygenation event, however, hexavalent uranium was dissolved during weathering and transported as uranyl complexes. The development of these aqueous solutions permitted the extensive formation of **unconformity-associated uranium** deposits in the period 1 800–1 200 Ma. Unlike the late Archaean palaeoplacers and the Phanerozoic sandstone uranium-type deposits, the Proterozoic unconformity-associated deposits contain extremely high concentrations of uranium, which makes them a much more attractive mining proposition. The time-bound nature of uranium placer deposits explains the lack of detrital uranium in the mid-Proterozoic, Witwatersrand-like, Tarkwaian gold placers of Ghana. As well as lacking uraninite, these have no detrital pyrite, but do carry detrital hematite, which may indicate that an oxidizing atmosphere was by then well developed. (Alternatively it could be pyrite oxidized *in situ*, or hematite derived from Superior-type B.I.F.!)

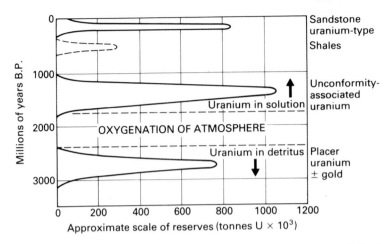

Fig. 30.1 The time-bound character of some major types of uranium deposit. The time period assigned on this figure to the oxygenation of the atmosphere must be considered as approximate. Experts in this field have postulated different time limits. The position of the upper line also varies according to whether we are concerned with the atmosphere and surface waters alone, or include the deep ocean waters, which will have taken longer to become significantly oxygenated.

30.3.2 Sedimentary manganese deposits

With the increasing development of cratonization shallow water, marine environments, within which large manganese deposits could form, were present. The early Proterozoic Kalahari Field, R.S.A. in the Transvaal Supergroup (2 500–2 000 Ma) rocks is enormous and estimates of about 7 500 Mt of plus 30% ore have been published. This, in terms of tonnage, is probably the **largest terrestrial manganese field in the world**. Many of the major Proterozoic manganese deposits were formed between 2 100 and 1 700 Ma ago; these include the West African deposits of Ghana, the Ivory Coast, other neighbouring countries and the large and important Moanda Mine, Gabon.

30.3.3 Sediment-hosted stratiform lead–zinc deposits

By about 1 700 Ma ago, the CO_2 content in the hydrosphere had reached a level that permitted the deposition of **thick dolomite sequences**. In a number of localities these **host syngenetic base metal sulphide orebodies**, such as those of the Balmat-Edwards and Franklin Furnace districts, U.S.A. In other sedimentary hosts, varying from dolomitic shales to siltstones, there are deposits such as McArthur River (Pb–Zn–Ag) and Mount Isa (Pb–Zn–Ag and separate Cu orebodies) in Australia and Sullivan, Canada. Various exhalative and biogenic origins have been suggested for these ores.

30.3.4 The chromium–nickel–platinum–copper association

The presence of small crustal plates permitted the development of large scale fracture systems and the intrusion at this time of giant dyke-like layered bodies, such as the **Great Dyke of Zimbabwe**, and enormous layered stratiform igneous complexes like that of the **Bushveld** in South Africa. Many of these **layered intrusions** were formed in the period 2 900–2 000 Ma ago. These are the repositories of enormous quantities of chromium and platinum, with other important by-products (see Sections 12.1 and 12.2). Though similar intrusions occur in other parts of the world, the great concentration of chromium is in southern Africa, and this has led some workers to postulate the presence of **chromium-rich mantle** beneath this region.

30.3.5 Titanium–iron association

About the middle of the Proterozoic, many **anorthosite plutons** were emplaced in two linear belts that now lie in the northern and southern hemispheres when plotted on a pre-Permian continental drift reconstruction. A number of these carry ilmenite orebodies, which are exploited in Norway and Canada (see Section 12.3). This was a unique magmatic event that has not been repeated. It suggests the gathering of reservoirs of magma (in the top of the mantle) that were able to penetrate upwards along deep fractures in the crust. The strength of the crust and the thermal conditions seem to have reached the point where magma could accumulate at the base of the crust in this extensive manner, rather like the accumulation of magma beneath present day rift valleys.

30.3.6 Diamonds

Diamondiferous kimberlites appear for the first time in the Proterozoic. This suggests that the geothermal gradients had decreased considerably, permitting the development of thick lithospheric plates, because diamonds, requiring extreme pressure for their formation, cannot crystallize unless the lithosphere is at least 120 km thick.

30.3.7 Banded iron formation

The greatest development of B.I.F. occurred during the interval 2 500–1 900 Ma ago. Although this rock type is important in the Archaean, it could not be developed on the large scale seen in the early Proterozoic because stable continental plates were not generally present. Following the development of **stable lithospheric plates**, B.I.F. could be laid down synchronously over very large regions; this took place possibly in intra-plate basins and certainly on continental shelves. The weathering of basic volcanics in the greenstone belts would have yielded ample iron and silica. **If the atmosphere was essentially CO_2-rich**, the iron could have travelled largely in ionic solution. It is now suspected that iron-precipitating bacteria may have played an important part in depositing the iron and oxidizing it to the ferric state, as modern iron bacteria are able to oxidize ferrous iron at very low levels of oxygen concentration. Although B.I.F. appears at later times in the Proterozoic, its development is very restricted compared with that in the early Proterozoic and this fall-off in importance has been correlated by some workers with the evolution of an oxidizing atmosphere. In the Phanerozoic, the place of B.I.F. is taken by the Clinton and Minette ironstones.

30.4 Mid to late Proterozoic

30.4.1 High grade linear belts

It has been suggested that a supercontinent existed through much of Proterozoic time and the trends of major high grade linear belts plotted on this supercontinent show that they lie on small circles having a common point of rotation. These linear belts affect middle to late Proterozoic as well as older rocks and include shear belts, mobile belts and linear zones of transcurrent displacements of magnetic and gravity anomaly patterns. They contain some deep dislocations that penetrate right down to the mantle and formed channelways for uprising magma. **Nickel mineralization** occurs in some of these belts, e.g. the Nelson River Gneissic Belt of Manitoba.

30.4.2 Sedimentary copper

There are anomalously high concentrations of copper in some late Proterozoic sediments in many parts of the world. These represent the oldest large sedimentary copper accumulations. Examples include the Katanga System of Zambia and Shaba [see Section 16.2.3(b)] and the Belt Series of the northwestern U.S.A. These, like the Upper Palaeozoic examples, correlate closely with widespread desert sedimentary environments.

30.4.3 Sedimentary manganese deposits

A second important period of manganese deposition occurred during the late Proterozoic, and manganese-rich sediments were laid down on or along the **margins of cratonic blocks**. The most important deposits are in

Brazil, central India, the R.F. and Namibia. Many of these, such as Morro do Urucum, Brazil (c. 900 Ma), are interbedded with B.I.F. On the other hand, the most characteristic late Proterozoic B.I.F. — the Rapitan Group, Canada — does not contain any manganese deposits and the late Proterozoic manganese oxide deposits in the Penganga Beds, India have no associated iron formation.

30.4.4 Tin

Tin mineralization does not appear in major quantities in the crust until the late Proterozoic, where it is associated with **high level alkaline and peralkaline anorogenic granites and pegmatites**. This is particularly the case in Africa where these deposits lie in three north–south belts (see Fig. 29.1). Another belt passes through the Rhondônia district of western Brazil.

30.4.5 Evaporite deposits

Sulphate evaporites of significant thickness and extent make their appearance in the early to mid-Proterozoic, and are quite common in the late Proterozoic.

30.4.6 Coal

The earliest known coals are found in early to mid-Proterozoic rocks where they have formed from compacted algal deposits. The first land-plant derived coals occur in the late Devonian rocks of the northern R.F.

30.5 The Phanerozoic

Towards the end of the Proterozoic, **a new tectonic pattern** developed which gave rise to Phanerozoic fold belts formed by continental drift. There was large scale recycling of oceanic crust, which greatly increased the number and variety of ore-forming environments by producing long chains of island and continental margin arcs, back-arc basins, rift-bordered basins and other features. Consequently, some Archaean volcanic types and Proterozoic sedimentological types reappear together with a few types dependent perhaps on more evolved plate mechanisms or extreme geochemical evolution of siliceous magmas. The latter process, plus recycling of crustal accumulations, may account for the important development of **molybdenum, tin and tungsten ores** in the Phanerozoic.

Although a few large phosphorite deposits and a large number of small ones are known in the Proterozoic, **marked worldwide epochs of phosphorite formation** belong largely to the Phanerozoic and commence with a very important epoch around the Proterozoic–Cambrian boundary (Fig. 30.2). This may well be related to more frequent continental fragmentation and drift providing suitable sites for phosphorite development in low latitudes during the Phanerozoic (see Section 29.2.3). Obviously, the supercontinent which existed through much of Proterozoic time will not have provided many favourable sites for widespread phosphorite deposition, and in the Archaean extensive regions of cratonic crust were not present.

The Proterozoic–Cambrian, phosphorite-forming epoch produced a marked concentration of deposits in a belt extending from Kazakhstan through central Asia into southern China, Vietnam and Australia, but deposits of this epoch are present in all the continents. It is estimated that in 1982 the annual production of phosphorite from deposits of this age was 23.5 Mt and these phosphorite deposits are believed to represent a world resource of around 34 Gt.

Cyprus-type copper–pyrite deposits and **sandstone uranium deposits** first appear in the Phanerozoic, while **podiform chromite**, found in the Archaean but not in the Proterozoic, becomes much more common. **Lead** becomes increasingly important in the volcanic-associated massive sulphide deposits. Some of the largest epigenetic metal concentrations of the Phanerozoic are those of the **porphyry copper** and **molybdenum** deposits of the continental margin and island arcs. **Gold**, which shows a peak of epigenetic mineralization in the late Archaean (2.7–2.6 Ga ago), peaks again in the Mesozoic to Quaternary where virtually all the deposits

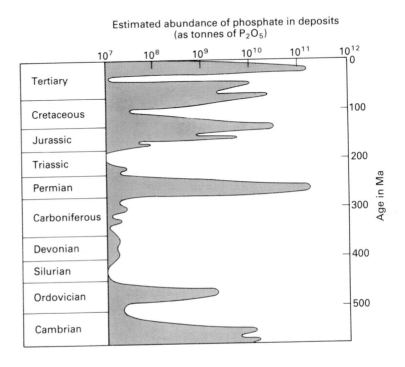

Fig. 30.2 Estimated abundance of phosphate in the world's known deposits of phosphorite (including reserves and resources) plotted against Phanerozoic time. Note that the tonnage scale is logarithmic to keep the peaks on the page!

occur in convergent margin settings. The Phanerozoic is also important for residual deposits, e.g. kaolin and bauxite deposits, many of which are Cretaceous–Recent in age.

Much of this Phanerozoic mineralization activity has been covered in the previous chapter. I hope that a study of these two brief chapters will reveal the importance of taking an account of plate tectonic settings, and the effect of continental drift and geological time in designing mineral exploration programmes.

I am I plus my surroundings and if I do not preserve the latter, I do not preserve myself.

José Ortega Y Gasset (1883–1955), Spanish essayist and philosopher.
Meditations on Quixote, To the Reader (1914).

There is not in the wide world a valley so sweet
As that vale in whose bosom the bright waters meet.

Thomas Moore (1779–1852), Irish poet.
Irish Melodies, The Meeting of the Waters.

Appendix 1: Formulae of Some Minerals Mentioned in the Text

Mineral	Formula
Acanthite	Ag_2S
Alabandite	MnS
Almandine	$Fe_3Al_2(SiO_4)_3$
Alunite	$KAl_3(SO_4)_2(OH)_6$
Amblygonite	$(Li,Na)Al(PO_4)$-(F,OH)
Anatase	TiO_2
Anglesite	$PbSO_4$
Anhydrite	$CaSO_4$
Apatite	$Ca_5(F,Cl)(PO_4)_3$
Arsenopyrite	$FeAsS$
Baddeleyite	ZrO_2
Baryte	$BaSO_4$
Bastnäsite	$(Ce,La)(CO_3)F$
Bertrandite	$Be_4Si_2O_7(OH)_2$
Beryl	$Be_3Al_2Si_6O_{18}$
Boehmite	$AlO(OH)$
Bornite	Cu_5FeS_4
Braunite	$(MnSi)_2O_3$
Carnotite	$K_2(UO_2)_2(VO_4)_2 \cdot 3H_2O$
Cassiterite	SnO_2
Celestite	$SrSO_4$
Chalcocite	Cu_2S
Chalcopyrite	$CuFeS_2$
Chromite	$FeCr_2O_4$
Cinnabar	HgS
Coffinite	$(USiO_4)_{1-x}(OH)_{4x}$
Columbite-tantalite	$(Fe,Mn)(Nb,Ta)_2O_6$
Copper	Cu
Covellite	CuS
Diamond	C
Diaspore	$AlO(OH)$
Digenite	Cu_9S_5
Enargite	Cu_3AsS_4
Eucryptite	$LiSiAlO_4$
Fluorite	CaF_2
Francolite	$Ca_5(PO_4,CO_3)_3(OH,F)$
Galena	PbS
Garnierite	$(Ni,Mg)_3Si_2O_5(OH)_4$
Gibbsite	$Al(OH)_3$
Gold	Au
Graphite	C
Gypsum	$CaSO_4 \cdot 2H_2O$
Hematite	Fe_2O_3
Illite	$(K,H_3O)(Al,Mg,Fe)_2(Si,Al)_4O_{10}[(OH)_2H_2O]$
Ilmenite	$FeTiO_3$
Jarosite	$KFe_3(SO_4)_2(OH)_6$
Kaolinite	$Al_2Si_2O_5(OH)_4$
Lepidolite	$KLi_2Al(Si_4O_{10})(OH)_2$
Limonite	Brown hydrous iron oxides
Löllingite	$FeAs_2$
Magnesite	$MgCO_3$
Magnetite	Fe_3O_4
Manganite	$MnO(OH)$
Marcasite	FeS_2
Microlite	$(Na,Ca)_2Ta_2O_6(O,OH,F)$
Molybdenite	MoS_2
Monazite	$(Ce,La,Y,Th)PO_4$
Montroseite	$VO(OH)$
Muscovite	$KAl_2Si_3O_{10}(OH)_2$
Nepheline	$NaAlSiO_4$
Parisite	$(Ce,La)_2Ca(CO_3)_3F_2$
Pentlandite	$(Fe,Ni)_9S_8$
Petalite	$LiAlSi_4O_{10}$
Pitchblende	UO_2
Pollucite	$CsSi_2AlO_6$
Pyrite	FeS_2
Pyrochlore	$NaCaNb_2O_6F$
Pyrolusite	MnO_2
Pyrophyllite	$Al_2Si_4O_{10}(OH)_2$
Pyrrhotite	$Fe_{1-x}S$
Quartz	SiO_2
Rhodochrosite	$MnCO_3$
Rutile	TiO_2

Scheelite	$CaWO_4$	Talc	$Mg_3Si_4O_{10}(OH)_2$
Sericite	Fine-grained muscovite	Tantalite	$(Fe,Mn)(Ta,Nb)_2O_6$
		Tetrahedrite-tennantite	$(Cu,Fe,Ag)_{12}$-$(Sb,As)_4S_{13}$
Serpentine	$Mg_3Si_2O_5(OH)_4$		
Siderite	$FeCO_3$	Trona	$Na_2CO_3 \cdot NaHCO_3 \cdot$
Silver	Ag		$2H_2O$
Smectite (montmorillonite)	$(Na,Ca)_{0.33}$-$(Al,Mg)_2Si_4O_{10}$-$(OH)_2 \cdot nH_2O$	Uraninite	UO_2
		Wad	Black manganese oxides and hydroxides
Spessartite	$Mn_3Al_2(SiO_4)_3$		
Sphalerite	$(Zn,Fe)S$	Wolframite	$(Fe,Mn)WO_4$
Spodumene	$LiAlSi_2O_6$	Wurzite	ZnS
Stannite	Cu_2FeSnS_4	Xenotime	YPO_4
Strontianite	$SrCO_3$	Zircon	$ZrSiO_4$

Appendix 2: Locations of Some Places Mentioned in the Text

1 Åheim, Norway
2 Allard Lake, Quebec, Canada
3 Almaden, Spain
4 Altenberg, Germany
5 Argyle Mine, Western Australia
6 Ambrosia Lake Field, New Mexico, U.S.A.
7 Ardlethan, N.S.W., Australia
8 Aswan High Dam, Egypt
9 Athabasca Uranium Field, Saskatchewan, Canada
10 Awash Valley, Ethiopia
11 Baku, Azerbaijan
12 Beijing, China
13 Bikita, Zimbabwe
14 Bingham Canyon, Utah, U.S.A.
15 Blind River, Ontario, Canada
16 Broken Hill, N.S.W., Australia
17 Bushveld Complex, R.S.A.
18 Butte, Montana, U.S.A.
19 Carlin Mine, Nevada, U.S.A.
20 Casapalca, Peru
21 Cerro Bolivar, Venezuela
22 Chiaturi, Georgia
23 Chernobyl, Ukraine
24 Climax and Henderson, Colorado, U.S.A.
25 Colorado Plateau, U.S.A.
26 Cornwall, England
27 Cornwall, Pennsylvania, U.S.A.
28 Crowdy Head, N.S.W., Australia
28 Cudgen, N.S.W., Australia
29 Donets Basin, Ukraine
30 Duluth, Minnesota, U.S.A.
31 East Alligator River, Northern Territory, Australia
32 East Alpine Ore District, Austria
33 Ekofisk Oilfield, North Sea
34 El Indio Mine, Chile
35 Erzgebirge, Czech Republic
   Felbertal, Austria, see 32

36 Fosdalen, Norway
37 Ganges Delta, Bangladesh
38 Glensandra Quarry, Scotland
39 Great Bear Lake, Northwest Territories, Canada
40 Great Dyke, Zimbabwe
41 Groote Eylandt, Northern Territory, Australia
42 Guadalajara, Mexico
43 Hamersley Iron Ore Region, Western Australia
44 Hishikari, Kyushu, Japan
45 Homestake Mine, South Dakota, U.S.A.
46 Imini, Morocco
   Jabiluka Mine, see 31
47 Kalgoorlie, Western Australia
47 Kambalda, Western Australia
48 Kempirsai, Kazakhstan
49 Khibina, R.F.
50 Kimberley, R.S.A.
51 King Island, Tasmania
52 Kirkland Lake, Ontario, Canada
53 Kiruna, Sweden
54 Kolar Goldfield, India
55 Kovdor, R.F.
56 Laisvall, Sweden
57 Lake Kariba, Zambia–Zimbabwe
58 Lake Magadi, Kenya
59 Lake Mead, Arizona–Nevada, U.S.A.
60 Lake Titicaca, Bolivia–Peru
61 Landoy Mine, Spain
62 Largentière, France
63 Lesotho
64 Lihir Island, P.N.G.
65 Llallagua, Bolivia
66 Luanshya Mine, Zambia
67 Lynn Lake, Manitoba, Canada
68 Madsen Mine, Ontario, Canada
69 Mansfeld Mine, Germany
70 Memé Mine, Haiti

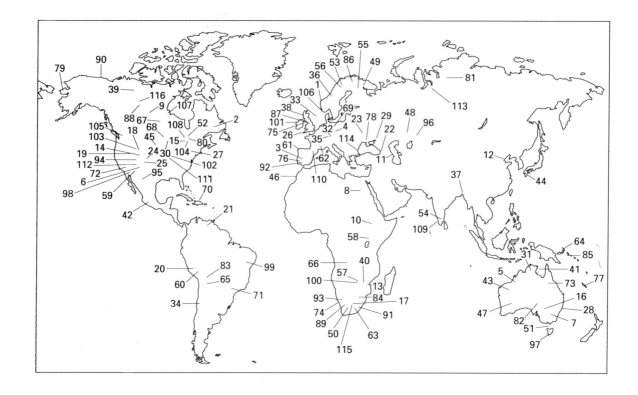

71 Morro Vehlo, Brazil
72 Mountain Pass, California, U.S.A.
73 Mount Isa, Queensland, Australia
    Mufulira Mine, Zambia, see 66
74 Namibian Diamond Fields,
    Namibia
75 Navan, Ireland
76 Neves-Corvo, Portugal
77 New Caledonia
78 Nikopol, Ukraine
79 Nome, Alaska
80 Noranda, Quebec, Canada
81 Noril'sk, R.F.
82 Olympic Dam, South Australia
83 Oruro, Bolivia
84 Palabora, R.S.A.
85 Panguna, P.N.G.
86 Pechenga, R.F.
87 Pennine Orefields, U.K.
88 Pine Point, Northwest Territories,
    Australia
89 Postmasburg, R.S.A.
90 Prudhoe Bay, Alaska
91 Richards Bay, R.S.A.
92 Rio Tinto, Spain
93 Rössing, Namibia

94 San Manuel-Kalamazoo, Arizona,
    U.S.A.
95 Santa Eulalia, Mexico
96 Sarbai, Kazakhstan
97 Savage River, Tasmania
98 Searles Lake, California, U.S.A.
99 Serra de Jacobina, Brazil
100 Sherugwel (Selukwe), Zimbabwe
101 Silvermines, Ireland
102 South-east Missouri, U.S.A.
103 Stillwater, Montana, U.S.A.
104 Sudbury, Ontario, Canada
105 Sullivan, British Columbia, Canada
106 Tellnes, Norway
107 Thompson Belt, Manitoba, Canada
108 Timmins, Ontario, Canada
109 Travencore and Quilon, India
110 Trimouns Mine, Luzenac, France
111 Tri-State Orefield, U.S.A.
112 Twin Buttes, Arizona, U.S.A.
113 Urengoi Gasfield, R.F.
114 Varentsi, Bulgaria
115 Witwatersrand Basin, R.S.A.
116 Yellowknife, Northwest Territories,
    Canada Zambian Copperbelt, see 66

Index